PLUMBING IS FUN...REALLY

Your plumbing system may not score very high on a glamour meter, but it's one of the most important systems in your home. Getting water in and out of the house without spilling it all over the place is essential to a family's quality of life.

The plumbing system brings luxuries easy to take for granted, such as hot showers, garbage disposers, and dishwashers—all of which make daily life easier and better.

The downside is that eventually something will fail. Toilets leak, pipes burst, and faucets drip. Somebody has to fix them and that somebody might as well be you.

Plumbers make a good living because some people—even experienced do-it-yourselfers—are a little nervous about the idea of releasing streams of water into their kitchens while they're hooking up a new sink. In fact, if they take it step-by-step, installing a sink is well within the skill range of most homeowners. Of course, some plumbing jobs are best left to professionals, and it's important to understand that you should not attempt projects beyond your level of skill and experience.

What plumbers know is that success under the sink is a matter of logic, common sense, and following the instructions—and that's what ***Plumbing 1-2-3*™** is all about. On the following pages you'll find more than 90 step-by-step repair, replacement, and installation projects. Each project has been reviewed by the experts at **The Home Depot®** to ensure that it's accurate and easy to follow so you can get the job done right the first time. You'll also find plenty of plumbing tips and good advice that will save you time and money, and make every job easier.

There can be a lot of satisfaction in taking care of and improving the plumbing systems in your home. It gets even better when you think of the savings that will come from doing the job yourself.

Plumbing 1-2-3™

Meredith® Books Development Team
Project Editor: John P. Holms
Art Director: John Eric Seid
Writer: Jim Stepp
Illustrator: Jim Swanson
Outline Development: Charlie Wing
Copy Chief: Terri Fredrickson
Managers, Book Production: Pam Kvitne, Marjorie J. Schenkelberg
Contributing Copy Editor: Steve Hallam
Contributing Proofreaders: Dan Degen, Beth Lastine, Ron Lutz, Debra Morris Smith, Shari Stelling
Contributing Designer: Tim Abramowitz
Indexer: Donald Glassman
Electronic Production Coordinator: Paula Forest
Editorial Assistants: Renee E. McAtee, Karen Schirm

Meredith® Books
Editor in Chief: James D. Blume
Design Director: Matt Strelecki
Managing Editor: Gregory H. Kayko
Executive Editor, Home Depot Books: Benjamin W. Allen

Director, Retail Sales and Marketing: Terry Unsworth
Director, Sales, Special Markets: Rita McMullen
Director, Sales, Premiums: Michael A. Peterson
Director, Sales, Retail: Tom Wierzbicki
Director, Book Marketing: Brad Elmitt
Director, Operations: George A. Susral
Director, Production: Douglas M. Johnston

Vice President, General Manager: Jamie L. Martin

Meredith Publishing Group
President, Publishing Group: Stephen M. Lacy
Vice President, Finance & Administration: Max Runciman

Meredith Corporation
Chairman and Chief Executive Officer: William T. Kerr

Chairman of the Executive Committee: E. T. Meredith III

The Home Depot®
Senior Vice President, Marketing and Communications: Dick Sullivan
Marketing Manager: Nathan Ehrlich

Image Studios / Image I.T.
Account Executive: Lisa Egan
Set Building: Rick Nadke
Primary Photography: Bill Rein
Contributing Photography: John von Dorn
Assistant: Max Hermans
Layout: Ron Jape
Production Manager: Jill Ellsworth
Account Rep.: Cher King
Technical Consultant: Richard Baron, Appleton, WI

Library of Congress Catalog Control Number: 2001132321
ISBN: 0-696-21451-2
Distributed by Meredith Corporation

Note to the Reader: Due to differing conditions, tools, and individual skills, Meredith Corporation and The Home Depot assume no responsibility for any damages, injuries suffered, or losses incurred as a result of following the information published in this book. Before beginning any project, review the instructions carefully, and if any doubts or questions remain, consult local experts or authorities. Because codes and regulations vary greatly, you always should check with authorities to ensure that your project complies with all applicable local codes and regulations. Always read and observe all of the safety precautions provided by any tool or equipment manufacturer, and follow all accepted safety procedures.

The editors of *Plumbing 1-2-3™* are dedicated to providing accurate and helpful do-it-yourself information. We welcome your comments about improving this book and ideas for other books we might offer to home improvement enthusiasts.
Contact us by any of these methods:

1. Leave a voice message at **800/678-2093**
2. Write to **Meredith Books, Home Depot Books, 1716 Locust Street, Des Moines, IA 50309–3023**
3. Send e-mail to **hi123@mdp.com**. Visit The Home Depot website at **homedepot.com**

Canadian EDITION

Plumbing 1-2-3™

Install, Upgrade, Repair, and Maintain Your Home's Plumbing System

Plumbing 1-2-3™ TABLE OF CONTENTS

Chapter 1
PLUMBING 1-2-3
PAGE 5

Chapter 2
THE PLUMBER'S TOOL KIT
PAGE 10

Chapter 3
REPAIR AND MAINTENANCE
PAGE 14

HOW TO USE THIS BOOK

Professional plumbers and store associates from The Home Depot® stores across the country created *Plumbing 1-2-3*™ to give homeowners a comprehensive and easy-to-follow guide to the most common home plumbing tasks. Their expertise and years of experience guarantee successful completion of every plumbing project, from basic repair and maintenance to installation of major appliances and new systems. Clear instructions and step-by-step photography make every project accessible and easy to understand.

Chapter 1: Plumbing 1-2-3 gets you started by introducing you to basic plumbing skills (see page 6). Picking projects within your skill range will help you gain confidence so when you move on to more complicated jobs, you know you'll be able to complete them successfully. The chapter starts with the basics—repairs and maintenance—then moves on to more complicated projects.

Chapter 2: The Plumber's Tool Kit will show you the tools required for most plumbing jobs. Refer to it when you're planning a project or when you want to know what something is for or how to use it.

Chapter 3: Repair and Maintenance covers common repair and maintenance jobs, such as fixing a leaky faucet, replacing a showerhead, or adjusting a toilet. Included are tips on maintenance and extending the life of major appliances, such as gas and electric water heaters. Troubleshooting tips from plumbing experts help you attack the messy subject of clogged fixtures and pipes.

Chapter 4: The Supply System deals with getting water where you want it to go and **Chapter 5: The DWV System** (drain, waste, and vent) covers carrying water and waste out of the house.

We introduce installation in **Chapters 6**, **7**, **8**, **9**, and **10**, taking you through the process of putting in **Tubs, Showers, Sinks and Lavatories, Toilets,** and **Appliances**. Getting things right the first time will prevent a lot of expensive headaches later on.

Finally, in **Chapter 11: Outdoor Systems**, you can take your plumbing expertise outside and learn how to design and install an underground irrigation system for your lawn and garden, as well as a hose bib and sump pump.

As you use this book, you'll learn what tools you need and how to plan a project. You'll see the big picture when tackling a remodeling project. Installing a tub may require removing other fixtures; it will also require carpentry, tiling, and electrical skills.

If local codes require a professional installation or if you want to hire a contractor, learning about the projects will give you an overview that will be useful when you're negotiating a contract or approving finished work.

Everyone knows the importance of working safely in the home, so be sure to read the **Safety Alerts** that are included with the projects. Pay close attention to red type and safety tips marked with stop signs; they're there to protect the safety and health of you and your family.

And finally, the experts at home centers and hardware stores in your area are great resources for advice on projects and problems. Take advantage of their experience and skills.

TRICKS OF THE TRADE

Tips from the pros at The Home Depot® are scattered throughout this book. Their expert advice will help you successfully complete the projects in *Plumbing 1-2-3*™.

Make smart work choices.

Prevent unsafe situations.

TOOL TIP

Use specialty tools to their best advantage.

Save time and gas mileage.

Understand all the details.

Save time and money.

Info you need to know before you begin.

Fix common mistakes. (Not that you'll make any.)

Avoid common mistakes.

Make sure your work is legal.

Learn new ways to work with old stuff.

Select the best materials.

CHAPTER

1 PLUMBING 1-2-3

Archaeologists speculate that the first plumbing system was probably installed on the isle of Crete around 4,000 years ago. It's a pretty safe bet that the first leak was noticed about the same time and that some ancient plumber made the first house call.

A modern homeowner would actually be familiar with the layout. Four separate drainage systems emptied waste into stone sewers. Terra-cotta pipe ran under the floors, providing hot and cold running water to fixtures and fountains. There's even evidence of the first working toilet—complete with a wooden seat and a small reservoir.

Technology and materials have definitely improved over the centuries, but the basics of getting water in and waste out of our homes—along with the associated problems—haven't changed (right down to finding the leaks and getting them fixed).

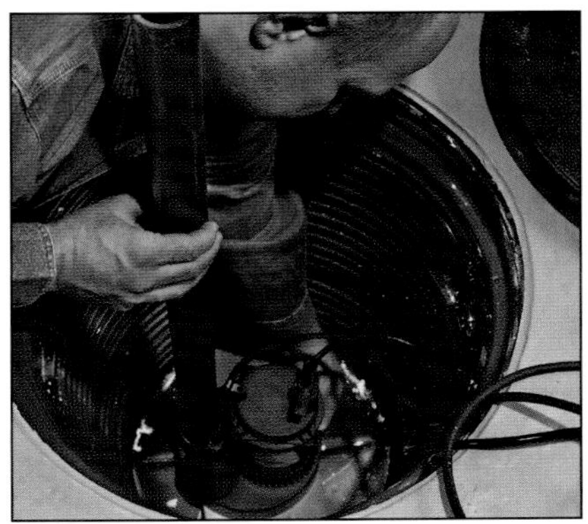

PLUMBING ISN'T SCARY

In the old days veteran plumbers joked that plumbing was easy—all you had to know was that "water flowed downhill and payday was on Friday." Actually, plumbing in those days was much more complicated. Working with lead pipe and cast iron required experience, a level of skill, and an investment in time beyond what most do-it-yourselfers were interested in tackling.

Even with all the advances in materials and technology, the average homeowner may still think of plumbing as mysterious and difficult. (Everyone's afraid of flooding their home.) The truth is that most home plumbing projects are accessible to anyone who's interested in doing things themselves and saving money.

SIMPLIFIED INSTALLATION

Manufacturers have simplified installations and provided detailed instructions along with technical support for their products. Home centers and hardware stores have knowledgeable staff who will answer questions and provide resources. Plumbing codes in general are becoming more standardized so materials are easier to find and installation procedures are more universal.

BETTER—AND SAFER

Today's plumbing systems use materials that are longer lasting, safer, and more user-friendly. A few basic tools and a little patience while learning the ropes are all you need. Some plumbing jobs, like unclogging a drain line or fixing a leaking pipe or toilet, may not be exciting work, but look at it another way. Most plumbing materials are inexpensive in relation to other do-it-yourself jobs—it's the labor that drives up the final cost. The money you will save by knowing a few basic plumbing skills can stay in your pocket or be applied to higher-quality fixtures than you might otherwise have budgeted.

EVOLUTION.
Early pipe was constructed of wood or earthenware. Lead became popular because of its durability. In fact, the Latin word "plumbus" means "lead." Lead was difficult to work with and proved to be a health hazard, so other materials were developed to carry water and waste, making installation easier and safer.

QUALIFYING YOURSELF

Assess your basic skills accurately and honestly so you can pick projects that you can complete successfully and with confidence. Store associates at The Home Depot qualify customers for plumbing projects by asking some basic questions to get a sense of their abilities.

LEARNING BY DOING
The best way to develop your plumbing skills is to start with a simple project such as repairing a leaking pipe, then move on to more demanding projects as your skills develop.

QUALIFYING QUIZ

■ **Do you mind getting your hands dirty?** Some plumbing projects can get messy; that's just the nature of the job.

■ **How about doing physical labor?** Some projects such as replacing a toilet can require a certain amount of heavy lifting. Enjoying the challenge of physical labor is an important attribute.

■ **Do you like working with tools?** Most plumbing projects require some basic tools. If you don't know a screwdriver from a pair of pliers or a plunger from a snake, now's the time to learn. The best introduction to using a tool is to watch it in action. **Chapter Two** will give you basic information on the most common plumbing tools, and you'll find more tool details in the projects themselves.

■ **Are you a tinkerer?** If you're good at taking things apart and putting them back together, you can learn plumbing skills.

■ **Are you willing to research projects and make a plan?** Doing your homework to develop an understanding of the process and scope of a project is essential. Learn the skills you'll need and explore all the safety issues before you start.

■ **Do you enjoy working on your house?** If you don't enjoy maintaining and improving your home, you may not want to get involved in replacing a sink or fixing a leak.

■ **Do you know your limitations?** It's OK to admit that a particular project is a little beyond your current skill level. It's better to pay a professional to do a job you're not comfortable with rather than paying extra for one to fix your mistake.

OK, YOU'RE QUALIFIED

Once you've answered these questions and qualified yourself, you're ready to begin picking the right level of projects for you.

"REPAIRING A CARTRIDGE FAUCET," (See page 20)

"TROUBLESHOOTING A LEAKING TOILET," (See page 40)

"ADJUSTING A POP-UP DRAIN," (See page 36)

PICKING PROJECTS

You can't always pick your projects. Plumbing failures and leaks can happen at any time and have to be dealt with quickly and efficiently.

You do, however, have choices when you're remodeling or upgrading; you can pick projects within your skill level. You have the time to do your homework and make a good plan. Become familiar with the materials and equipment. Ask yourself: "Can I do this and do I have the time?" If the answer is yes, get started!

TOOLS OF THE TRADE

Each project has a list of tools and materials you'll need to complete the job. Review the list for the project you select and make sure you understand how each tool is used. Ask the salespeople at your local home center how to safely and properly use the tool.

KNOW YOUR PLUMBING SYSTEM

A working knowledge of your plumbing system will help you understand why something isn't working properly. The chapters on the supply and drainage systems, pages 82–101, will tell you how each system operates and give you some troubleshooting tips. Make a map showing how the plumbing in your home is installed. Know where the water meter and the main shutoff valve are located. Find out where water comes into the house and where the drainpipes and traps go out.

STARTER PROJECTS

Begin with easy projects. Most repair and maintenance jobs are fairly simple. Check through the house and see if repairs are needed. Does a faucet drip? Is the toilet acting up?

Or maybe it's time to finally address a slow-draining sink or dishwasher. You'll find repair and maintenance projects in Chapter 3.

INTERMEDIATE JOBS

Getting comfortable with your plumbing skills? Consider tackling more complex projects, such as remodeling or updating fixtures and appliances. **Installing Sinks and Faucets, Installing Toilets,** and **Installing Appliances** will guide you through each upgrade.

ADVANCED REMODELING

Do you need to replace that tub or shower? Or maybe you'd like a shower stall downstairs. These projects require planning, and you'll need other skills besides plumbing. Before you start, draw a detailed diagram. List the tools and materials you'll need for the job. Estimate the amount of time from start to finish and ask yourself if you can spare the time to do the job. Ask yourself not only if you have the plumbing skills required, but also if you have the electrical, tiling, painting, and carpentry skills needed. If the answers are all yes and you're comfortable with them, check out **Installing Tubs** and **Installing Showers**.

WORKING OUTDOORS

Plumbing is not just an indoor activity. Installing a below-ground irrigation system and placing a hose bib in a more convenient location are both outdoor plumbing projects. **Installing an Irrigation System**, page 180, requires solid planning, some advanced plumbing skills, a little electrical savvy, and the willingness to use a shovel.

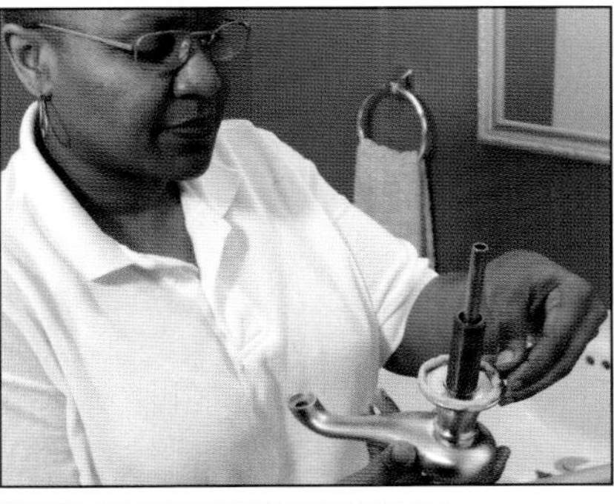

"INSTALLING WIDESPREAD FAUCETS," (See page 138)

"UNCLOGGING DRAINS AND WASTE LINES," (See page 80)

GROWING YOUR SKILLS

The key to growing your skills is learning to do things correctly from the start. Looking for shortcuts before you understand the basics can lead to very costly mistakes.

Plumbing is a process. It helps to know how something works before you try to fix it. Pick a project you're comfortable with, assemble all the tools and materials you're going to need, and review the entire process by reading all of the instructions before you start. Once everything is in place, all you have to do is follow the project through to completion step-by-step.

Your skills will grow in proportion to your willingness to do things right.

ESTIMATING JOB COSTS

The cost of a project is based on materials and labor. If you're doing the work yourself, you'll only be concerned with materials, but if you're hiring a plumber, you'll have to figure in his or her rate. Before you start, know what you want to do and list what you need to complete the project. Shop around and compare prices. Compare low-, middle-, and high-end fixtures. You may find that the cost difference between good and best isn't prohibitive.

Purchase products and materials from businesses that back what they sell with good service, and have a clear return policy if problems arise. They have a presence in the community and a reputation to maintain. Older hardware stores and secondhand shops are potential sources for antique fixtures. Consider places that sell discontinued items, but keep in mind it may be very difficult if not impossible to find replacement parts.

When estimating costs, add 10 to 15 percent to the total bill so you can cover unexpected expenses without blowing a hole in your budget.

DEALING WITH PLUMBING CODES

Plumbing codes—both local and national—are standards and guidelines intended to make sure plumbing installations are safe and will work properly. Together, they codify plumbing techniques and proper use of materials throughout Canada.

IMPORTANT NOTE: The projects in this book follow common national guidelines, but local codes are often more strict than national standards. Also be aware that codes can vary from location to location in both materials and installation methods for the same project. Check with your local building department for requirements and to schedule inspections.

Do your research before you start actual work to make sure you're meeting code. If the inspector doesn't approve your work, you'll have to do the work all over again and that gets expensive. When applying for permits, present precise drawings of your plans that are easy to read. Include a complete list of materials to be used in the project.

Installations need to be seen to be inspected. Don't put up the drywall or cover pipes until the inspector has signed off.

HIRING A PLUMBING CONTRACTOR

Great word of mouth is the sign of a good contractor. Get at least three different quotes. The bids should include a list of materials or "specs" of everything that will be installed and a work schedule. Payments should be made in at least three phases: no more than half of the money up front, a payment during the course of the job, and a final payment after the job is completed to your satisfaction. Be aware that not all types of plumbers will handle all types of work. A contractor should be responsible for getting necessary permits as well as scheduling and dealing with inspectors.

Make sure the contractor's license is valid in your area and that he or she has valid insurance to protect you from liability for the duration of the job. Ask for references and visit the contractor's finished sites as well as some works in progress. Check with the Better Business Bureau if you have any concerns about a contractor's reliability.

Question everything. Ask the contractor to explain what he or she is doing and why. And finally, don't accept shoddy work.

GENERAL PLUMBING SAFETY

If you use common sense and follow the instructions, you can do any plumbing project safely and with confidence. But there are safety issues to consider, such as pipes that carry potentially hazardous liquids and gases in and out of the house. It's essential to have a proper working environment.

Remember that safety is always your number one priority.

PLUMBING AND ELECTRICITY

Water and electricity don't mix. Use proper precautions and follow all manufacturer's instructions when combining plumbing and electrical work.

RESPECTING NATURAL GAS

Gas is very safe, but sometimes there are leaks. If you suspect a gas leak, open windows and doors for ventilation. Don't touch electrical switches.

If the odor is faint and it's near a gas appliance, check the pilot light. If the pilot light is on and you're certain the gas is coming from the appliance, turn off its gas supply. If the smell persists, leave your home immediately. Call the gas supplier's emergency phone number from a neighbor's house.

HANDLING EMERGENCIES

▶ No matter how carefully a system was installed, eventually pipes will burst, a toilet or tub will overflow, a joint will break, or a drain line will crack. Remaining cool, calm, and collected in such emergencies is the key to minimizing damage and making cost-effective repairs.

▶ Take a walk around your home and map out your plumbing system. Know where the main supply valve is and know how to shut it off. Map all individual shutoffs near appliances and fixtures and install new ones where they'd be useful. (See "Installing Valve Stops and Shutoffs," page 140.)

▶ Learn how to use plungers and drain augers beforehand so you're not reading instructions during an emergency. (See "Unclogging the Clogs," pages 72–73.)

SAFETY EQUIPMENT

- **Wear eye protection.** Safety goggles are an absolute must whenever you're cutting, pounding, soldering, or working above your head.They protect against splashing liquids and airborne particles. Get used to wearing them when you're working.
- **Wear a respirator.** Airborne particles can damage your lungs. Use dust masks when cutting materials and respirators specified by manufacturers when you're working with any chemicals.
- **Wear gloves**. Sturdy gloves are essential for handling chemicals, sharp objects, or heavy, awkward loads. Special gloves are designed for different types of protection.
- **Wear good-quality work clothes.** Protect yourself from physical injury and the potentially high bacteria content of wastewater and debris by wearing heavy clothing—especially long sleeves and long pants. Launder clothing immediately after each use.
- **Clean your hands.** Always clean your hands with an antibacterial soap.
- **Use the right tool.** Use your tools for the job for which they were designed. A screwdriver isn't a cold chisel and an adjustable wrench isn't a hammer. Improper tool use can cause injuries and damage fixtures and equipment.
- **Respect ladders.** If you're working from a ladder, read and follow the precautions printed on the side. Never stand on the top or higher than the third step from the top. A helper should always be present to steady the ladder, and all four of the ladder's feet should be seated firmly on level ground.

VENT GASES PROPERLY

Carbon monoxide (CO) is odorless, colorless, and tasteless. Any fuel-burning appliance can be a source of CO. Vent properly. Follow codes and manufacturer's instructions. CO detectors can also be valuable additions to your home safety list.

Keep a list of utility, fire, police, and medical emergency numbers by the telephone for quick reference.

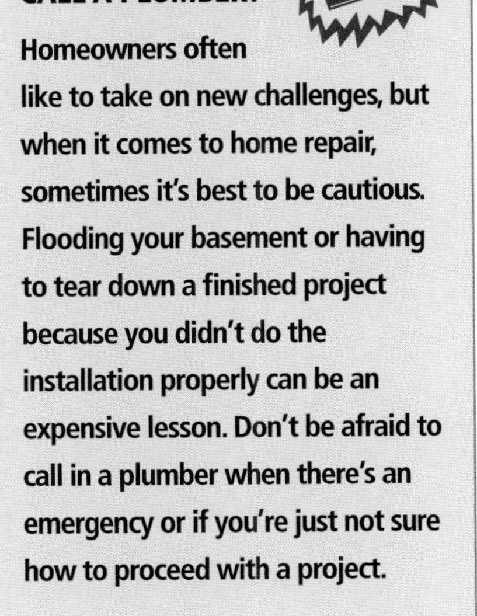

KNOW WHEN TO CALL A PLUMBER!

Homeowners often like to take on new challenges, but when it comes to home repair, sometimes it's best to be cautious. Flooding your basement or having to tear down a finished project because you didn't do the installation properly can be an expensive lesson. Don't be afraid to call in a plumber when there's an emergency or if you're just not sure how to proceed with a project.

CHAPTER 2

THE PLUMBER'S TOOL KIT

Here are some of the tools you'll use for the plumbing projects in this book. While many projects can be completed with a few common hand tools, some tasks require specialized items designed to make the job easier. As you expand your plumbing skills, you will also expand your tool kit.

If you do carpentry or electrical work around the house, you'll already have a lot of plumbing tools on hand. Purchase specialty items as you need them for projects. Consider renting tools that are very specialized or that you'll only need once or twice. Friends and neighbors are often good sources for specialized items, but if you borrow tools, treat them carefully, clean them, and return them promptly.

Always use tools as they're intended to be used; an adjustable wrench is not a hammer. Study the manuals and instructions that come with the tools so you can handle them safely and maintain them properly.

Buy the best you can afford; good tools will last a lifetime and can be passed down through generations. The old adage remains true—take care of your tools and they'll take care of you.

TOOLS ARRANGED ALPHABETICALLY, TOP TO BOTTOM, IN COLUMNS LEFT TO RIGHT.

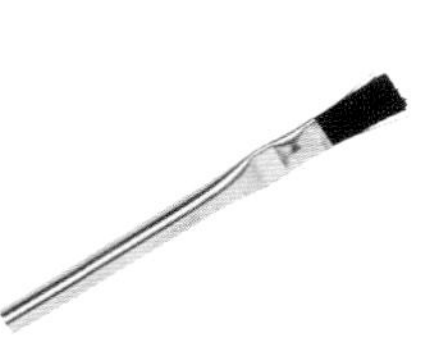

ACID/FLUX BRUSH

Used to apply flux to copper fittings.

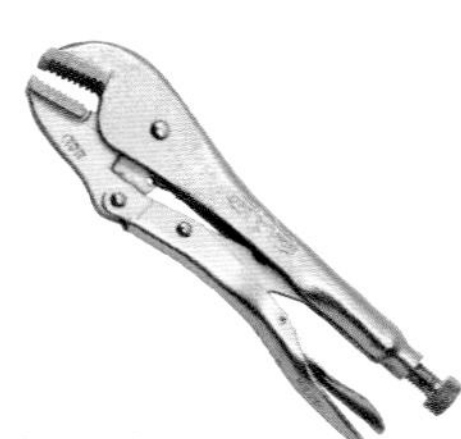

ADJUSTABLE LOCKING PLIERS

Adjustable pliers with jaws that lock in place to clamp securely on an object.

ADJUSTABLE WRENCH

Movable jaw allows adjustment of wrench to fit a variety of bolt heads and nuts.

BASIN WRENCH

Tightens or removes hard-to-reach faucet mounting and coupling nuts.

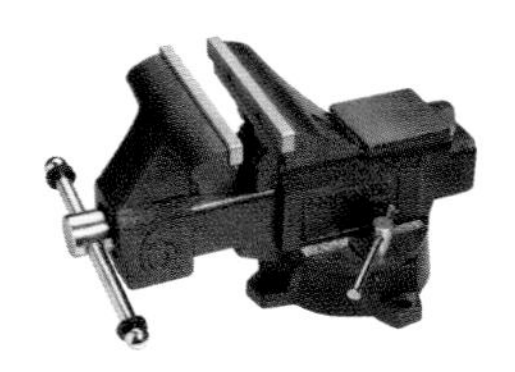

BENCH VISE

Lightweight vise for clamping objects for cutting or assembly.

BLOW BAG

Builds up water pressure then releases a burst of water for pushing out clogs in floor drains and waste lines. Check local codes for approval before using.

CAST-IRON SNAP CUTTER

Allows cutting of tough cast-iron pipe.

CAULKING GUN

Delivers a bead of sealant or glue from a disposable tube. Caulk seals seams and plugs holes.

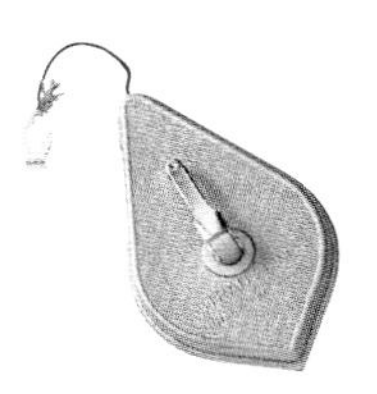

CHALK LINE

Marks a line out with chalk for cutting or reference.

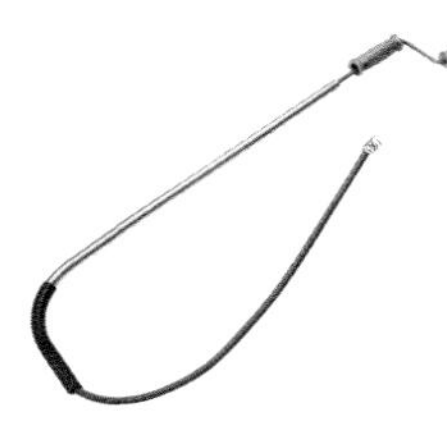

CLOSET AUGER

Designed to clear clogs in toilets without damaging the porcelain surface.

COIL SPRING TUBING BENDER

Fits over copper tubing, allowing bending without crimping the pipe.

COLD CHISEL

Marks or breaks hardened metals such as cast-iron pipe.

COMBINATION TOOL, 4-IN-1

A number of tools in one for deburring and cleaning the exterior and inside of copper pipe.

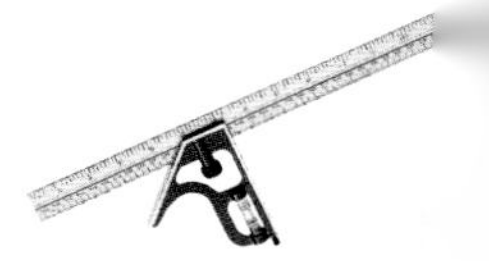

COMBO SQUARE

Used for marking common angles.

COPPER FITTING BRUSH

Deburrs and cleans debris from the interior surface of copper pipe and fittings.

COPPER TUBING DEBURRER

Deburrs and cleans debris from the exterior surface of copper pipe.

CORDLESS REVERSIBLE ⅜-INCH DRILL

Makes it easier to drive screws and fasteners.

CORDLESS REVERSIBLE SCREWDRIVER

Drives as well as removes screws and fasteners.

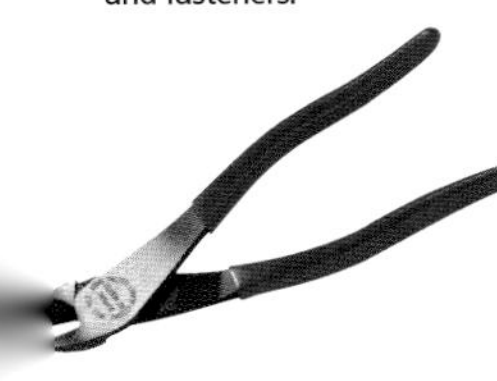

DIAGONAL CUTTERS

(Left- and right-handed)

Prevents distortion of dryer pipe when cutting.

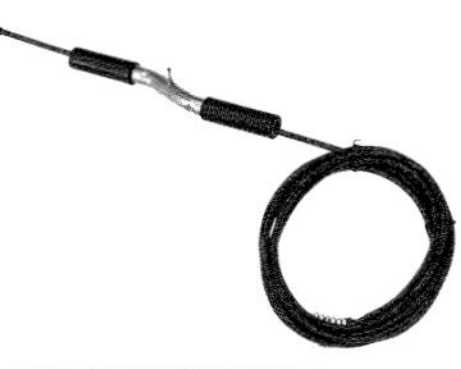

DRAIN SNAKE OR DRAIN AUGER

A slender tube with a handle on one end that clears blocked or slow-draining waste and drain lines.

ELECTRIC ELEMENT SOCKET

Makes the job of removing the electrical element in a water heater easier.

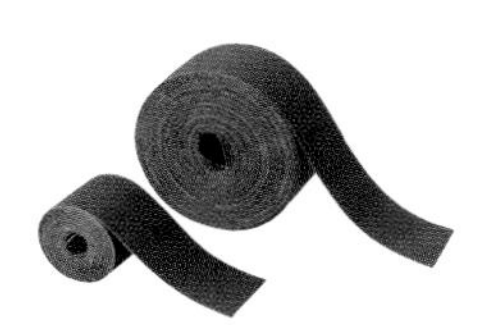

EMERY CLOTH

A very finely gritted sandpaperlike cloth used to remove burrs and clean debris from copper supply pipe.

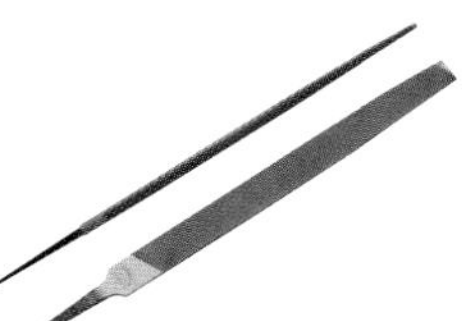

FILES—ROUND AND FLAT

Smooths metal, wood, or plastic.

FLAME PROTECTION

A fireproof sheet of material to place between a propane torch flame and a flammable object such as wood while soldering.

FLANGED PLUNGER

Folds up into a cup during use, then discharges a forceful burst of water to clear clogs in sinks and toilets.

FLARING TOOL

Fits over the end of copper pipe and flares out the end as the screw is tightened so end will seal properly with the gas line fitting.

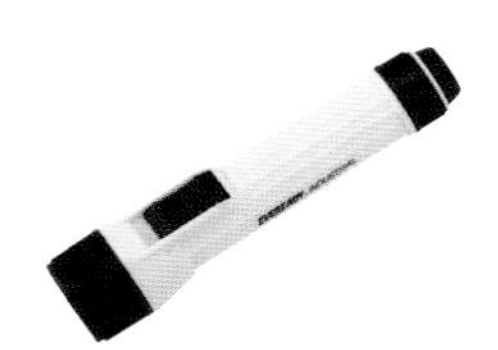

FLASHLIGHT

A handheld light that directs light into hard-to-see places.

HACKSAW

Cuts metal, metal piping, and plastic piping.

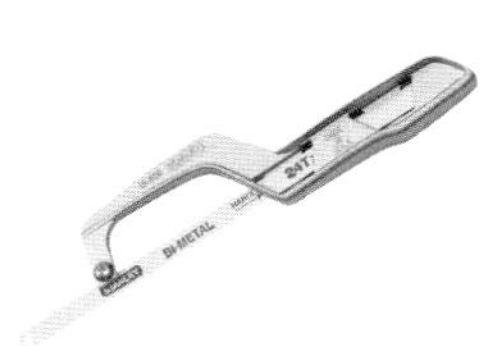

HACKSAW—MINI

Cuts pipe and metal in hard-to-reach places.

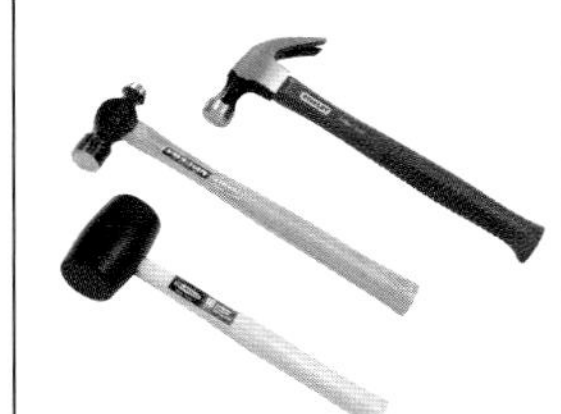

HAMMER

Drives nails and is used for striking metallic objects.

HAND AUGER

A handheld device with a crank and a cylinder containing a coiled tube that is fed out by turning the crank to clear clogs.

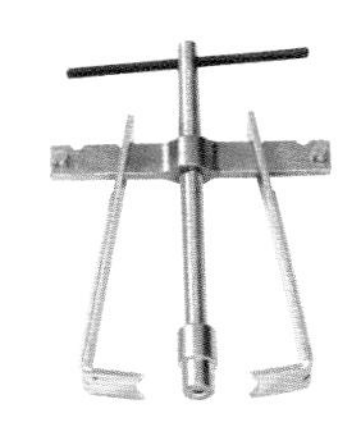

HANDLE PULLER

Removes stubborn handles without damaging them.

HEX KEY OR ALLEN WRENCH SET

Tightens and loosens set screws.

HOLE SAW

A bit that fits on the end of a power or hand drill for cutting holes in wood.

INSPECTION MIRROR

A telescoping handle with an adjustable mirror at the end that allows seeing into tight locations.

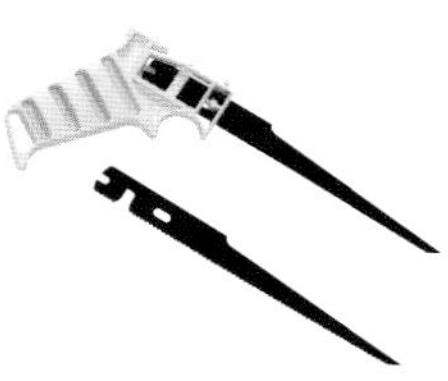

KEYHOLE SAW

A handheld saw with a thin, replaceable blade for cutting in tight places.

LEAD-FREE FLUX AND SOLDER

Flux cleans the surface of copper pipe; solder joins the pipe to the fitting with a tight seal.

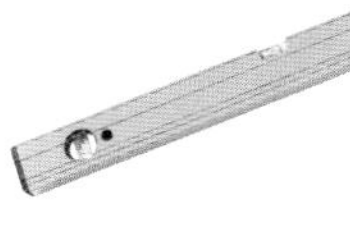

LEVEL—CARPENTER'S

Determines vertical and horizontal level; used for determining slope of pipe runs.

LEVEL—LINE

A level that sits on a string to find slope for running pipe.

THE PLUMBER'S TOOL KIT

LEVEL–PLUMB BOB
Determines vertical leveling of an object.

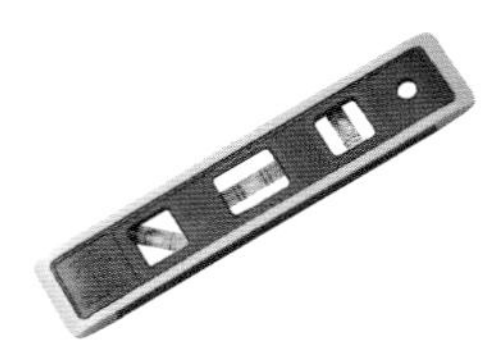

LEVEL–TORPEDO
A small level that allows easy access to tight places. Often comes with a magnetic edge, which makes it a good choice for leveling metal appliances.

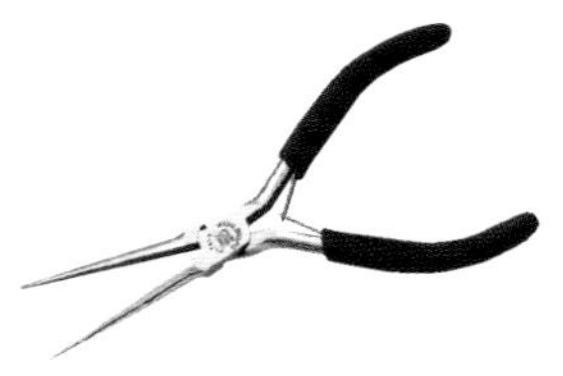

LONG-NOSED PLIERS
Grips wires and small objects while holding them in place in tight locations.

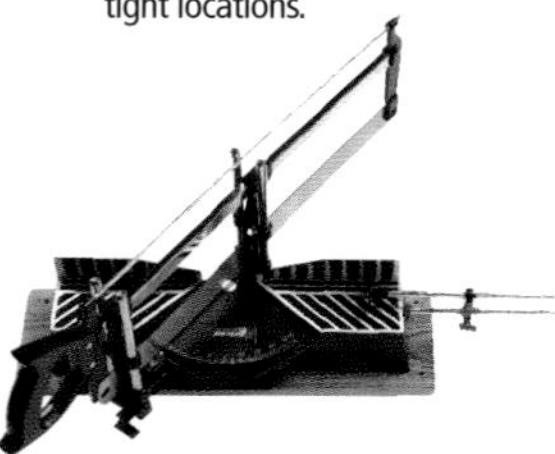

MITER BOX
Cuts pipe while holding it in place for square, even cuts.

NEEDLE NOSE PLIERS
Grips small objects while holding them in place in tight locations.

PENETRATING OIL
Loosens "frozen" nuts, bolts, and threaded pipe by penetrating the threads.

PIPE JOINT COMPOUND
Ensures a perfect seal for threaded pipe.

PIPE WRENCH
Tightens, loosens, and clamps pipes and large fittings.

PLASTIC PIPE CEMENT
Solvent that cements plastic to plastic plumbing fittings. Available in all-purpose formulas.

PLASTIC PIPE PRIMER
Primer prepares PVC and CPVC plastic pipe for cement when connecting to a fitting.

PLASTIC TUBING CUTTER
Cuts plastic piping, providing a clean and square cut. Important when seating the pipe in a fitting.

PLUMBER'S PUTTY
Provides a watertight seal for fixtures. Make sure the type is compatible with materials and surfaces.

POWER CIRCULAR SAW
Use different blades to cut wood and plastic.

POWER REVERSIBLE 3/8-INCH DRILL
Used for driving screws and fasteners and drilling holes. No battery means unlimited use.

PROPANE TORCH
Supplies fuel from a cylinder and through a torch fitting to produce a flame for soldering copper pipe.

PRY BAR
A lever that allows lifting or separation of heavy objects.

PUTTY KNIFE
Used to apply to or scrape putty and caulk from surfaces.

RATCHET-TYPE PVC CUTTER
A tool for cutting plastic pipe that guarantees a square, smooth cut.

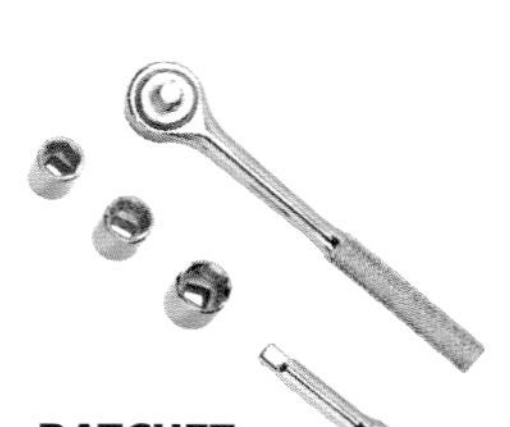

RATCHET WRENCH AND SOCKET SET
Easily tightens and removes bolts and nuts.

RECIPROCATING SAW
Cuts wood, metal, or plastic with the proper blade.

RETRACTABLE STEEL TAPE MEASURE
Measuring device. Purchase one that is at least 16 feet long.

RIGHT-ANGLE DRILL
Drills easily in hard-to-reach places.

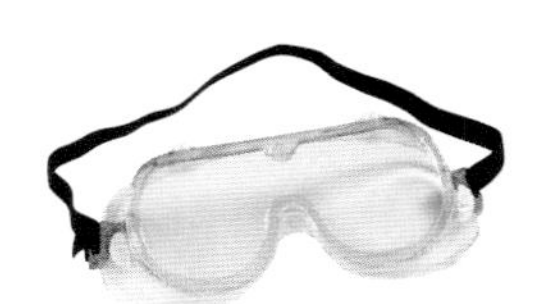

SAFETY GOGGLES
Protects the eyes from flying particles or splashing fluids.

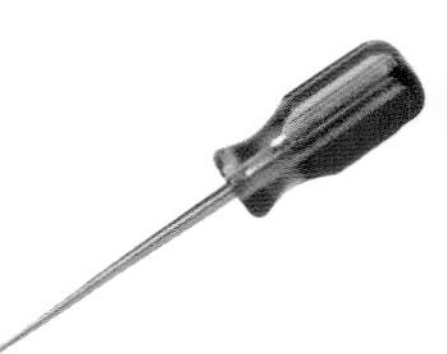

SCRATCH AWL
Marks metal by etching the surface.

SEAT DRESSING TOOL
A specialized tool that smooths and resurfaces nonremovable seats.

THE PLUMBER'S TOOL KIT

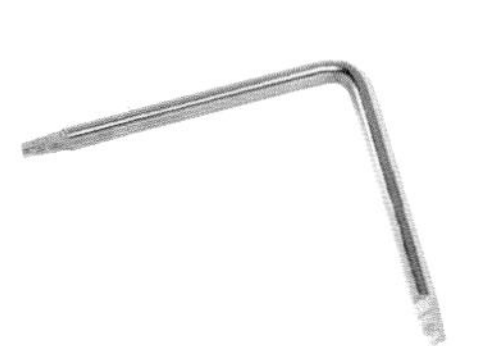

SEAT WRENCH

A specialty tool designed to reach into the faucet stem to remove or tighten removable seats.

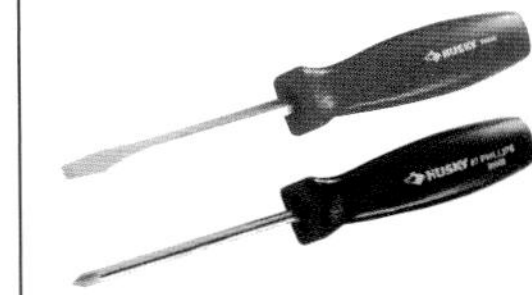

SLOTTED AND PHILLIPS SCREWDRIVERS

Drives and removes screws by hand.

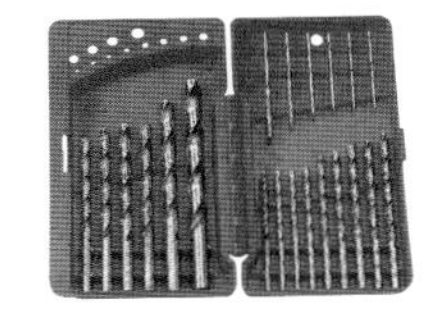

SPEED BITS

High-speed bits designed for drilling through metal, plastic, or wood.

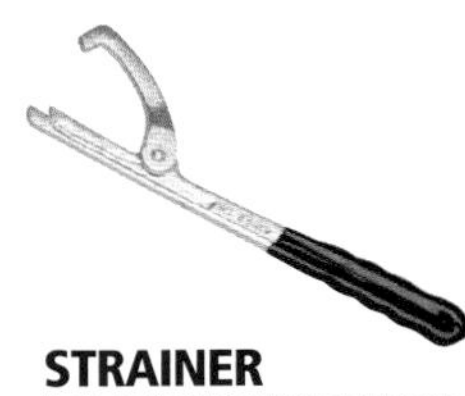

STRAINER LOCKNUT WRENCH

A specialty wrench with a locking jaw used to remove the basket strainer on a sink drain.

TWEEZERS

Small hand pliers that grasp, remove, and hold small objects.

SELF-ADJUSTING PLIERS

Provides an instant, positive grip for removing or tightening pipe.

SMALL WIRE BRUSH

Cleans debris and mineral deposits from pipe and fittings.

SPIRAL SAW

A high-speed cutting tool with a drill-like blade for cutting holes in countertops, drywall, and other soft surfaces.

STRAP WRENCH

Allows a secure hold to remove or tighten pipe or fittings without marring them.

UTILITY KNIFE

Cuts and trims a variety of materials.

SHOWER STEM SOCKET

Deep-throated sockets used to remove compression stems embedded in shower walls.

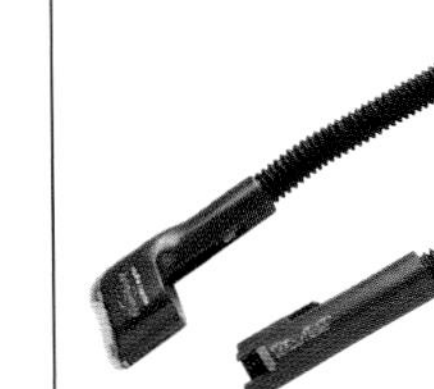

SNAKE FLASHLIGHT

A flexible flashlight that can be wrapped and positioned to direct light into hard-to-see places, freeing hands to work.

SPUD WRENCH

Specially designed tool to remove and tighten very large (2- to 4-inch) nuts by grabbing onto the lugs for increased leverage.

TEFLON TAPE

A ribbon of Teflon that wraps around threaded pipe, providing a secure seal.

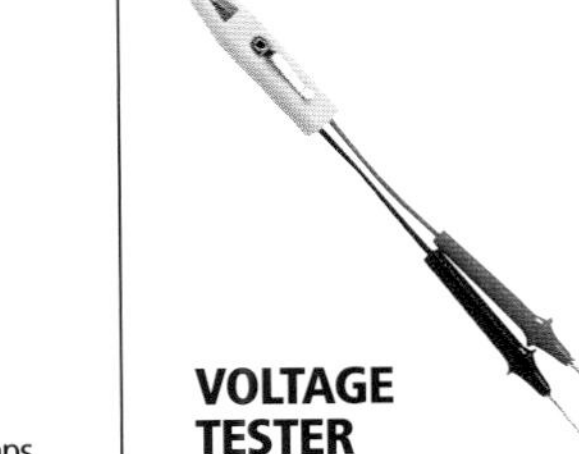

VOLTAGE TESTER

Checks for voltage across an electrical circuit.

SILICONE GREASE

Lubricates faucet and valve assemblies.

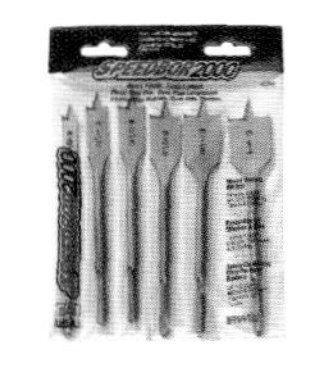

SPADE BITS

Specifically designed for wood. Bits range from ¼ to 1¼ inches. Use when running pipe through studs or joists.

SQUARE–CARPENTER'S

Used to lay out right angles and check corners for "square" (90 degrees).

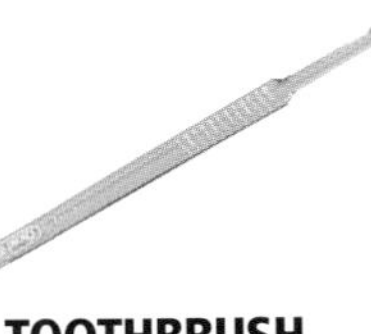

TOOTHBRUSH

Cleans screens of mineral deposits and debris.

WATER-PUMP PLIERS

A multipurpose gripping tool with ridged jaws for secure contact. Tape jaws to prevent damage to fragile fixtures.

SLIP PLIERS

The jaws adjust to two settings by slipping from a smaller setting to a larger one.

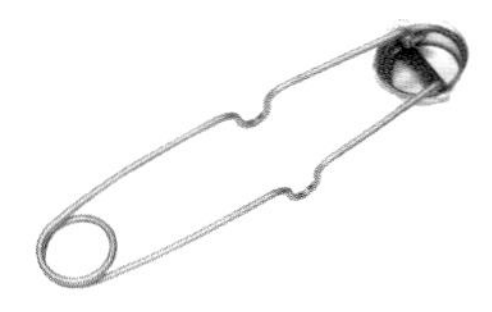

SPARK LIGHTER

A tool that produces a spark to light a propane torch for soldering copper pipe.

STEEL WOOL

Smooths and cleans metal surfaces. Graded by numerals with 000 being the finest.

TUBING CUTTER

Creates straight, smooth cuts in plastic or metal pipe.

CHAPTER

3 REPAIR AND MAINTENANCE

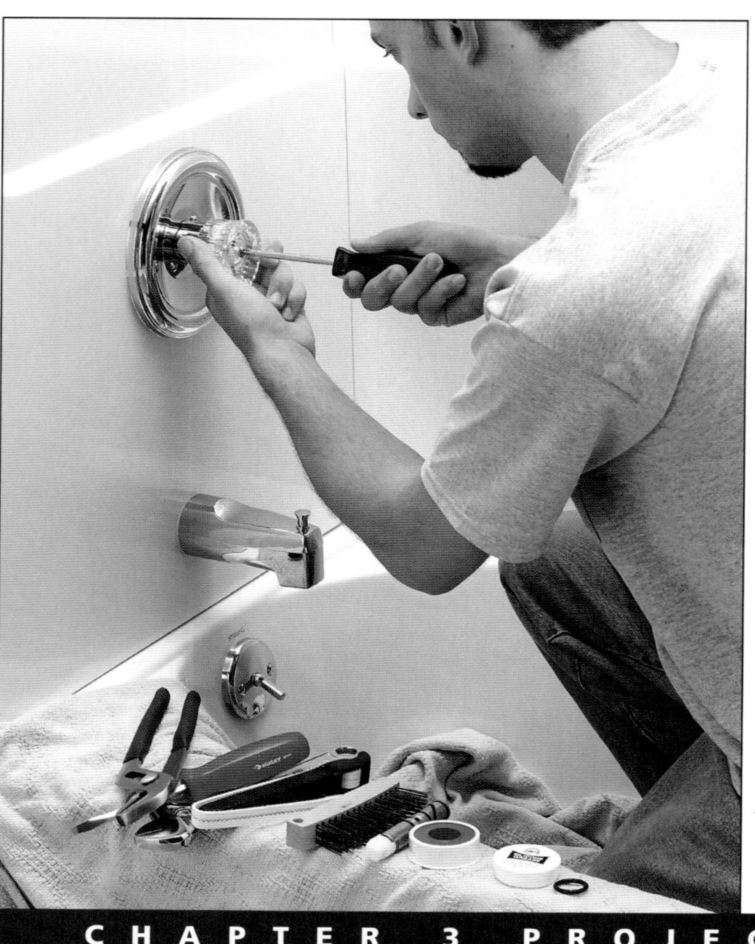

The repair and maintenance projects in this chapter will show you how to fix everything from sink and shower drains to jammed disposers and leaky toilets. Whatever your skill level, you'll find projects you can handle. As you solve basic plumbing problems, you'll gain the experience and confidence to tackle bigger, more complicated jobs. Plunge right into simple fixes such as adjusting a pop-up drain, curing sweating pipes, or replacing a toilet flapper. Repair a faucet, replace a sink strainer, or silence noisy pipes. Replace a water heater thermostat, clear clogged pipes, and fix slow-filling appliances.

Review "Emergencies—Quick Fixes for Leaking or Burst Pipes," page 68, so you'll be ready when it happens!

CHAPTER 3 PROJECTS

12 TIPS FOR GETTING THE JOB DONE RIGHT THE FIRST TIME.

1. Try the simplest solution first and then work up to the hard stuff. Don't take the toilet apart if you just have to use a plunger.
2. Study the project thoroughly before you dive in. Going step-by-step through each phase of the job will save time and give you the result you want. Prepare a shopping list so you won't have to run to the store at 9:30 on a Sunday evening.
3. Wear safety glasses and gloves. Taking the proper safety precautions when you do plumbing work is essential to you and your family's well-being.
4. Map your plumbing layout. Knowing where everything is will speed up your reaction time in an emergency and may save on expensive repairs and replacements.
5. Build your kit one tool at a time. Adding specialty tools to your basic kit will allow you to tackle more complicated projects, which will save you time and money.
6. Keep a repair and maintenance log. If you know what you've done and when you did it, you'll keep your home running smoothly and you'll eliminate some big surprises along the way.
7. Stick with code. Building and plumbing codes aren't in place to make your life miserable; they're there to make the systems in your home operate safely and efficiently. Follow them carefully.
8. Get advice from experts. Home centers and plumbing suppliers are more than places to buy washers and P-traps; take advantage of the years of practical experience walking around on the sales floor.
9. Buy the best fixtures and materials you can budget. It's a fact that you get what you pay for—the better the fixture, the longer it will last and the less time you'll spend maintaining it.
10. Don't put off minor repairs. What could have been a simple fix like replacing a leaking faucet can become a major (and expensive) headache over time.
11. Call in a pro when the water's over your head. Know your skill level and don't attempt jobs you're not prepared for.
12. Rent the big tools on a job-by-job basis. There's no point in spending the money on a tool like a commercial drain auger if you're only going to use it once.

CLEANING A FAUCET AERATOR, SINK SPRAYER, AND SHOWERHEAD

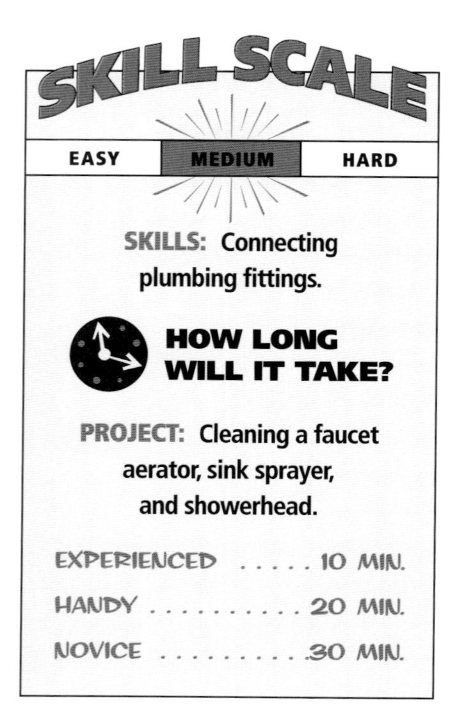

SKILL SCALE
EASY
MEDIUM
HARD
SKILLS: Connecting plumbing fittings.
HOW LONG WILL IT TAKE?
PROJECT: Cleaning a faucet aerator, sink sprayer, and showerhead.
EXPERIENCED ..... 10 MIN.
HANDY .... 20 MIN.
NOVICE .... 30 MIN.

Hard water is rich in minerals that build up and eventually clog aerators, sink sprayers, and showerheads. Removing the deposits by soaking them in a solution of 1 part vinegar and 1 part water or lime-dissolving solution and cleaning them with a small brush or paper clip will usually restore water flow. If your shower is clean but isn't running at the pressure you might like, it's usually because flow restrictors are required on showerheads to conserve water. Flow restrictors are required by code and by law and cannot be removed from the showerhead.

✓ STUFF YOU'LL NEED

TOOLS: Water-pump pliers, small wire brush, toothbrush, paperclip, small bowl

MATERIALS: Masking tape, white vinegar solution or lime-dissolving solution

GOOD IDEA

Get in the habit of setting out the parts of a fixture in the order that you removed them so you can easily put them back together. You'll have lots of little pieces that are easy to lose.

Wrap masking tape around the jaws of pliers to protect against damaging the surface of the aerator.

1 REMOVE THE AERATOR. Use a pair of water-pump pliers to unscrew the aerator from the faucet spout. Protect the surface of the aerator by wrapping the jaws of the pliers with masking tape. Be careful as you apply pressure to the pliers when removing the aerator. Too much pressure can crush the aerator cylinder.

2 DISASSEMBLE THE AERATOR. Remove the internal parts of the aerator by pushing them out with your fingers. If the components are stuck, presoak the aerator in a vinegar solution. Use a pick or tweezers, but be careful not to damage any of the parts. Once they're removed, lay out the parts and inspect them for damage. Damaged components will have to be replaced. Soak reusable parts in a vinegar solution overnight to loosen mineral deposits.

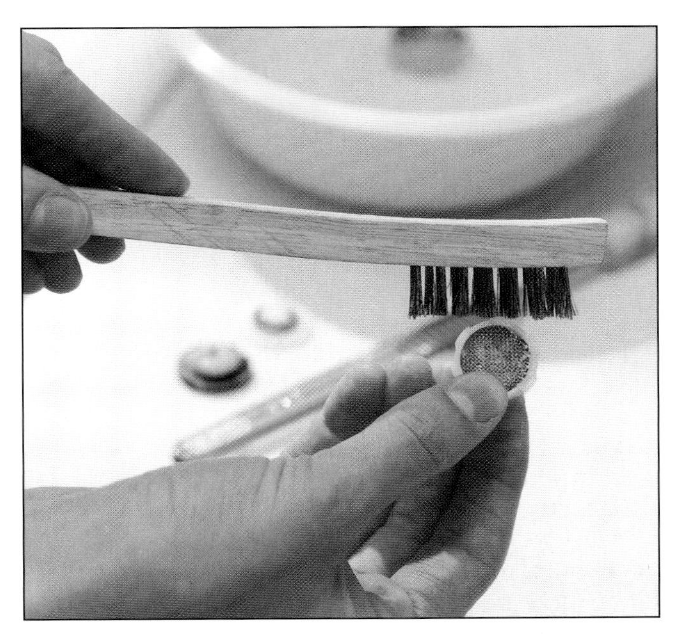

3 **CLEAN THE AERATOR PARTS.** Remove the parts from the vinegar solution and wipe them dry with a rag. Use a small wire brush to remove mineral deposits. It may be necessary to soak the parts again if they are not clean. Inspect all parts for damage after cleaning. Purchase a replacement kit or a new aerator if parts are damaged or if you are unable to clean any of the parts. Reassemble the aerator and screw it back into the spout hand-tight. Turn on the water and check for leaks. Tighten if necessary.

SHOWERHEAD

Flow restrictor
Swivel ball
Spray adjustment cam lever
Spray outlets
Collar
Washer
Body

CLEANING A SHOWERHEAD. Not all showerheads break down the same way, but what's shown above is a common example. Newer showerheads come with a water-conservation device called a flow restrictor that cannot be removed. Unscrew the swivel ball nut and remove the showerhead. Disassemble the internal parts. Soak overnight in white vinegar or a lime-dissolving solution. Use a small wire brush to clean mineral deposits. Use a paper clip to remove mineral deposits from holes in the disk. Inspect for damage and replace any damaged components. Reassemble the showerhead. Apply silicone grease to the shower arm threads and install the showerhead. Turn on the water and inspect for leaks.

SINK SPRAYER

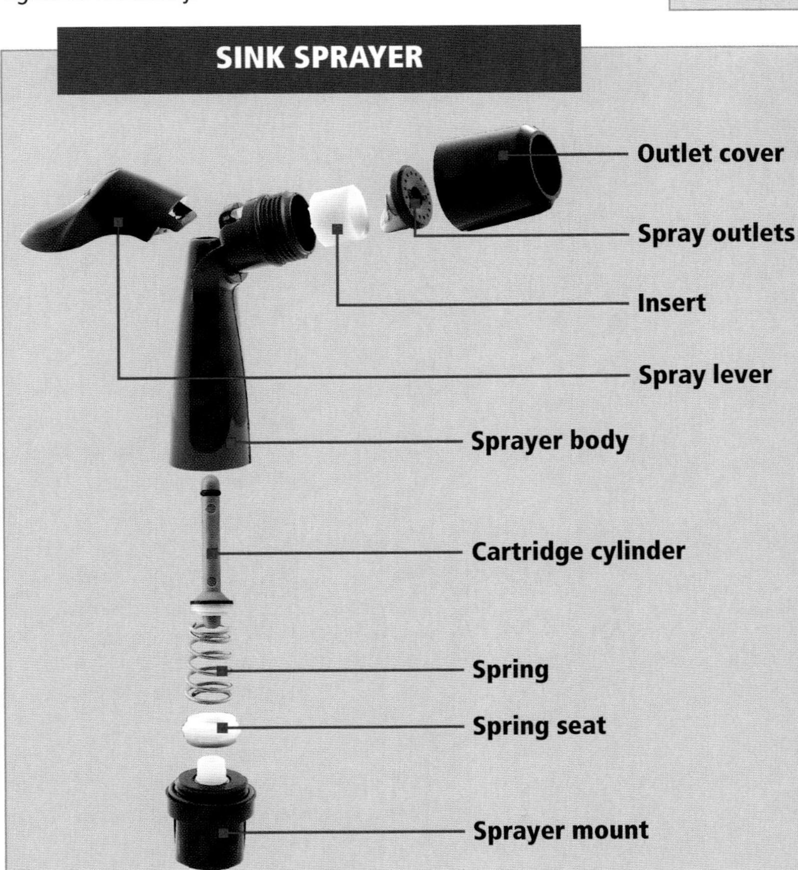

CLEANING A SINK SPRAYER. Pry off the sprayer head cover and remove the perforated disk. Soak the disk overnight in white vinegar or a lime-dissolving solution. Use a paper clip to clean out clogged holes and a small wire brush to remove remaining mineral deposits. Inspect for damage and replace any damaged parts. Reassemble.

REPAIRING A ROTARY BALL FAUCET

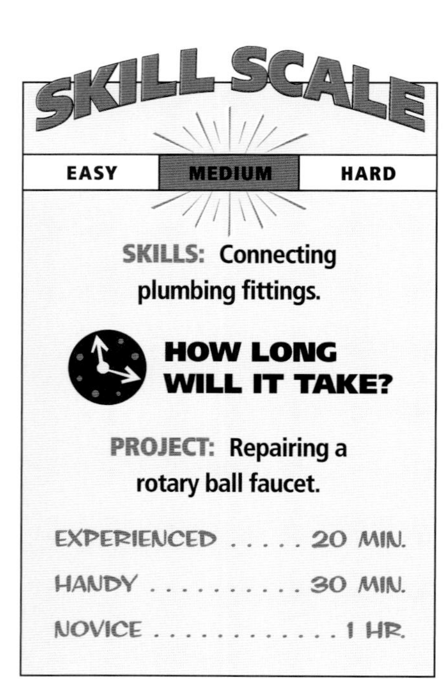

SKILL SCALE
EASY
MEDIUM
HARD
SKILLS: Connecting plumbing fittings.
HOW LONG WILL IT TAKE?
PROJECT: Repairing a rotary ball faucet.
EXPERIENCED ..... 20 MIN.
HANDY .... 30 MIN.
NOVICE .... 1 HR.

STUFF YOU'LL NEED

TOOLS: Hex set key wrench, water-pump pliers, tweezers, utility knife

MATERIALS: Masking tape, silicone grease, rotary ball repair kit, rotary ball faucet replacement ball if necessary

Rotary ball faucets have a smooth plastic or metal ball which rotates in the socket of the faucet body when the faucet handle is moved. Grooves in the ball pass over or away from the water inlet valve seats that control the amount and mixture of hot and cold water that is supplied to the spout.

Leaks at the spout or the handle can be easily fixed by tightening the housing or adjusting the ring. If this doesn't work, buy a replacement kit. If you aren't sure of the faucet type, take the parts with you to the store.

GOOD IDEA Don't throw away any old parts until you're finished with the job; you may want them for reference.

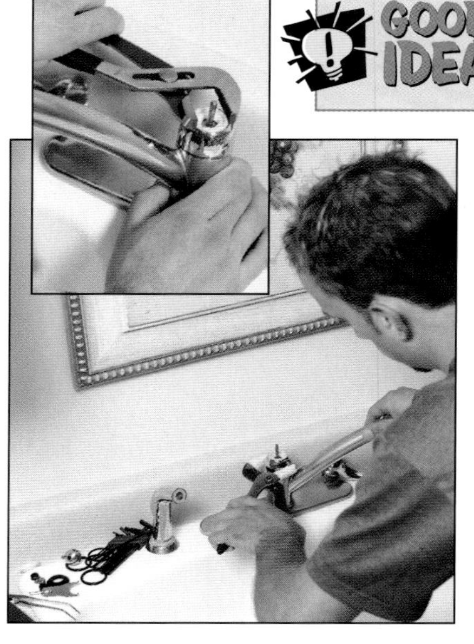

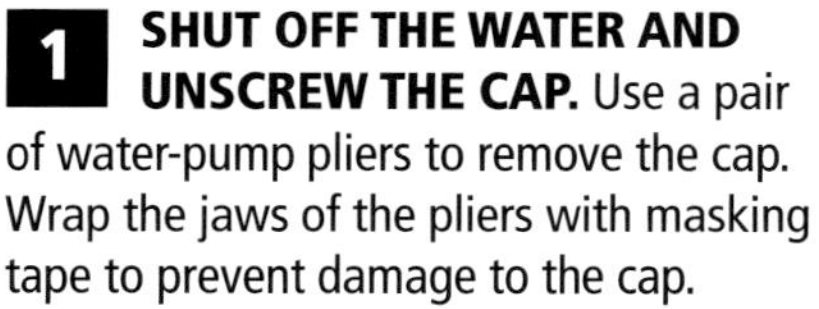

1 SHUT OFF THE WATER AND UNSCREW THE CAP. Use a pair of water-pump pliers to remove the cap. Wrap the jaws of the pliers with masking tape to prevent damage to the cap.

2 REMOVE THE CAM. Lift off the cam housing, seal, and ball. Inspect each part for damage. Replace damaged parts.

3 LIFT OUT THE SEATS. Use tweezers to remove the valve seats and springs. The springs are cone-shaped, with one end larger than the other. Note how the springs are installed before you lift them out—you must install the replacement springs in exactly the same order you remove them or the faucet won't work properly. Remove the spout by twisting and lifting at the same time.

To work properly, the springs need to be installed per manufacturer's instructions. Put the springs on a piece of tape in the order you took them out so you know exactly how they go back into place.

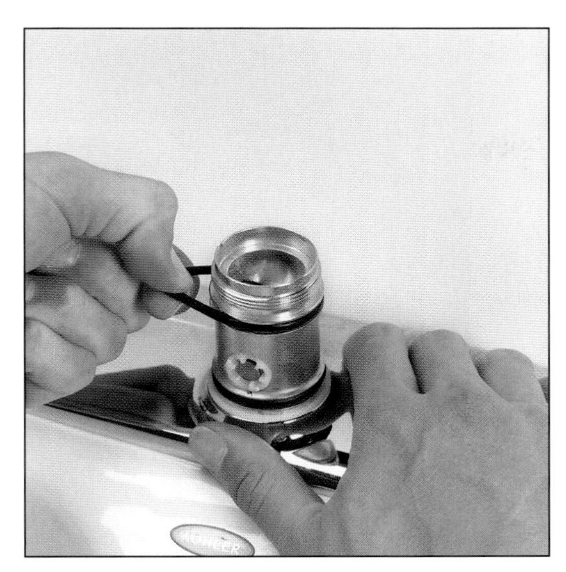

4 **PEEL OFF (DON'T CUT) THE O-RING.** Slip the O-ring from its groove and peel it from the housing. It's important to keep the ring whole so you can find an exact replacement at the store. Use the tip of a screwdriver to help release it if necessary. If you can't peel it off, pry it out with a screwdriver and cut the ring with a utility knife.

5 **REASSEMBLE THE FAUCET.** Take the old parts to your local home center to find the right replacement. Coat the new O-ring with silicone grease and seat it in the faucet housing groove. Push the spout over the O-ring and faucet housing. Install the new valve seals and springs. Make sure the seats and springs are installed correctly—they will only work correctly if installed properly, so read the installation instructions carefully. Place the cam ball in the housing. The tab on the cam fits into the notch in the body. Screw the cam housing onto the threads of the housing using the wrench that comes with the kit. Screw on the cap and install the handle. Turn the water on and check for leaks. Tighten the adjusting ring firmly to prevent leaks.

ROTARY BALL FAUCET

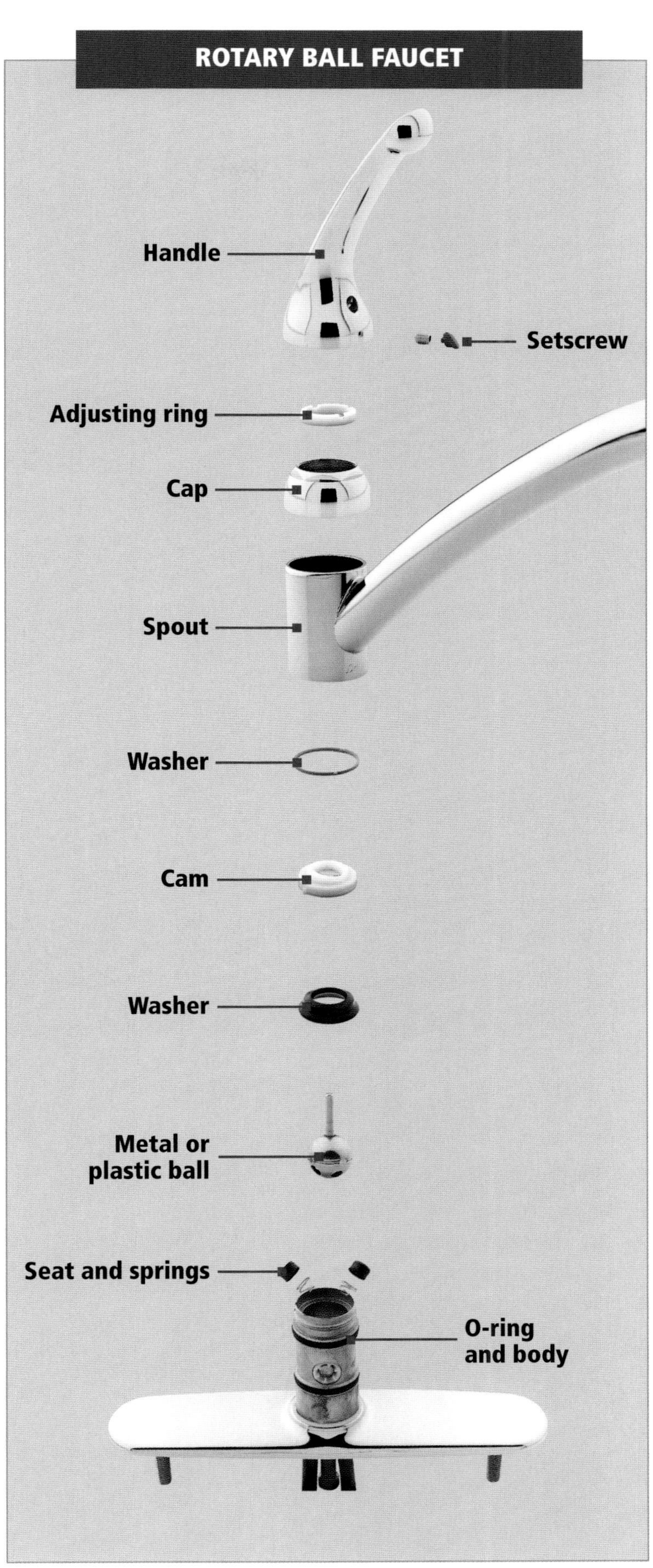

REPAIRING A CARTRIDGE FAUCET

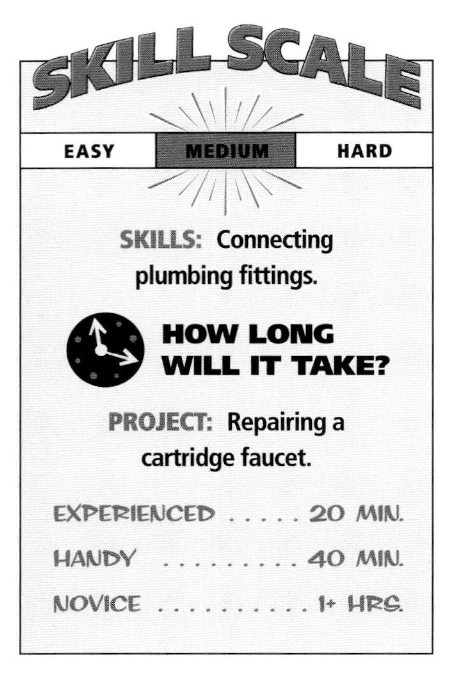

SKILL SCALE
EASY
MEDIUM
HARD
SKILLS: Connecting plumbing fittings.
HOW LONG WILL IT TAKE?
PROJECT: Repairing a cartridge faucet.
EXPERIENCED ..... 20 MIN.
HANDY .... 40 MIN.
NOVICE .... 1+ HRS.

TOOLS: Screwdriver, water-pump pliers, needle nose pliers, utility knife

MATERIALS: New cartridge, O-ring, silicone grease

Cartridge faucets control the flow by channeling water through passageways inside the cartridge. Depending on how the faucet is rotated, the cartridge exposes an opening for incoming water and allows water to flow to the spout with the desired mix of hot and cold. To stop the flow, the cartridge is rotated so an opening does not face the water supply. O-rings seal the interior housing of the cartridge body. Leaks occur when an O-ring is worn.

The hardest part of replacing a cartridge faucet may be finding the correct replacement kit. Even though the operating principle is the same, faucet designs vary. (See "Closer Look," below.)

CARTRIDGE FAUCET

Cap
Screw
Handle
Retaining ring
Clip
Cartridge
Spout
O-ring

CLOSER LOOK

THE CARTRIDGE JUNGLE.

Just in case you thought it was easy, cartridges come in many shapes and sizes. Take the one you want to replace to the store so you can get an exact match and save yourself an extra trip.

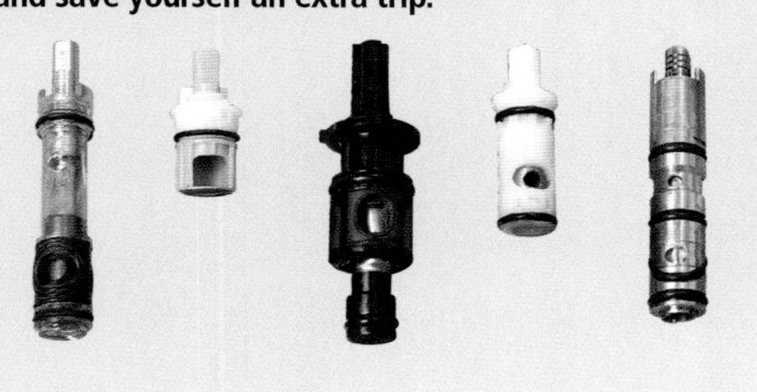

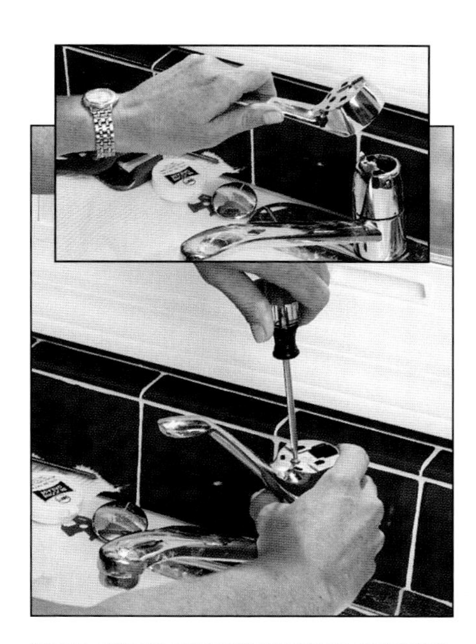

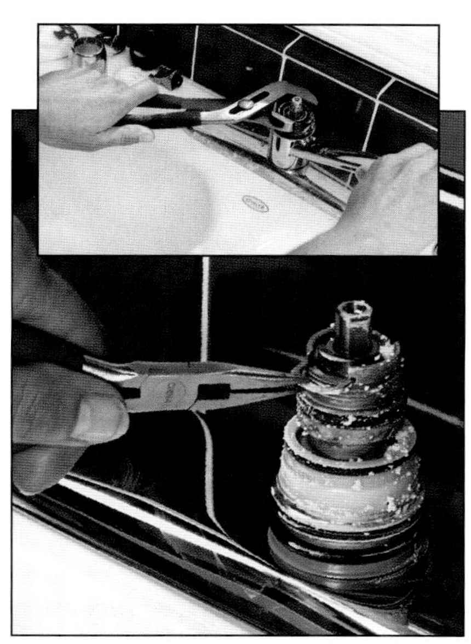

1 **TURN OFF THE WATER SUPPLY AND REMOVE THE FAUCET HANDLE.** Turn off the water at the shutoff valves or the main valve. Pry off the handle cap. Unscrew the faucet handle screw using a screwdriver. Lift the handle from the faucet assembly (See inset.)

2 **UNSCREW THE RETAINING NUT AND REMOVE THE RETAINING CLIP.** Spin the faucet out of the way and use water-pump pliers to remove the plastic retaining nut. (See inset.) Pull out the retaining clip that is just beneath it with needle nose pliers. Lift the faucet spout straight up from the faucet body and remove.

3 **PEEL OFF (DON'T CUT) THE O-RING.** Slip the O-ring from its groove and peel it from the housing. It's important to keep the ring whole so you can find an exact replacement at the store. Use the tip of a screwdriver to help release it if necessary. If you can't peel it off, pry it out with a screwdriver and cut the ring with a utility knife.

4 **PULL OUT THE CARTRIDGE STEM.** Grip the exposed end of the cartridge stem. Pull it straight up and out of the faucet body. Some brands of faucets may require a cartridge puller especially made for the job. Check the manufacturer's instructions. If that's the case, don't use pliers to remove the cartridge, or you might damage or destroy it.

5 **CLEAN AND REASSEMBLE THE FAUCET.** Clean the faucet body to remove debris. Coat the new O-ring with silicone grease to lubricate it, then seat it into the faucet body O-ring groove. Insert the new stem cartridge. Replace the faucet spout and reassemble. Turn on the water. Check the hot and cold water to make sure they are not reversed. If reversed, disassemble the faucet and rotate the cartridge 180 degrees, then reassemble.

LOOK BEFORE YOU LEAP!

There are many little pieces in a faucet, and they all need to be in exactly the right place for the faucet to work. It's easy to lose track of the assembly order and even easier to lose the parts themselves. Read the instructions carefully before you begin and arrange the parts on a flat surface as you remove them.

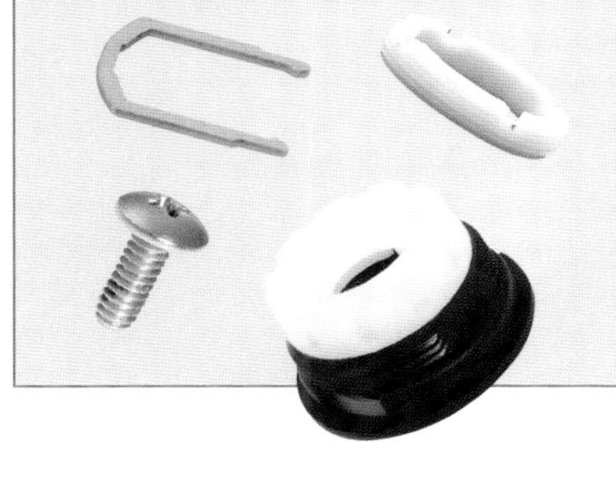

REPAIRING A CERAMIC DISK FAUCET

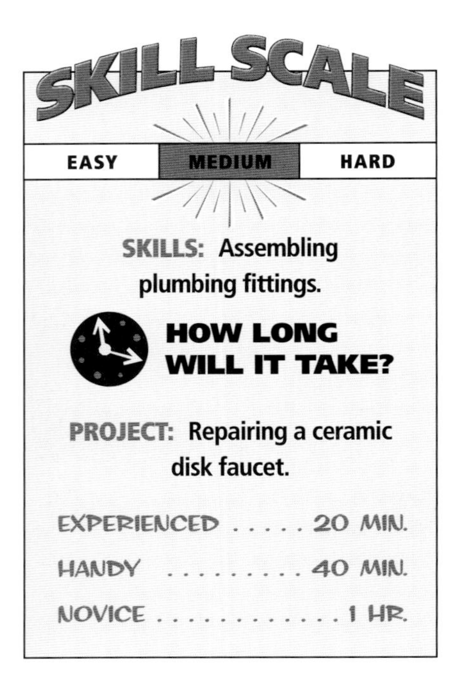

SKILL SCALE
EASY
MEDIUM
HARD
SKILLS: Assembling plumbing fittings.
HOW LONG WILL IT TAKE?
PROJECT: Repairing a ceramic disk faucet.
EXPERIENCED . . . . . 20 MIN.
HANDY . . . . 40 MIN.
NOVICE . . . . 1 HR.

STUFF YOU'LL NEED

TOOLS: Screwdriver or hex key set

MATERIALS: Ceramic disk faucet replacement kit with seals, abrasive pad, silicone grease

No shutoff valves below the sink? Think about installing them now. See "Installing Valve Stops and Shutoffs," pages 140–141.

Disk faucets rely on replaceable neoprene seals to ensure a tight water seal. The seals are set in the bottom of the disk. If water drips from the spout or pools around the top of the faucet during use, replace the seals.

To work on the faucet, first shut off the water supply. Turn on a faucet farther down the line to help drain water before you begin.

Most ceramic disk faucets come with a lifetime warranty for some of the parts. Check with the manufacturer for information on how to order.

CERAMIC DISK FAUCET

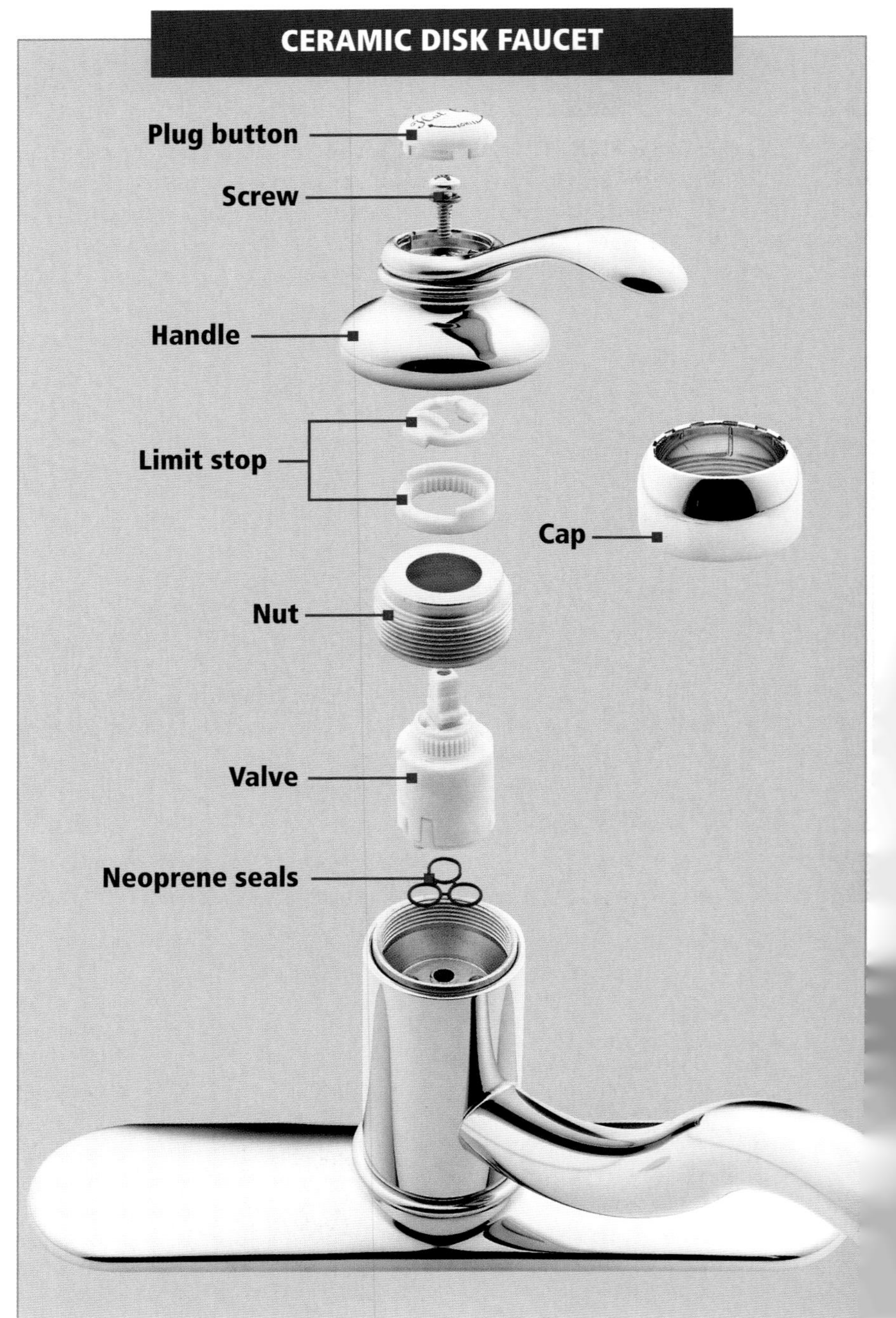

1 TURN OFF THE WATER AT THE SUPPLY OR MAIN SHUTOFF AND REMOVE THE FAUCET HANDLE. Use a hex key set or screwdriver to loosen the setscrew. Lift off the handle and dome housing.

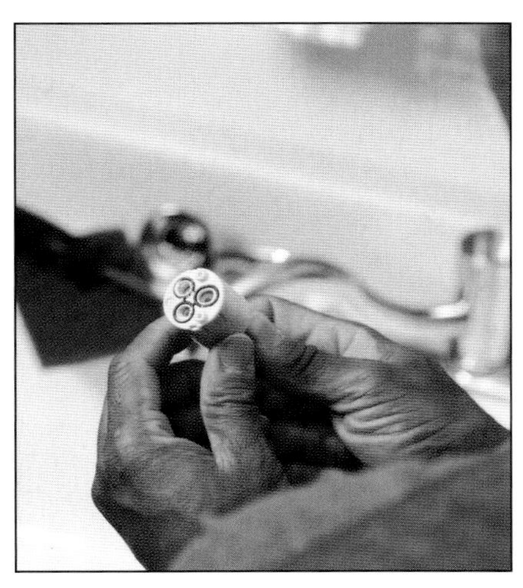

2 LIFT OUT THE DISK. Unscrew the disk cartridge screws with a screwdriver. Lift the disk up and out. Inspect the disk for cracks—replace the disk, if damaged.

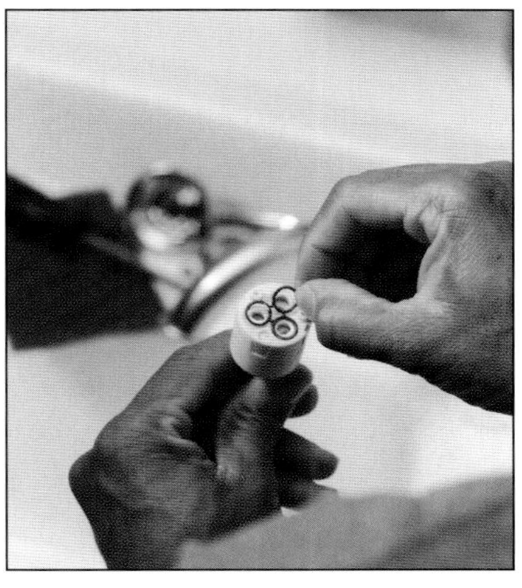

3 TAKE OUT THE SEALS. Take the disk and seals to your local home center to find the right replacement parts.

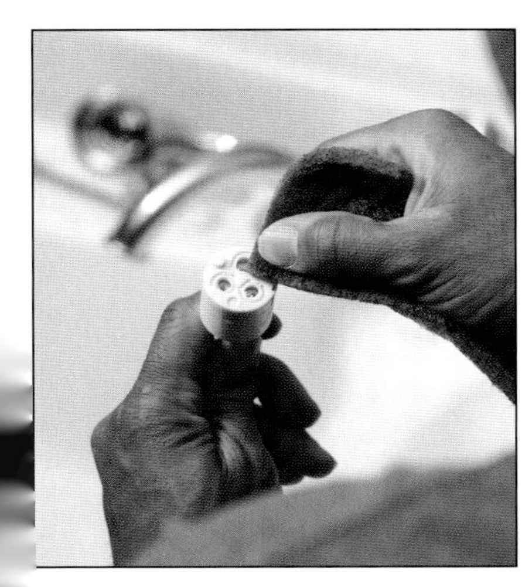

4 CLEAN SEAL SEATS. Use an abrasive pad to clean the seal seats.

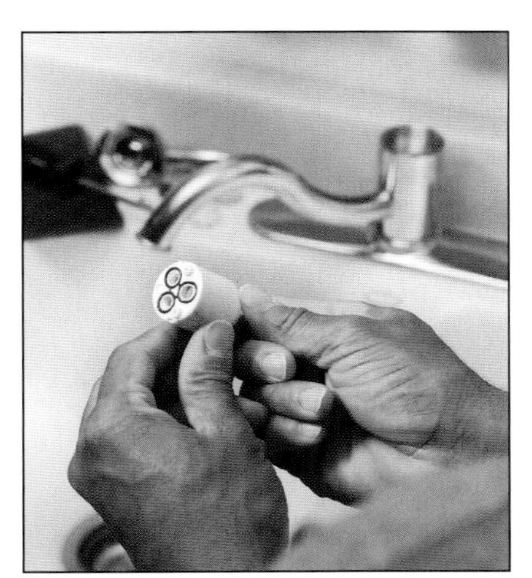

5 INSTALL NEW SEALS. Line up the disk seals with the faucet body and install the disk.

6 ASSEMBLE THE FAUCET. Install the escutcheon cap and handle, and tighten the setscrew. Remove air from the line before opening the supply valves fully. (See "Work Smarter," below.)

DON'T CRACK THAT CERAMIC DISK!

Air rushing through a ceramic disk can crack it. First, open the faucet in the center position to balance the flow of water, then gradually open the shutoff valves to bleed out the air. Don't turn off the faucet until water flows freely and all the air is out.

REPAIRING A COMPRESSION FAUCET

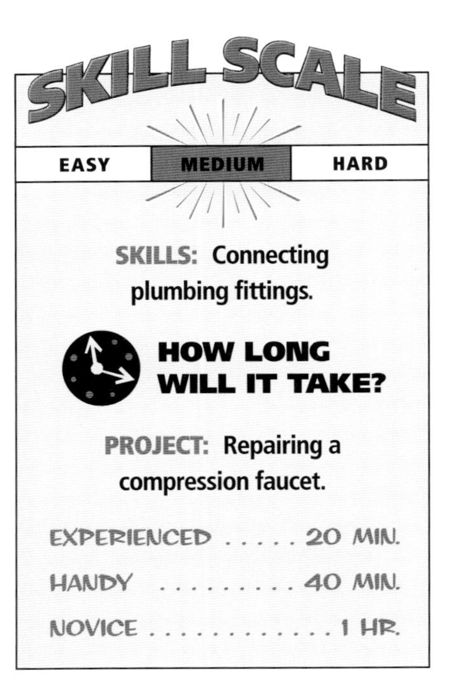

SKILL SCALE
EASY
MEDIUM
HARD
SKILLS: Connecting plumbing fittings.
HOW LONG WILL IT TAKE?
PROJECT: Repairing a compression faucet.
EXPERIENCED ..... 20 MIN.
HANDY .... 40 MIN.
NOVICE .... 1 HR.

Compression faucets have been around for more than a century. Rubber washers are compressed into the valve seat to stop the flow of water. Damaged seats and worn washers cause compression-stem faucets to leak from the spout.

Compression faucets are designed with separate hot and cold water valves. This flexibility allows for many design sizes and shapes.

Repair kits with everything you'll need are available at home centers. Take the parts to your local home center so you can buy the exact replacement.

✓ STUFF YOU'LL NEED

TOOLS: Screwdriver, water-pump pliers, utility knife

MATERIALS: Universal washer kit, O-ring, silicone grease

Never reuse an old washer, no matter how good it looks.

You may find packing string rather than an O-ring in some compression assemblies. If that's the case, you'll need to replace the packing string before you reassemble the faucet.

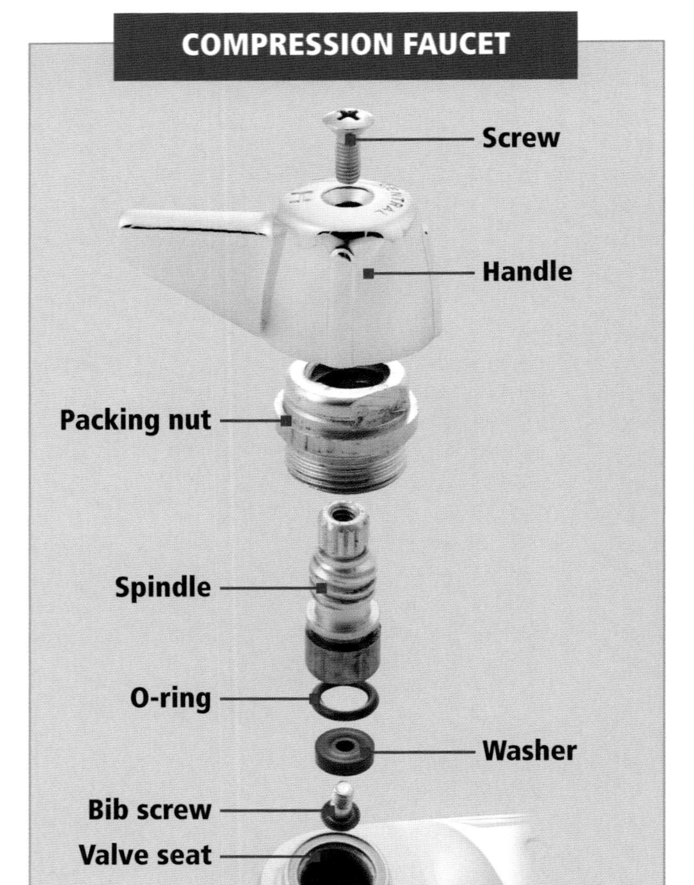

1 TURN OFF THE WATER SUPPLY AND UNSCREW THE STEM ASSEMBLY. Pry off the handle cap. Remove the handle screw with a screwdriver. Lift it up and off the handle. To remove corroded handles, you may have to use a handle puller (see "Closer Look" on page 25). Unscrew the packing nut from the faucet body with water-pump pliers.

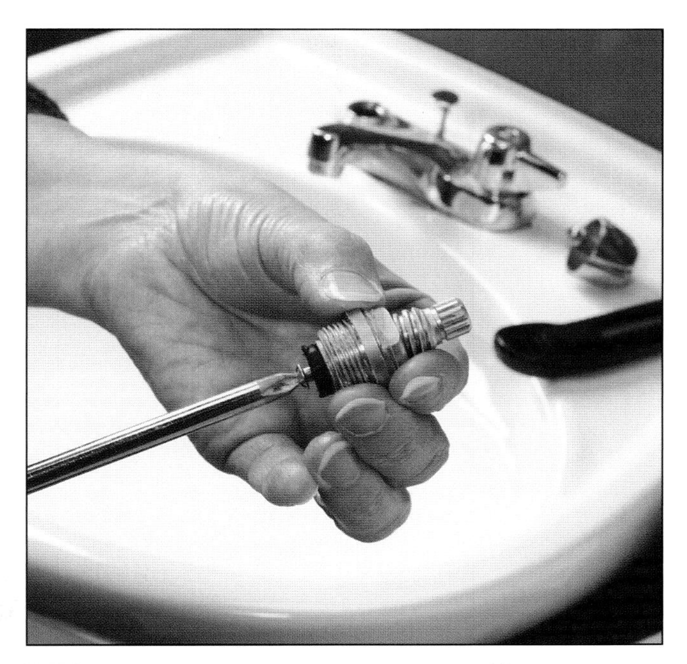

2 **TAKE OFF THE WASHER.** Use a screwdriver to remove the bib screw that holds the washer in place. Pry out the worn washer and discard it. The bib screw can deteriorate and should be replaced when a faucet is repaired. While the valve is out, examine the valve seat (see Compression Faucet, page 24) by touch. If you feel any roughness, replace it. (See "Resurfacing a Worn Valve Seat," page 27.)

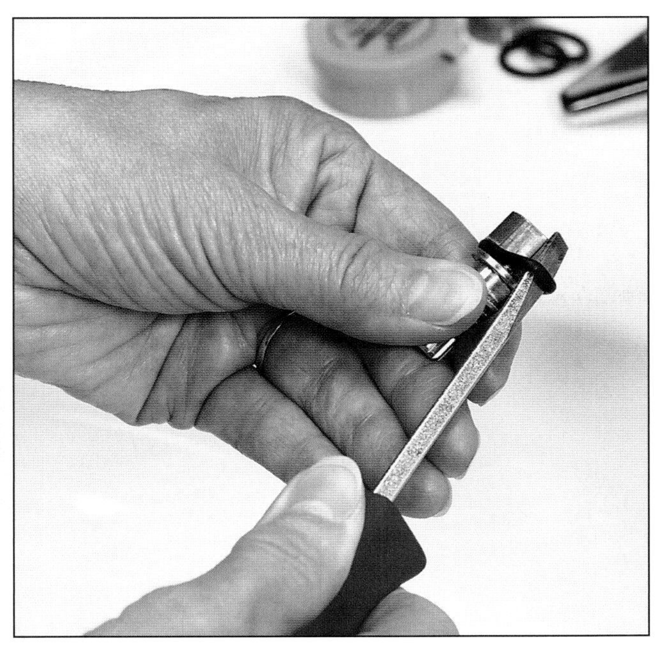

4 **PEEL OFF (DON'T CUT) THE O-RING.** Slip the O-ring from its groove and peel it from the housing. It's important to keep the ring whole so you can find an exact replacement at the store. Use the tip of a screwdriver to help release it if necessary. If you can't peel it off, pry it off with a screwdriver and cut the ring with a utility knife.

3 **REMOVE THE STEM FROM THE RETAINING NUT.** Inspect the threads for damage and replace the stem if necessary.

CLOSER LOOK

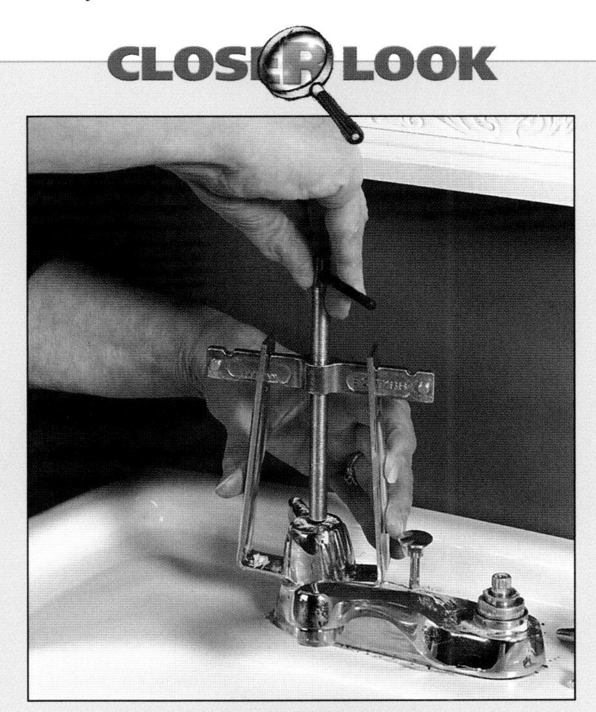

REMOVING A STUBBORN HANDLE.

Not able to budge that old handle? A handy tool to remove corroded handles is a handle puller. Clamp the side extensions of the handle puller beneath the handle. Thread the puller into the faucet stem. Continue to tighten the puller until the handle is free.

REPLACING A WORN VALVE SEAT

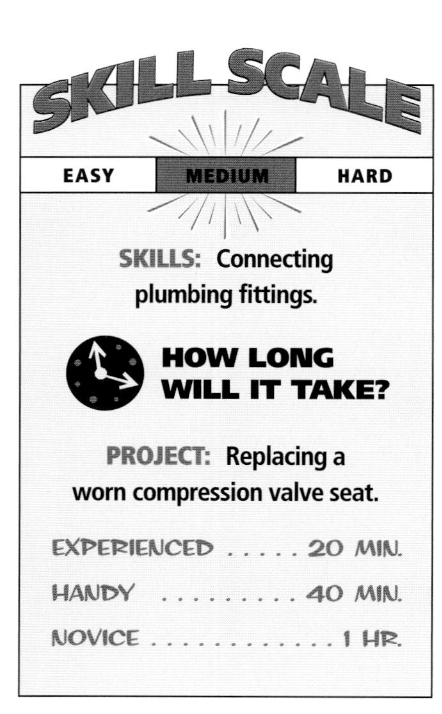

SKILL SCALE
EASY
MEDIUM
HARD
SKILLS: Connecting plumbing fittings.
HOW LONG WILL IT TAKE?
PROJECT: Replacing a worn compression valve seat.
EXPERIENCED ..... 20 MIN.
HANDY .... 40 MIN.
NOVICE .... 1 HR.

Only compression valves have valve seats. When repairing a compression valve, check to see if the valve seats need to be resurfaced. Poke your finger down into the faucet body to feel if the valve seat is rough. A rough seat will quickly damage a new washer. You should either replace or resurface the valve seat at the same time you are replacing the faucet washer. (See "Resurfacing a Worn Valve Seat," page 27.)

TOOLS: Screwdriver, water-pump pliers, seat wrench, seat cutter

MATERIALS: Valve seat, silicone grease, pipe dope

If you're replacing one valve seat, you might as well do the other. And since you've already got the assembly apart, replace the washers, bib screws, and O-rings on the valves.

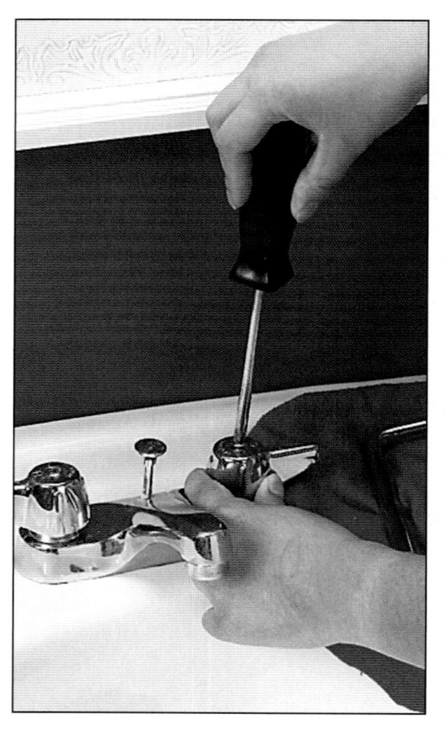

1 TURN OFF THE WATER SUPPLY AND DISASSEMBLE THE FAUCET VALVE. Lower the sink stopper and cover it with a cloth so loose parts won't fall into the drain. Pry off the handle cap. Remove the handle with a screwdriver.

2 REMOVE THE VALVE. Loosen the compression valve with water-pump pliers and remove the valve. Keep any washers or O-rings with the valve and set it aside.

FLUSH OUT THE FAUCET BODY.

Resurfacing the valve seat was an easy job, but I didn't realize the debris I ground off was still in the faucet body. There was enough inside the faucets to ruin the washer I had just replaced. Flush the system before you reassemble the faucet. Cover the hole with a rag and turn on the water gently to remove any debris or other gunk. You'll make a longer-lasting repair.

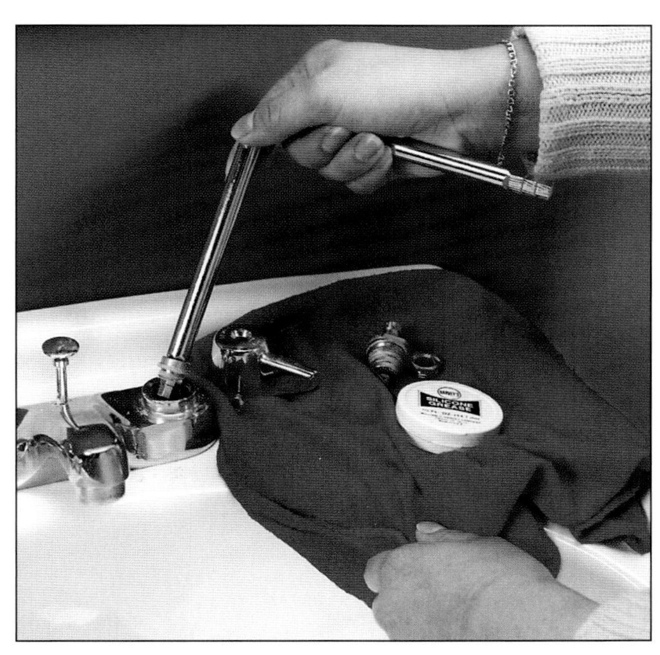

3 **BACK OUT THE SEAT.** Use a seat wrench to remove the valve seat. Select an end that will fit snugly into the seat. Insert the end of the wrench into the seat and tap the top to seat it firmly. The valve seat may be stuck, so the first turn should be quick and firm to release it without stripping the threads. Once it's loose, turn the wrench counterclockwise and remove it. Take the old seat to your local home center to be sure you replace it with the correct part.

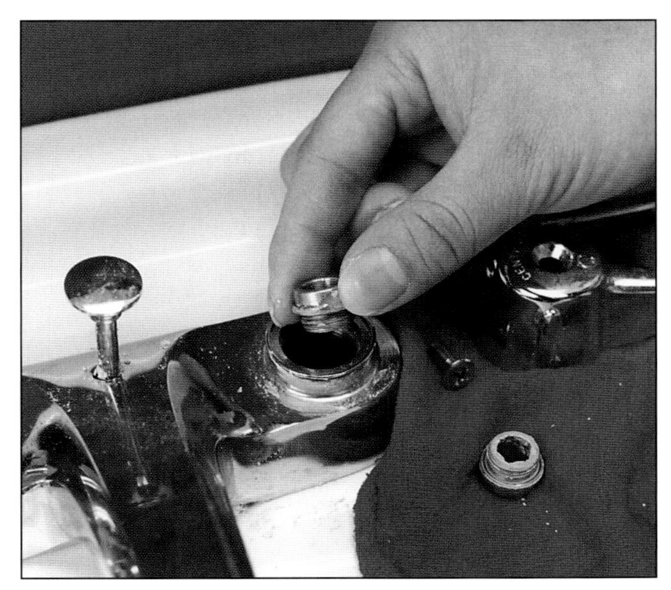

4 **REPLACE THE SEAT.** Install the new seat into the faucet. Apply pipe dope to the seat threads to seal them; insert the end of the seat wrench into the seat, and set the seat in place. Screw the valve into place. Assemble the compression valve, faucet handle, and handle cap. Turn on the water supply and check for leaks.

CLOSER LOOK

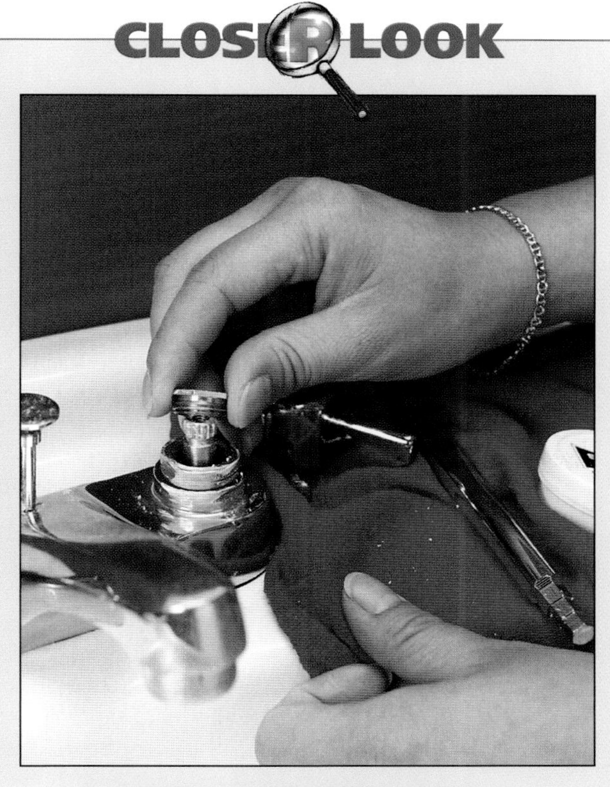

RESURFACING A WORN VALVE SEAT.

Unable to remove the valve seat? You don't have to replace it. Worn valve seats can be resurfaced using a seat cutter that can be purchased at a home center or hardware store. Select a seat cutter that will fit snugly inside the retaining nut. Slide the cutter and retaining nut over the threaded end of the valve seat dressing tool. Attach the locknut and cutter head to the shaft of the tool. Carefully screw the retaining nut into the faucet body. To resurface the seat, lightly press down on the handle while turning it clockwise two to three complete turns. This should be enough to resurface the seat. Remove the tool and inspect the seat by feeling it to make sure it's smooth. If smooth, reassemble the faucet.

REPAIRING A SINGLE-HANDLE TUB AND SHOWER FAUCET

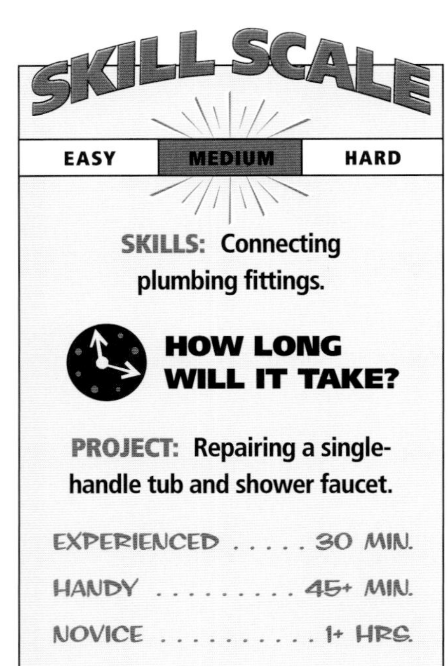

SKILL SCALE
EASY
MEDIUM
HARD
SKILLS: Connecting plumbing fittings.
HOW LONG WILL IT TAKE?
PROJECT: Repairing a single-handle tub and shower faucet.
EXPERIENCED . . . . . 30 MIN.
HANDY . . . . 45+ MIN.
NOVICE . . . . 1+ HRS.

Single-handle faucets for tubs can be of disk, cartridge, or rotary ball design. Whatever the style, one handle controls both the water flow and the amount of hot and cold water.

Diverter gates direct the flow of water to the spout or showerhead. The diverter gate seldom requires repair. When it does need repair, it's usually because the lever won't stay up or the lever has been broken. Don't fix it; replace the spout. (See "Installing a Tub Spout," page 115.)

In some cases it's easier to replace the entire cartridge than it is to repair it, and cartridges often come with lifetime warranties. Check with your home center or the manufacturer.

STUFF YOU'LL NEED

TOOLS: Screwdriver, strap wrench or taped water-pump pliers, adjustable wrench, small wire brush, plastic putty knife

MATERIALS: O-ring, silicone grease, thread dope or Teflon tape, replacement cartridge kit, silicone caulk

DEBRIS IN SUPPLY LINES IS A BAD THING.

It's a law of plumbing that dirt and gunk will find their way into any opening. Debris in your pipes can damage fixtures and cause blockages. Before you install a new faucet or fixture, turn on the water for a second to flush out the line. (You can cover the opening with your hand or a rag so you don't spray the room.) An alternative is to flush the toilet or turn on an outside hose bib. This will also bleed the system of dirt and air.

Know the locations of the main shutoff valve or the supply valves for each fixture in your home.

1 TURN OFF THE HOT AND COLD WATER SUPPLY VALVES OR THE MAIN SHUTOFF, AND REMOVE THE HANDLE. Pry off the handle cap. Remove the screw holding the handle in place. If there are screws holding the escutcheon cap in place, remove them as well.

2 SLIDE OFF THE ESCUTCHEON CAP. If the edge is sealed with caulking, pry off the escutcheon using a plastic putty knife. Clean the old caulking off the cap with the putty knife and set aside.

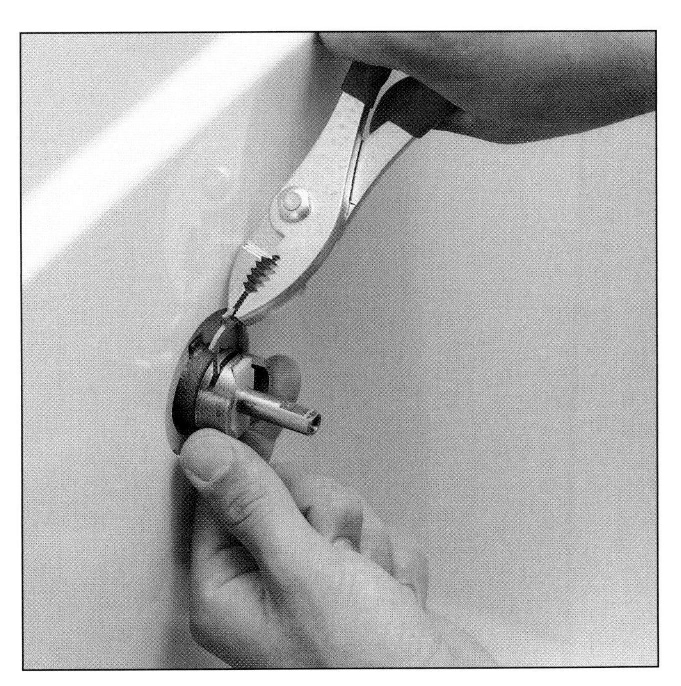

3 REMOVE THE RETAINING CLIP. You can't get the cartridge out without first removing the retaining clip. Remove the sleeve, grip the clip end with a pair of pliers, and pull it out.

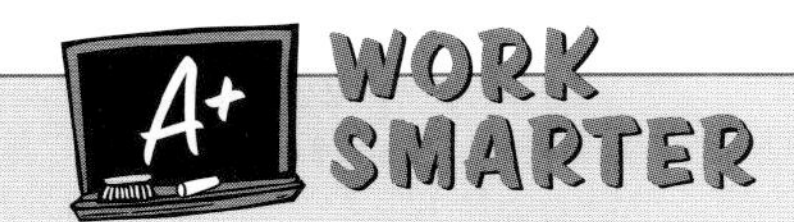

SAVE THE TUBS!
PREVENT LOSING PARTS!

Don't assume that a screw or a small part can't slip through a screen or drain opening, or that a pair of water-pump pliers won't chip the surface of your tub. Work smarter by:

1. **Protecting your tub from cracking or chipping by covering the surface with a piece of carpet, a blanket, or a towel.**
2. **Covering or blocking the drain opening with a rag or a piece of duct tape.**
3. **Closing any plunger-style drains and covering them with a screen.**

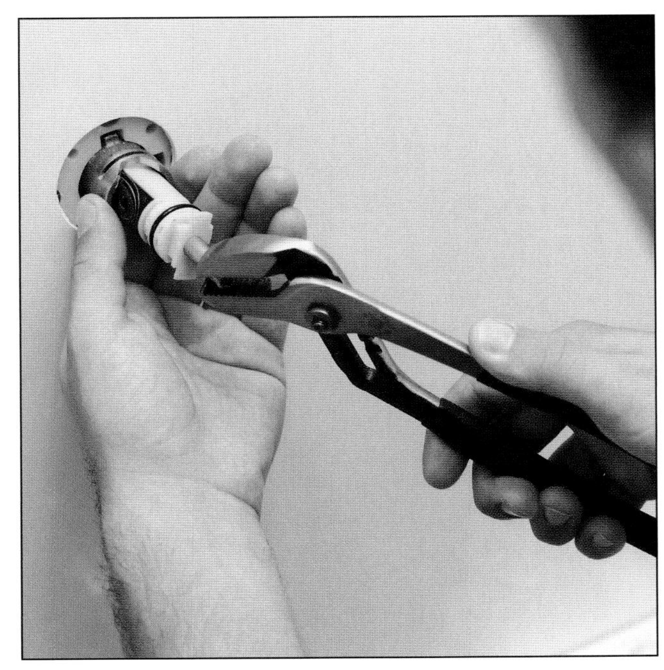

4 PULL THE CARTRIDGE STRAIGHT OUT. Grab the end of the faucet cartridge with a pair of pliers. Pull it straight out of the housing. If the cartridge gets stuck, apply white vinegar with a small brush to dissolve lime. Take the cartridge with you to your local home center so you can purchase an exact replacement.

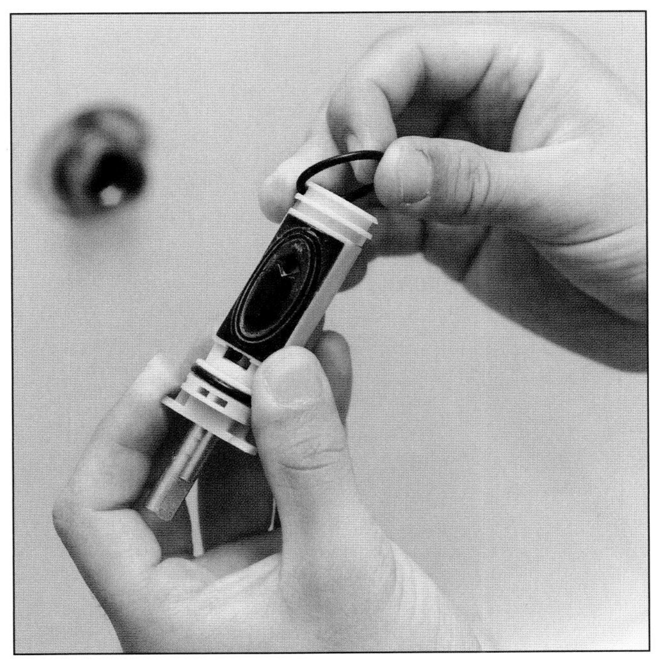

5 REPLACE THE O-RING. Slide off the old O-ring. Coat the new O-ring with silicone grease. Install it on the cartridge. Reassemble and test the faucet. The hot feed should be on the left; if it isn't you've reversed the cartridge. Disassemble the faucet and turn the cartridge over 180 degrees. Don't forget to recaulk the escutcheon plate.

REPAIRING A TWO-HANDLE TUB AND SHOWER FAUCET

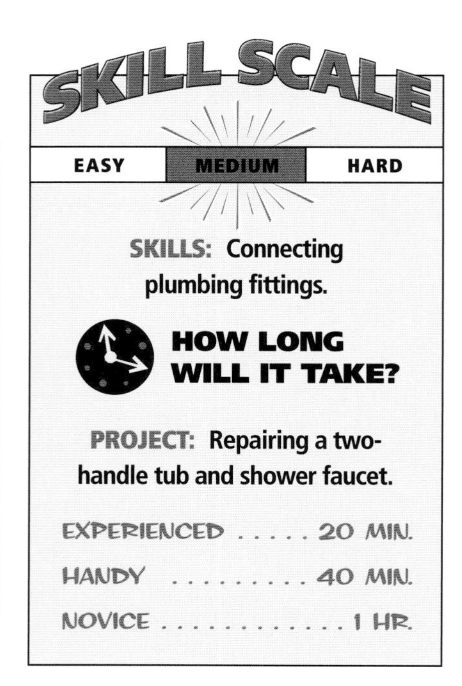

SKILL SCALE
EASY
MEDIUM
HARD
SKILLS: Connecting plumbing fittings.
HOW LONG WILL IT TAKE?
PROJECT: Repairing a two-handle tub and shower faucet.
EXPERIENCED ..... 20 MIN.
HANDY .... 40 MIN.
NOVICE .... 1 HR.

✓ STUFF YOU'LL NEED

TOOLS: Screwdriver, plastic putty knife, shower stem socket kit, pliers, utility knife, flashlight, allen wrench (for set screws)

MATERIALS: O-ring, cartridge replacement kit, silicone grease, white vinegar or lime-dissolving solution

Know the locations of the main shutoff valve or the supply valves for each fixture in your home.

A two-handle tub and shower faucet has separate handles to control the water flow and the mixture of hot and cold water. One handle controls the flow of hot water and the other controls the flow of cold water. The faucet can have either a compression or cartridge valve.

To divert the water flow from the spout to the showerhead, lift the lever for the diverter gate located on the spout. If the diverter gate doesn't work, replace the spout. (See "Installing a Tub Spout," page 115.)

Replacing one cartridge? Replace them both; you'll save time and effort.

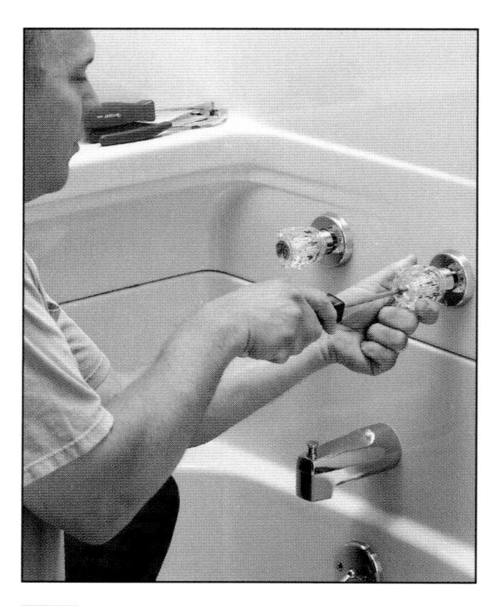

1 TURN OFF THE WATER SUPPLY AND REMOVE THE HANDLE. Pry off the handle cap. Use a screwdriver to remove the handle.

2 PRY OFF THE ESCUTCHEON. If there is caulking around the escutcheon, use a plastic putty knife to remove the caulk. Slide the escutcheon off the stem.

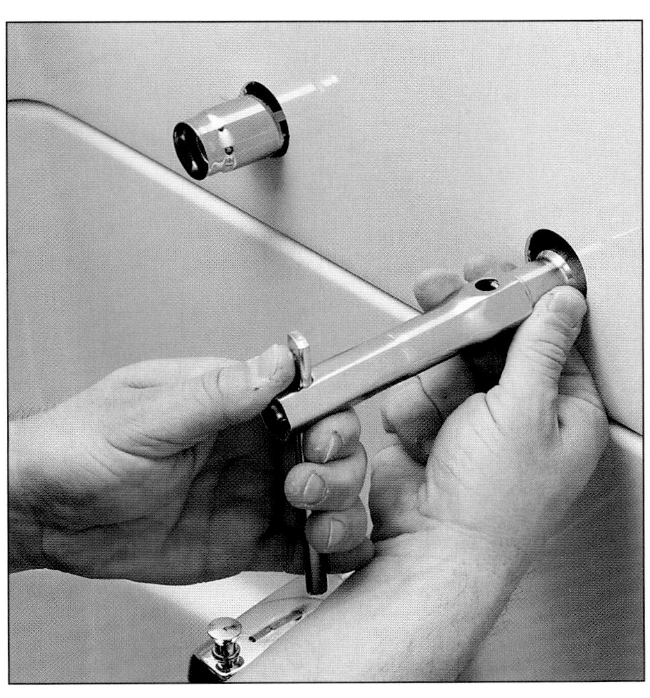

3 **UNSCREW THE RETAINING NUT.** Use a shower stem socket (available at home improvement centers and hardware stores) to remove the retaining nut.

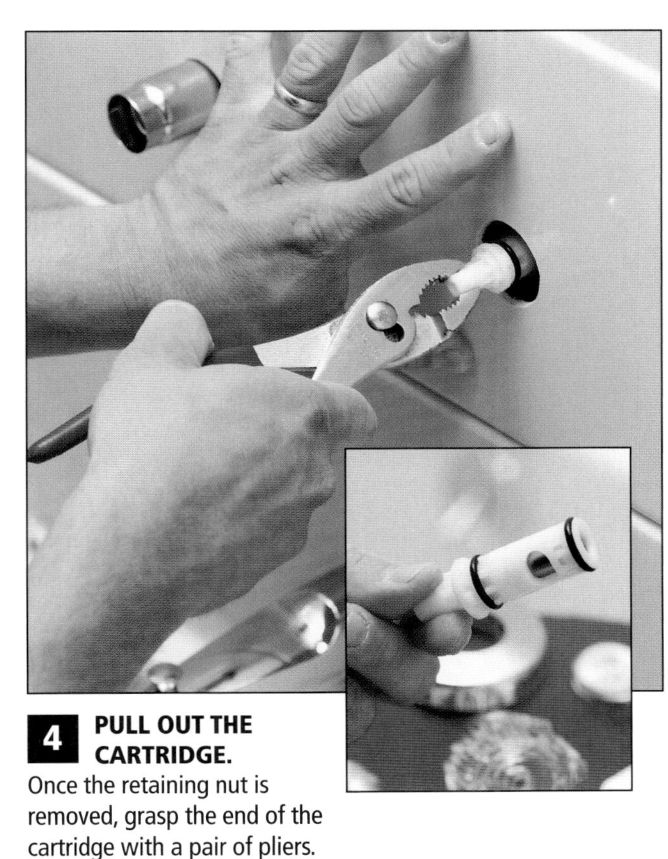

4 **PULL OUT THE CARTRIDGE.** Once the retaining nut is removed, grasp the end of the cartridge with a pair of pliers. Pull it straight out.

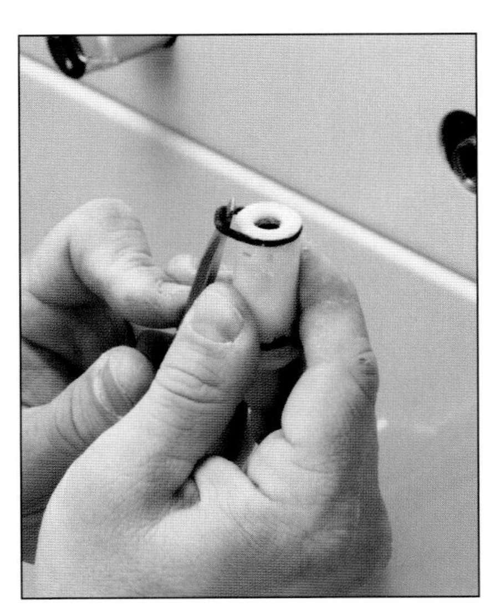

5 **PEEL OFF (DON'T CUT) THE O-RING.** Slip the O-ring from its groove and peel it from the housing. It's important to keep the ring whole so you can find an exact replacement at the store. Use the tip of a screwdriver to help release it if necessary. If you can't peel it off, pry it off with a screwdriver and cut the ring with a utility knife.

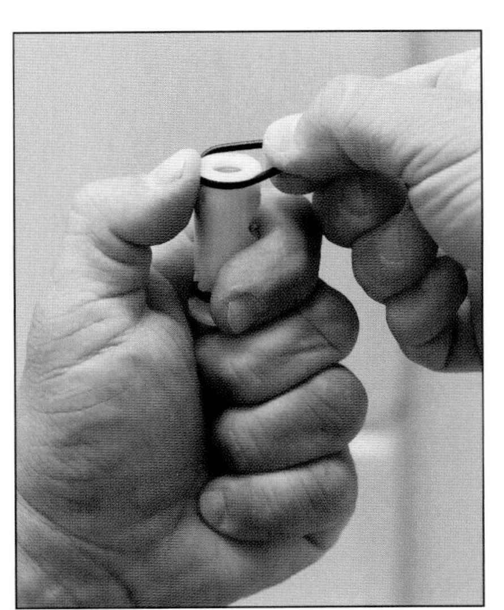

6 **SLIDE ON THE NEW O-RING.** Apply silicone grease to the new O-ring. Slide it over the cartridge, seating it into the O-ring groove.

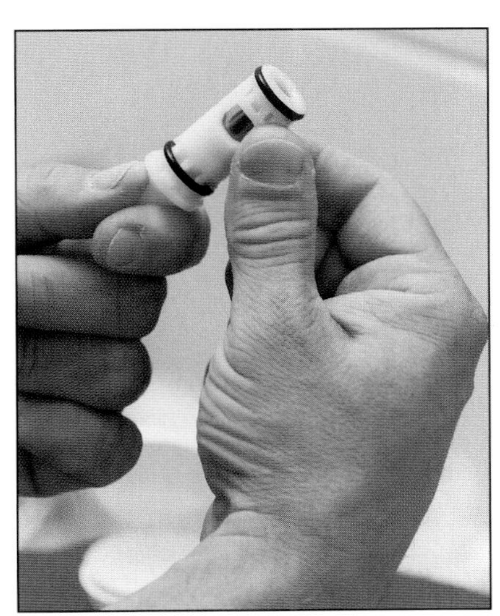

7 **ASSEMBLE THE FAUCET.** Turn on the water supply and test for leaks.

REPAIRING A THREE-HANDLE TUB AND SHOWER FAUCET

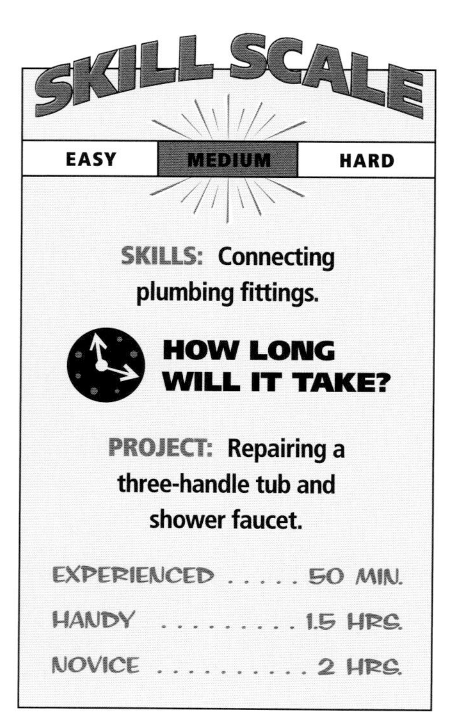

SKILL SCALE
EASY
MEDIUM
HARD
SKILLS: Connecting plumbing fittings.
HOW LONG WILL IT TAKE?
PROJECT: Repairing a three-handle tub and shower faucet.
EXPERIENCED ..... 50 MIN.
HANDY .... 1.5 HRS.
NOVICE .... 2 HRS.

A third or middle handle diverts the flow of the water from the spout to the showerhead. The diverter valve can be either a compression or cartridge type. Compression valves can be easily repaired. Cartridge valves that are worn or damaged will have to be replaced.

Separate handles control the water flow and the mixture of hot and cold water. As with the two-handle tub and shower faucet, the valves in the faucet can have either compression or cartridge valves.

The hot side is often the first to go so when pros do the job, they will repair or replace the hot, cold, and diverter while they've got everything apart. You should, too.

✓ STUFF YOU'LL NEED

TOOLS: Screwdriver, stem socket set, utility knife, water-pump pliers

MATERIALS: Compression replacement kit, silicone grease, white vinegar solution, and toothbrush

CLOSER LOOK

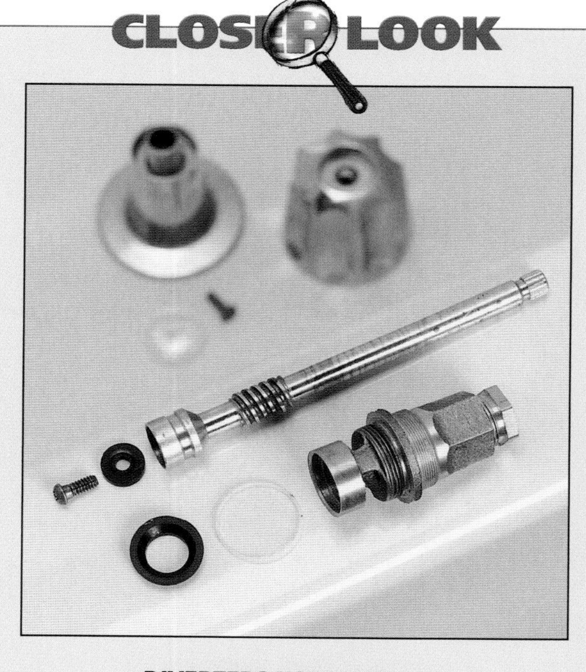

DIVERTERS USUALLY HAVE COMPRESSION FITTINGS.

Parts for a compression diverter include the handle cap, handle, escutcheon, stem, bonnet nut, washers, and screws.

1 REMOVE THE HANDLE AND ESCUTCHEON CAP. Place a blanket in the tub to protect it from dropped tools. Shut off the water supply. Pry the cap off the diverter handle. Remove the screw with a screwdriver and slide off the handle. The escutcheon may just slide off, or you may have to unscrew it to remove it. If using water-pump pliers or a strap wrench to remove the escutcheon, wrap the jaws of the pliers with masking tape.

2 DISCONNECT THE BONNET NUT. Use a shower stem socket set to remove the bonnet nut and stem.

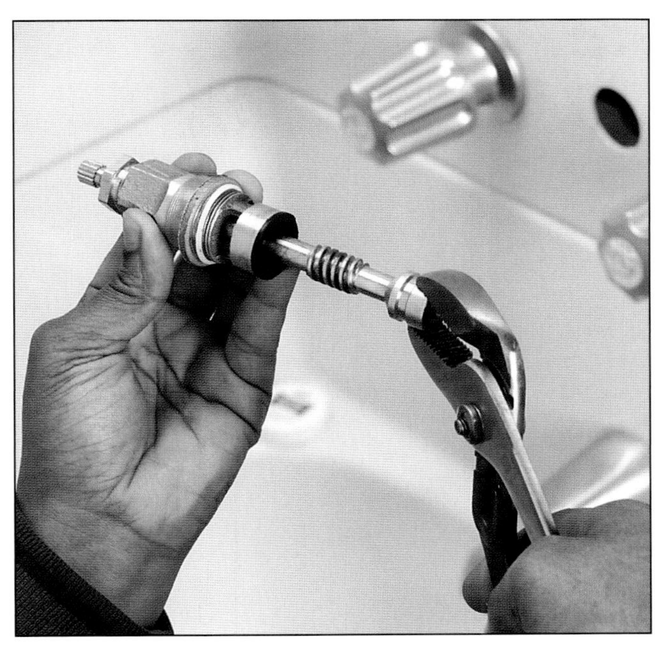

3 UNSCREW THE STEM FROM THE BONNET NUT. Hold the bonnet nut with one hand. Use your other hand to remove the stem from the bonnet nut with water-pump pliers. Inspect the threads of the stem for damage. If there is evidence of damaged threads, replace the stem. Check the valve seat with a flashlight to make sure it's OK. It may need resurfacing or replacement. (See "Resurfacing a Worn Valve Seat," page 27.)

Know the locations of the main shutoff valve or the supply valves for each fixture in your home.

Look carefully at the replacement washer. If it has markings on one side, put that side into the stem.

4 REMOVE THE BIB SCREW. Clean mineral deposits on the stem with a white vinegar solution or buy a solution to dissolve mineral deposits at your local home center. A toothbrush easily removes deposits.

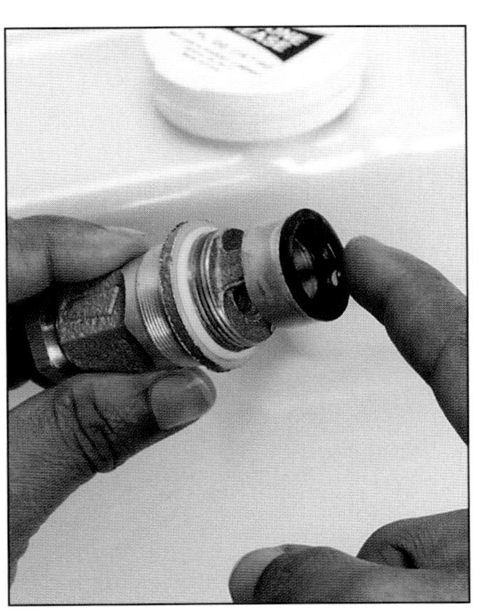

5 REPLACE THE OLD WASHER. Seat the new washer in the stem. Apply silicone grease to the washer so it seats properly. Replace the bib screw with a new one. Reassemble the faucet. Turn on the water supply and check for leaks and drips.

TILE CAN BE TRICKY!

If your shower is finished in tile—especially in an older home—replacing faucets can be a little more difficult. That's because the controls may be set way back in the wall so they are more difficult to reach. You may have to knock out enough tile to get the access you need to do the job. Replacing those tiles can become part of the project.

REPLACING A FAUCET AND SINK SPRAYER

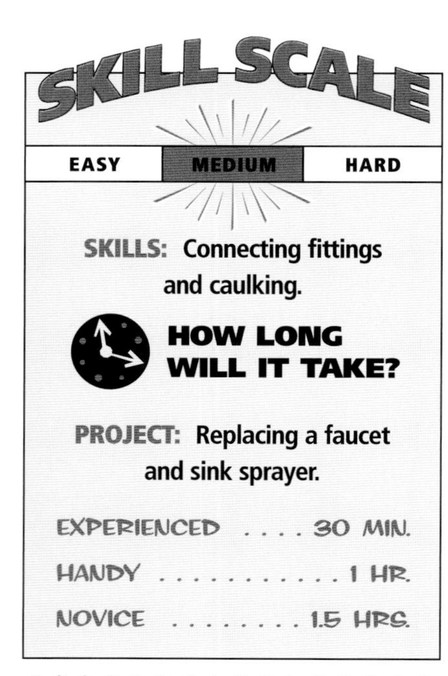

STUFF YOU'LL NEED

TOOLS: Flashlight, adjustable wrenches, basin wrench, utility knife

MATERIALS: New faucet, flexible supply lines (braided), plumber's putty, pipe compound or Teflon tape, silicone caulk, penetrating oil, bucket, pan, or rags

WORK SMARTER

USE THE RIGHT PIPE SEALANT.

When you're using a pipe or thread sealant like silicone, pipe compound, Teflon tape, or plumber's putty, read the label to make sure it's compatible with the materials in the sink or fixtures. Plumber's putty, for instance, will dissolve plastic parts and rubber fittings and discolor cultured marble sinks.

Living with a faucet that drips or constantly leaks is bad for the environment, your pocketbook, and, on top of that, it's annoying. Wear and tear or a kitchen remodel can prompt a faucet replacement. Anyone can replace a faucet and sink sprayer. The difficult part of the job often is removing the old fixtures. Corrosion can stiffen connections, and working under a sink isn't exactly convenient.

CHOOSE THE NEW FIXTURES

Kitchen faucet-sprayer systems come in many types and styles. Just make sure you select one that will fit your sink. Measure the distance between the centers of the sink openings, then measure the diameters of the openings.

PREPARE THE WORK AREA

Here's how to get ready to go to work:

✔ **Purchase all the fixtures and gather all the tools you'll need for the job.** Lay them out near the installation so you don't have to keep looking for what you need instead of staying on the job.

✔ **Clear all the detergents, rags, and kitchen gear from the cabinet.** Get under the sink with a flashlight and inspect the job site.

✔ **Turn off the hot and cold water supply lines at the shutoff valves beneath the sink.** If there are no shutoff valves, you'll need to turn off the main water supply. Install shutoffs as part of this project (See pages 140–141).

✔ **Place a bucket, pan, or rags below the work area.** Even with the water off, there will be some water in the lines.

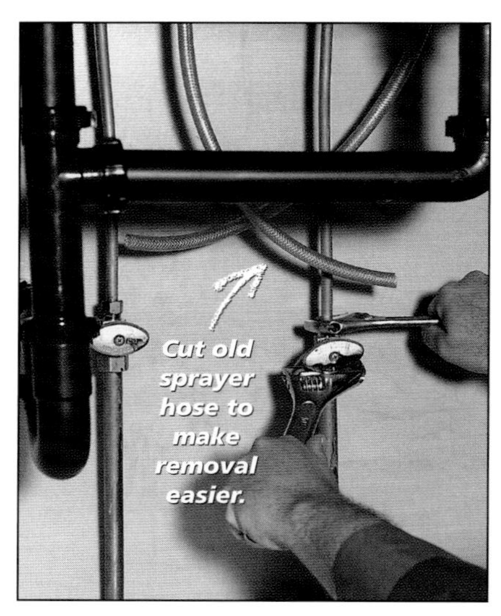

Cut old sprayer hose to make removal easier.

1 DISCONNECT THE SUPPLY LINES AND CUT THE SPRAYER HOSE. Disconnect the hot and cold supply lines using two adjustable wrenches. If you're replacing the sprayer hose, save a little time by cutting it with a utility knife so that you only have to disconnect it from the faucet body.

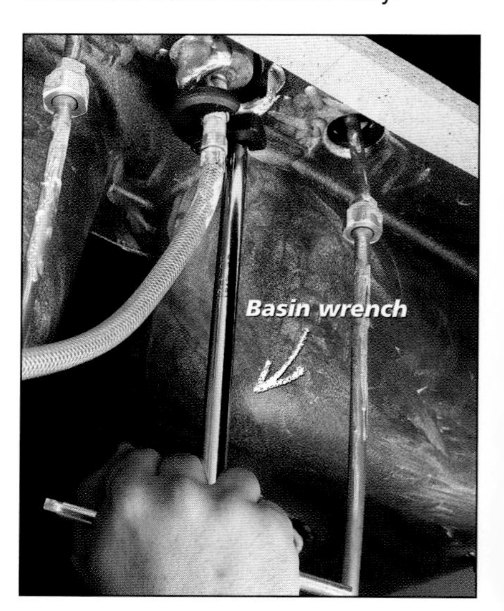

2 REMOVE THE RETAINING NUTS WITH A BASIN WRENCH. Getting to the tailpiece nuts can be tricky. A basin wrench will take a little practice, but it allows easier access. Lift out the faucet body and clean up old putty from the sink. If the old sprayer flange base is OK, leave it alone. If you're replacing it, run a bead of silicone caulk around the base and seat it in its hole in the sink. Thread the new sprayer and hose into the sink.

Use Teflon tape or pipe compound on the threads when you connect the supply to the faucet.

3 ATTACH THE SUPPLY LINES TO THE FAUCET BODY. It's much easier to work above the sink than below it, so plumbers will attach anything they can before they install the faucet. Label the hot and cold lines so you won't have to guess when you're hooking them up. Wrap the threads in Teflon tape before you connect the supply lines. Put a bead of plumber's putty on the faucet base before you thread the supply lines into the hole.

4 SEAT THE FAUCET BODY ON THE SINK. Press the faucet body into the plumber's putty to seat it. Square the faucet on the sink before you go underneath to attach it or you'll have to loosen the connections later to fix it.

Feeling for leaks often isn't enough; you must look for them, too. So keep a flashlight handy.

TOOL TIP

WHY BASIN WRENCHES ARE COOL.

After a few bruised knuckles you'll learn why a basin wrench is the right tool for working with those pesky tailpiece nuts. Unlike an ordinary adjustable wrench, there's an extension that allows you to reach up into tight places. The head swivels to tighten or loosen nuts, and the jaws are self-ratcheting so they will automatically tighten around the nuts. The right tool makes all the difference and helps avoid cuts and bruises. Basin wrenches with extensions for more flexibility and reach are also on the market.

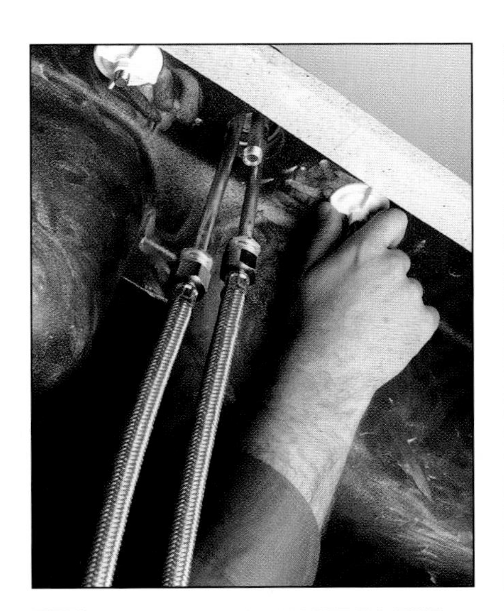

5 ATTACH THE FAUCET TO THE SINK AND CONNECT THE SPRAYER HOSE. Use the hardware that came with your faucet to attach the faucet to the bottom of the sink.

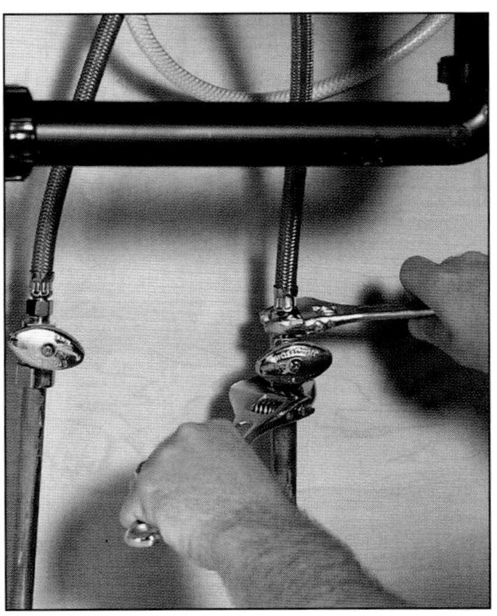

6 ATTACH THE HOT AND COLD SUPPLY LINES. Use two adjustable wrenches to attach the hot and cold supply lines to the correct compression fittings. Connect the sprayer hose to its nipple. Make sure you don't tangle the sprayer hose in the supply lines or it won't extend to its entire length.

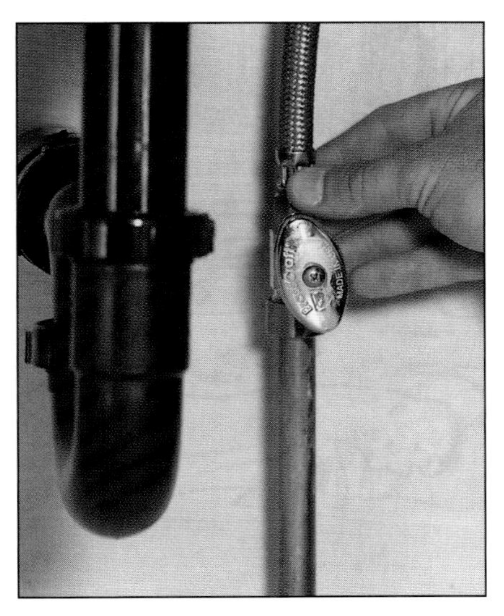

7 TURN ON THE WATER AND CHECK FOR LEAKS. Turn on the water and test the system for leaks by visually examining and feeling all the connections. Clean up any excess putty around the base of the faucet.

ADJUSTING A POP-UP DRAIN

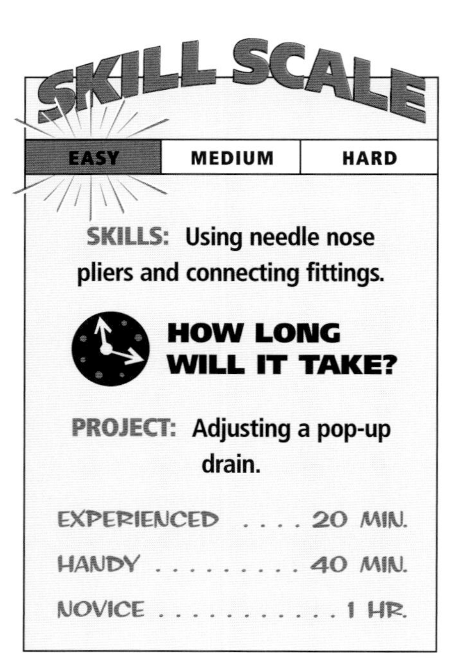

SKILL SCALE
EASY
MEDIUM
HARD
SKILLS: Using needle nose pliers and connecting fittings.
HOW LONG WILL IT TAKE?
PROJECT: Adjusting a pop-up drain.
EXPERIENCED .... 20 MIN.
HANDY .... 40 MIN.
NOVICE .... 1 HR.

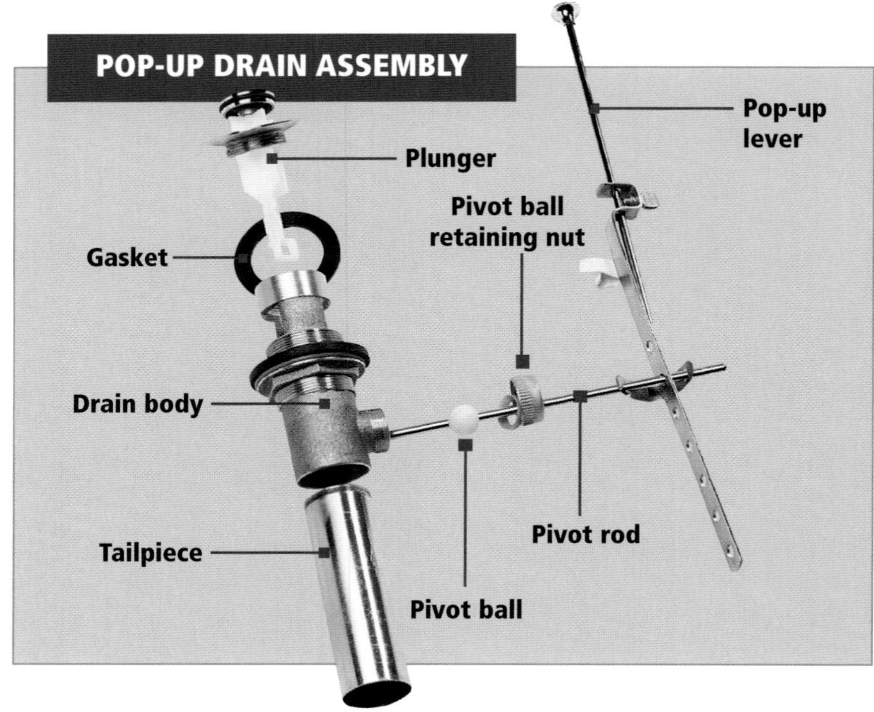

STUFF YOU'LL NEED

TOOLS: Needle nose pliers, water-pump pliers, small wire brush

MATERIALS: Vinegar

Four common problems occur with pop-up drains. The stopper won't lift out with the control, it won't remain in the closed or open position, the stopper doesn't hold water, or water leaks from the pivot ball. All are easy to fix:

1 Stopper won't stay open or closed. Tighten the pivot ball nut with water-pump pliers until the stopper maintains either an open or a closed position.

2 Water leaks from pivot. Unscrew the pivot ball retaining nut with water-pump pliers. Remove the pivot rod, ball, and gasket. Replace the gasket and reassemble.

3 Stopper won't lift out. (Below left)

4 Stopper won't hold water. (Below right)

ADJUST THE STOPPER WITH NEEDLE NOSE PLIERS. To adjust a stopper that won't lift, pull the clevis setscrew down until the stopper pops up. Tighten the setscrew once the stopper is at the proper height.

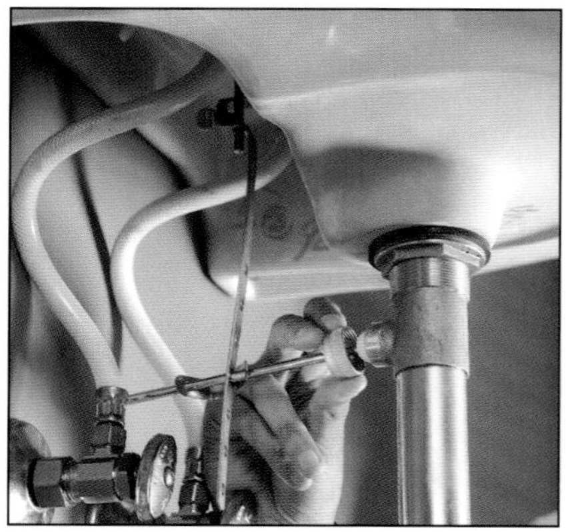

REMOVE THE PIVOT BALL RETAINING NUT. Pull out the rod and lift out the stopper. Clean the rubber seal by soaking it in vinegar and scrubbing with a small wire brush. Insert the stopper. Engage the pivot rod and replace the nut.

REPLACING A SINK TRAP

EASY	MEDIUM	HARD

SKILLS: Disconnecting and installing pipe fittings.

HOW LONG WILL IT TAKE?

PROJECT: Replacing a sink drain trap.

EXPERIENCED . . . . . 10 MIN.
HANDY . . . . 20 MIN.
NOVICE . . . . 30 MIN.

TOOLS: Water-pump pliers, bucket

MATERIALS: Replacement sink trap, washers

DANGER IN THE DRAIN!

If a chemical drain cleaner was used to try to clear the trap, you need to take some extra precautions when you open it up. Wear rubber gloves, a long-sleeved shirt, and safety glasses to prevent injury.

Sink traps may require periodic cleaning to dislodge clogs that prove too stubborn for plunging or removing with a plumber's snake. Replacement is the easiest and most direct way to solve this problem. Disconnect the trap with water-pump pliers, clean out the debris, and reinstall the trap.

PREPARING FOR THE JOB

A newer sink drain trap probably won't fail and need replacement because traps are constructed with more durable materials and are highly resistant to corrosion. Older sink drain traps may need replacing. You should replace yours if it is rusted or has developed leaks around the fittings.

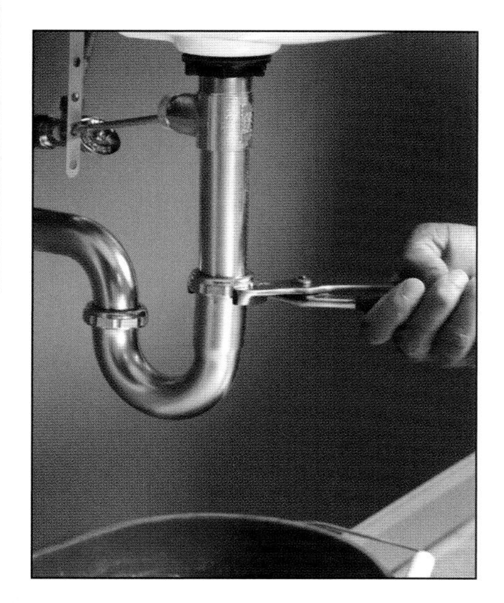

1 REMOVE THE SLIP NUTS. Place a bucket under the drain trap to catch debris and any wastewater remaining in the trap. If there is a clean-out plug, remove it and drain wastewater into the bucket. If there isn't, loosen both slip nuts with water-pump pliers and slide them out of the way.

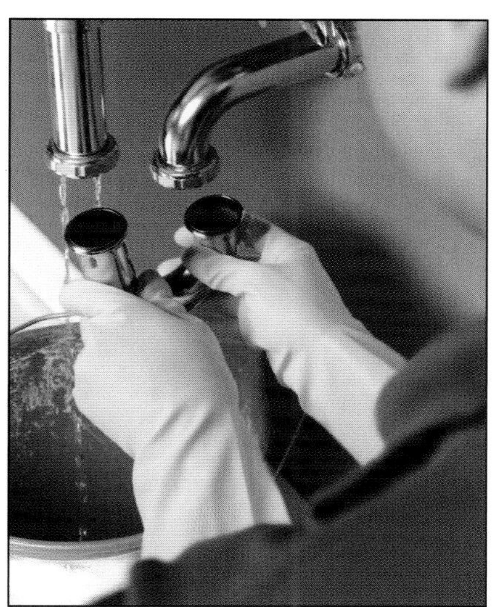

2 REMOVE THE TRAP. Pull down on the trap to dislodge it and remove the washers. Dump out any debris remaining in the trap or drain line. Take the trap and washers to your local home center so you can find the same size replacement parts. Slide on the new washers. Fit the new trap in place and tighten the slip nuts.

SMOOTH TRANSITION.

If you make a transition from a metal trap to PVC or ABS, use a rubber transition band that you can purchase at a home center. Finding an exact replacement can be difficult; the transition band provides a strong, effective solution.

REPAIRING OR REPLACING A SINK STRAINER

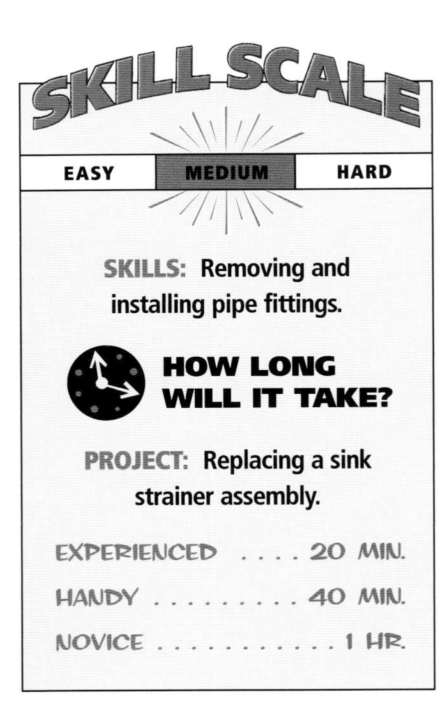

The sink strainer assembly connects the sink to the drain line. To fix a leak, you'll need to take it all apart. Remove and clean the sink strainer basket, then replace any worn washers and gaskets. If the seal where the strainer basket meets the lip of the drain line was not properly installed, it may leak.

Repairing does not mean replacing every part. If you don't mind reusing old parts, don't replace them. The sink drain body may be usable even though it may not shine like a new one; reuse it or any of the metal parts. The drain locknut can be reused because it is hidden below the sink. The only parts you should not reuse are washers and gaskets; they may not provide a proper seal. Besides, they are the least expensive parts to replace.

SINK STRAINER ASSEMBLY

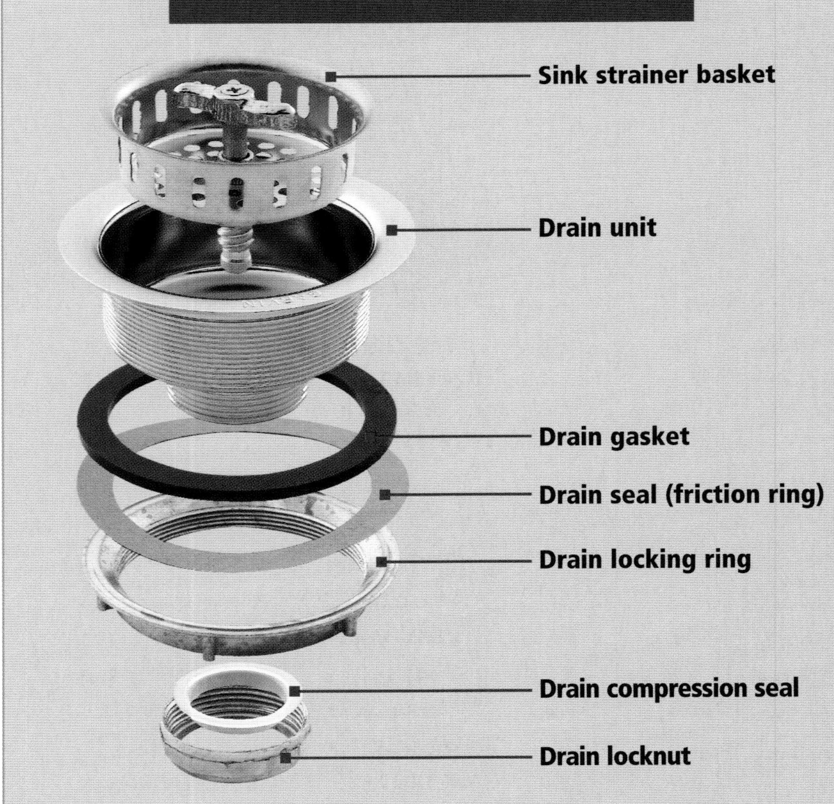

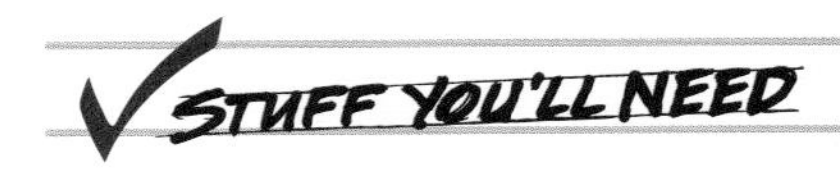

STUFF YOU'LL NEED

TOOLS: Water-pump pliers, hammer, basket strainer wrench, plastic putty knife, mini hacksaw, screwdriver

MATERIALS: Plumber's putty

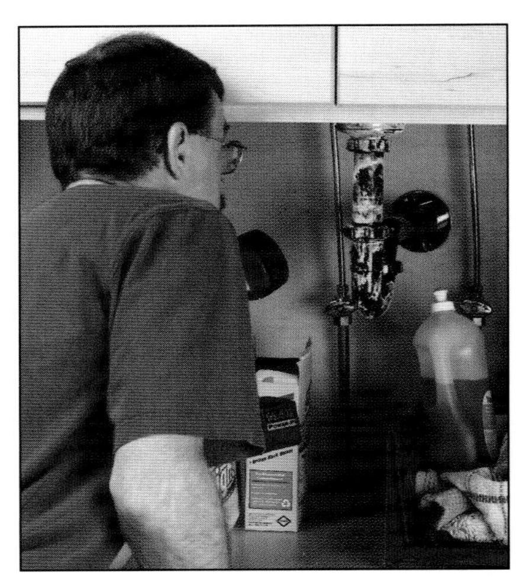

1 GIVE YOURSELF ROOM TO WORK. Inspect the area below the sink and remove any obstacles.

2 DISCONNECT THE SLIP NUTS. Use water-pump pliers to loosen the slip nuts and slide them out of the way. Remove the tailpiece.

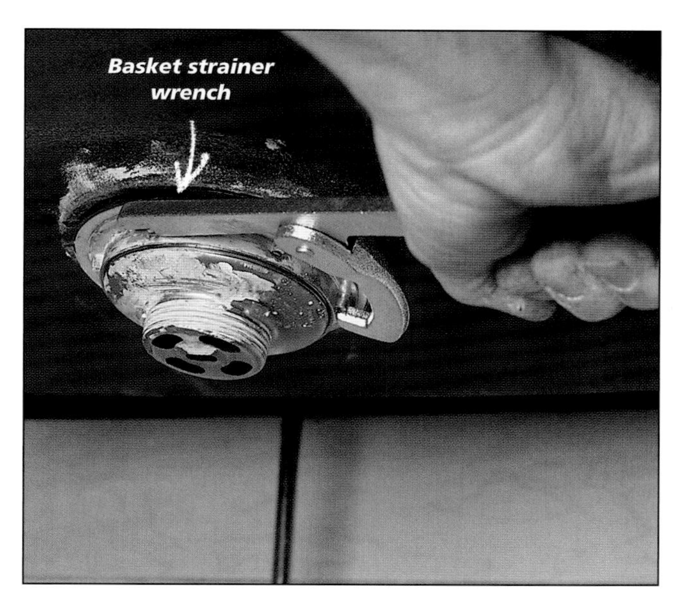

3 **UNSCREW THE LOCKING NUT.** Use a basket strainer wrench to remove the sink strainer assembly locknut. If the locknut will not budge, tap on the lug with a hammer and screwdriver to loosen it.

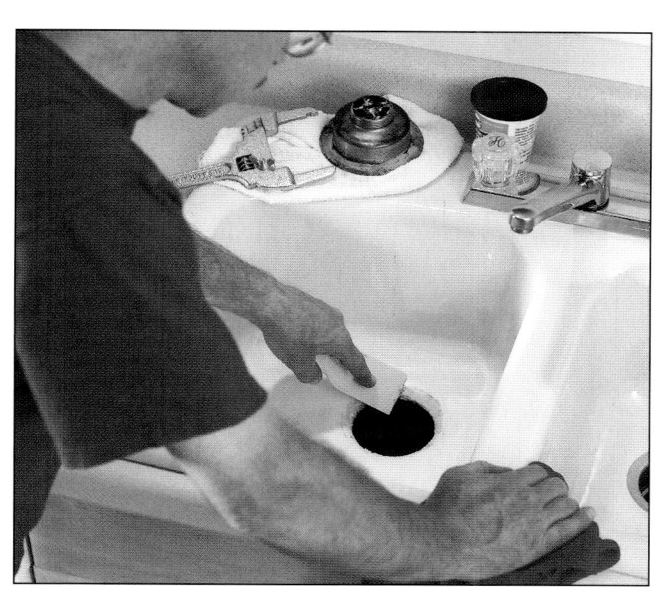

4 **SCRAPE OFF THE OLD PUTTY WITH A PLASTIC PUTTY KNIFE.** If you reuse the old strainer, remove any old putty under the flange. Do not reuse old gaskets and washers. Replace them with new ones.

5 **COAT THE FLANGE.** Apply a bead of plumber's putty to the underside of the drain unit. Insert the drain unit into the sink.

6 **REASSEMBLE THE STRAINER.** Install a new rubber gasket and friction ring. Hand-tighten the new locknut. Connect the tailpiece to the assembly body with slip nuts. Test with running water. Tighten the nuts if you see any leaks.

DIFFICULT NUT TO CRACK?

You may find the locking nut is very difficult to loosen. If all else fails, cut a groove in the nut at about a 30-degree angle with a mini hacksaw. Insert a screwdriver into the groove and twist or tap with a hammer until the nut breaks off.

OOPS!

Even experienced plumbers sometimes break plastic fittings. That final turn with the pliers may seem like a good idea, but it can result in a cracked fitting and a trip to the home center. Tighten all fittings hand-tight, then a quarter-turn at a time with the pliers until the leaking stops.

TROUBLESHOOTING A LEAKING TOILET

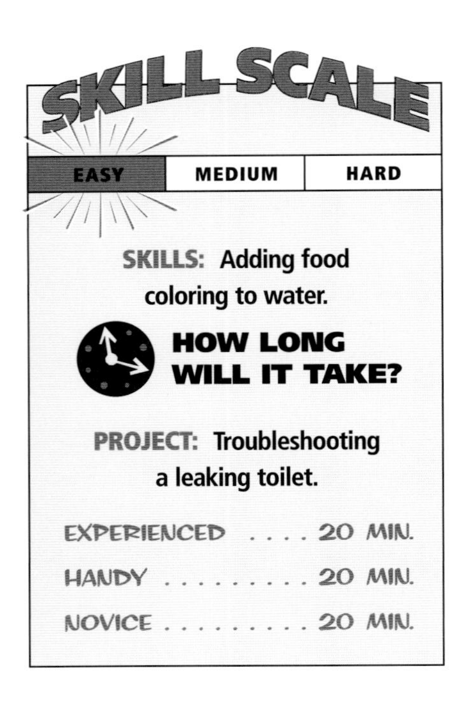

SKILL SCALE
EASY
MEDIUM
HARD
SKILLS: Adding food coloring to water.
HOW LONG WILL IT TAKE?
PROJECT: Troubleshooting a leaking toilet.
EXPERIENCED . . . . 20 MIN.
HANDY . . . . 20 MIN.
NOVICE . . . . 20 MIN.

STUFF YOU'LL NEED

MATERIALS: Red food coloring, paper towels, rags

Before you begin working on the toilet, put on a pair of rubber gloves and clean the base thoroughly with a disinfectant. It will make the job a little more pleasant!

Fix a leaking toilet immediately. A leak that is not repaired may become more than an annoyance. It can develop into structural damage not only to the floor, but also to the ceiling below the toilet. Both may require hiring a contractor for a repair which can be extensive and expensive.

TROUBLESHOOTING

Troubleshooting a toilet to locate a leak is easy. There are no tools required, just food coloring (red is recommended) and paper towels. The food coloring makes the leak easily visible so you can see what repairs are needed.

ANATOMY OF A TOILET

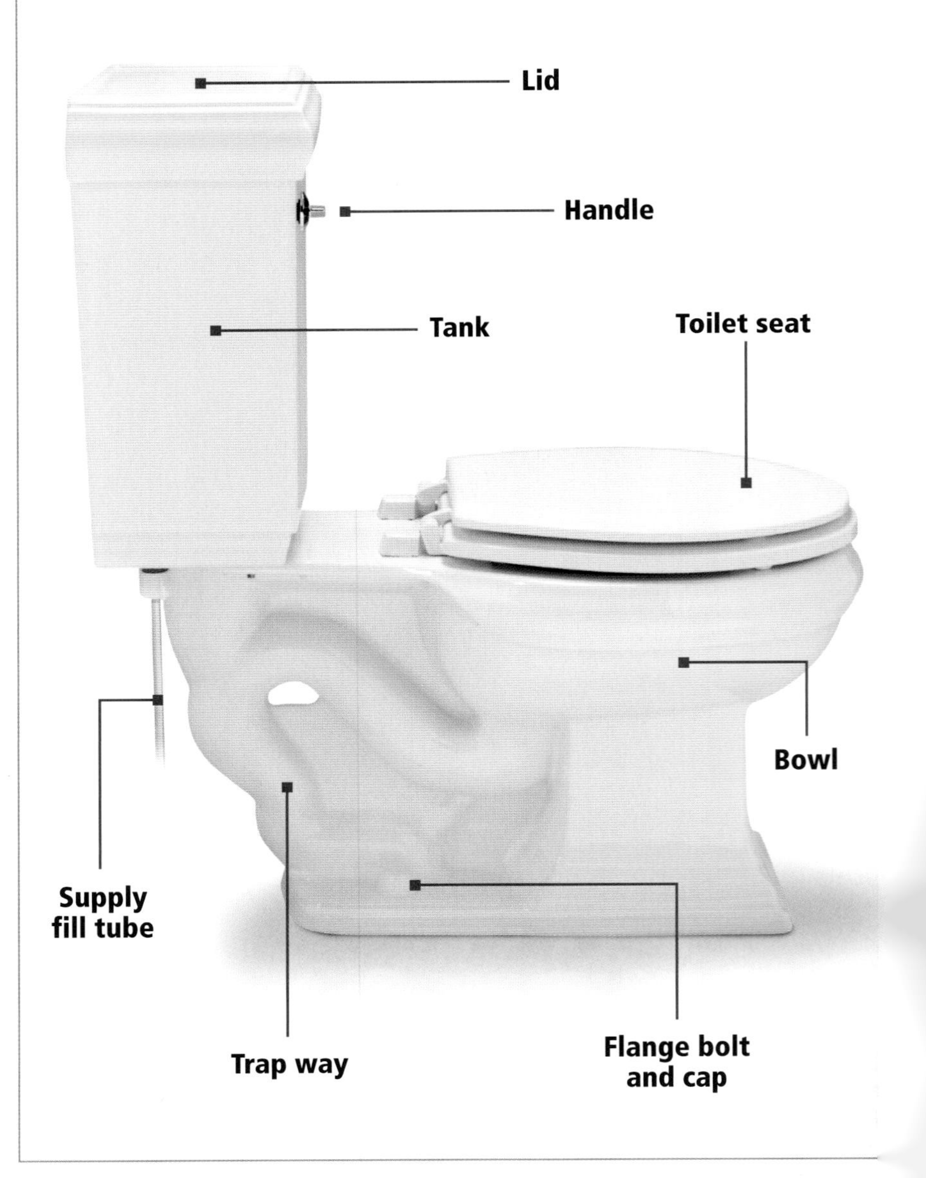

1 **CLEAN THE OUTSIDE OF THE TOILET.** Flush the toilet. Thoroughly dry all exterior surfaces. This will help you spot a leak easily.

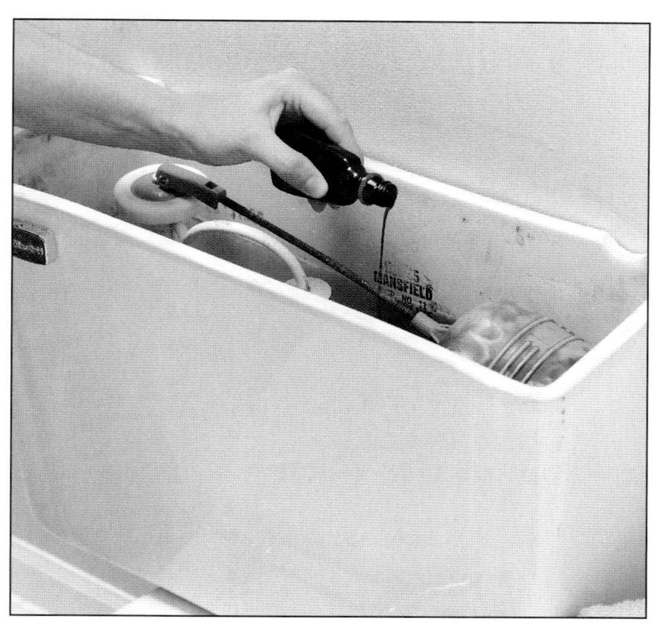

2 **ADD RED FOOD COLORING.** Pour about a teaspoon of red food coloring into the water in the tank and a teaspoon into the bowl. Wait an hour.

3 **CHECK FOR LEAKS.** Wipe a dry paper towel around the base and under the tank. Red coloring on the towel shows there's a leak. If the leak is around the base, you will need to replace the wax ring (see pages 44–45). Leaks under the tank could be either a leaking fill valve, bolt gasket, or spud washer (see pages 42–43).

SEEING RED?

If your paper towel turns red around the base of the toilet, you'll probably need to replace the wax ring. Red beneath the tank may mean the gasket between the bowl and tank needs to be replaced. A damp paper towel that shows no red is a sign that the tank is sweating and you may need to insulate the tank (see page 47).

REPAIRING A LEAKING TANK

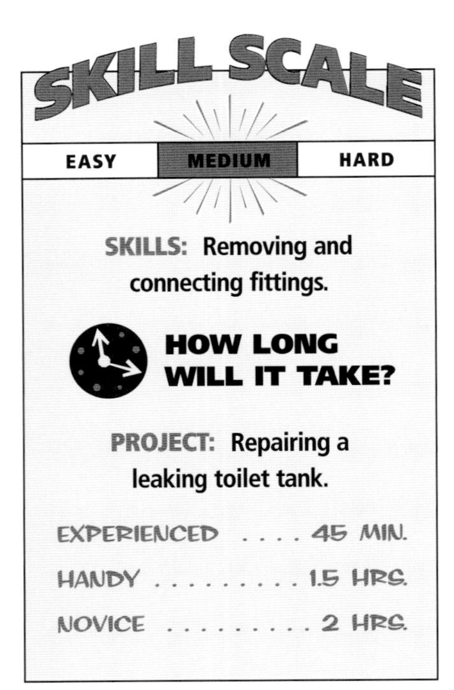

SKILL SCALE
EASY
MEDIUM
HARD
SKILLS: Removing and connecting fittings.
HOW LONG WILL IT TAKE?
PROJECT: Repairing a leaking toilet tank.
EXPERIENCED .... 45 MIN.
HANDY .... 1.5 HRS.
NOVICE .... 2 HRS.

Toilets keep running because either water is leaking into the toilet bowl or water is leaking out of the tank. Once you've figured out which is occurring, you are close to solving the problem.

THREE COMMON LEAKS

There are three common areas of the toilet that might need repair. The fill valve may be leaking, there may be a leak around a tank bolt, or the spud washer may have developed a leak. Here's how to fix them.

STUFF YOU'LL NEED

TOOLS: Two adjustable wrenches, screwdriver, helper, spud wrench, small wire brush

MATERIALS: Flush valve gasket, fill valve, gaskets, spud washer, plumber's putty, white vinegar

Don't worry about touching the water in the tank. Unlike the water in the bowl, it's clean.

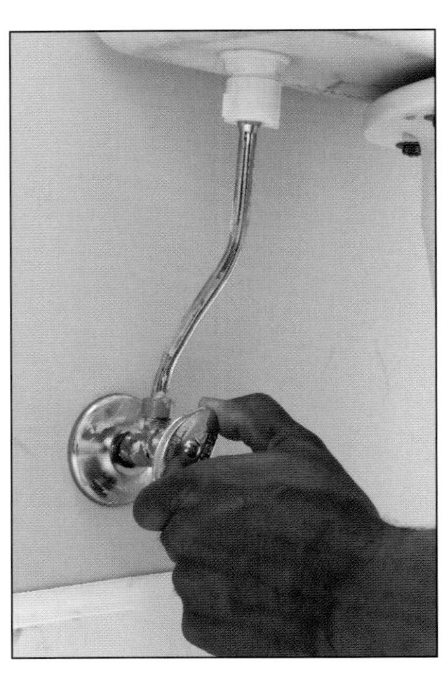

1 DRAIN AND CLEAN ALL SURFACES. Turn off the water supply valve. Flush the toilet. Disconnect the supply line from the tank and sponge the inside of the tank until it's dry.

2 DISCONNECT THE FILL VALVE. Use two adjustable wrenches to remove the fill valve. Remove the old fill gasket. Take the gasket with you to the home center so you'll be sure to get the right size part. If you're not replacing other parts, install the new fill valve and gasket. Turn on the supply line and check for leaks. If necessary, tighten a quarter turn. Otherwise, proceed to step 3.

WORK SMARTER

DON'T OVERTIGHTEN!
Overtightening the bolts that connect the tank to the toilet base can crack the base, the tank, or both. Tighten gently until snug and alternate from side to side so they seat evenly.

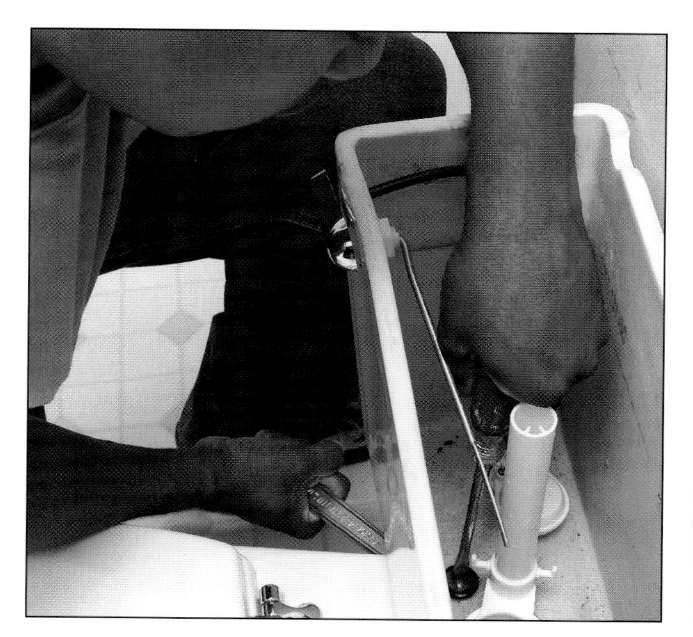

3 **UNSCREW THE TANK BOLT.** A screwdriver and adjustable wrench will remove most tank bolts. Remove the tank bolt, nut, and gasket. Clean the bolt and nut with white vinegar and a small wire brush. If you aren't replacing the spud washer, reinstall bolts and nuts with new gaskets. Alternate the tightening of the nuts to evenly draw the tank tight. If you are replacing the spud washer, continue to step 4.

4 **LIFT THE TANK** straight up and off the toilet base to remove it. Make sure you have a helper; toilet tanks are usually in an awkward place and are heavier than they appear to be. Set the tank upside down on the floor. It's best to set it on an old towel or rug because there may be some water left in the tank.

5 **REPLACE THE SPUD WASHER.** Take the spud washer to your local home center to find the right replacement. Place a new spud washer over the flush valve tailpiece. Lower the tank onto the base so the tank bolts go through the holes. Reinstall the tank bolts, gaskets, and nuts. Alternate tightening the nuts from side to side so they tighten evenly. Reinstall the supply tube coupling and fill valve. Turn on the water supply and check for leaks.

TIGHTEN BY DEGREES.

You might give the toilet a flush to test it and notice a slight leak. If so, turn off the supply valve. If there is a leak along the supply valve, tighten the fittings an additional quarter turn. If the leak is around the base of the tank, check the washers in the tank to make sure they're seated properly. If the washers appear to be properly seated, tighten the tank nuts another quarter turn.

REPAIRING A LEAKING TOILET BASE

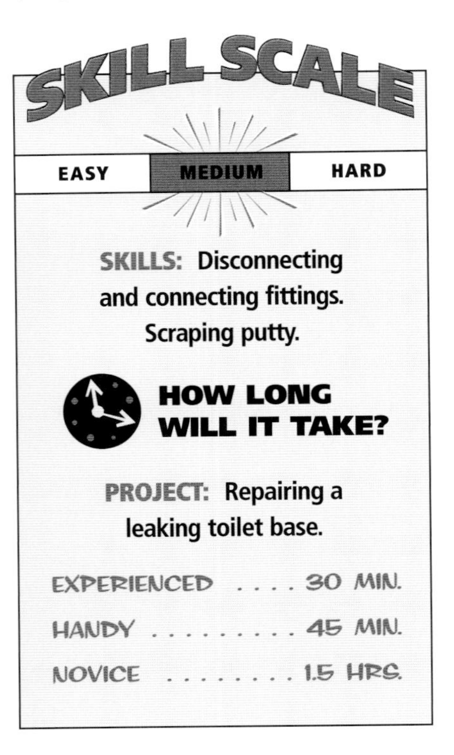

SKILL SCALE
EASY
MEDIUM
HARD
SKILLS: Disconnecting and connecting fittings. Scraping putty.
HOW LONG WILL IT TAKE?
PROJECT: Repairing a leaking toilet base.
EXPERIENCED .... 30 MIN.
HANDY .... 45 MIN.
NOVICE .... 1.5 HRS.

Plan ahead before you start this project. If the toilet is the only one in the house, you might want to repair it when everyone else is away for the weekend. See if neighbors will let you use theirs during the project—especially if it's time to collect on those favors you've given.

GIVE YOURSELF SOME ROOM

Bathrooms tend to be small and crowded with fixtures and furnishings. Before beginning, review the working space. Do you have enough room? You might have to remove the door, clothes baskets, and decorative items. This messy job requires room to work. Take the time—it will make the job easier.

TOOLS: Adjustable wrench, putty knife, caulking gun

MATERIALS: No-seep wax ring, flange, rag, silicone caulk

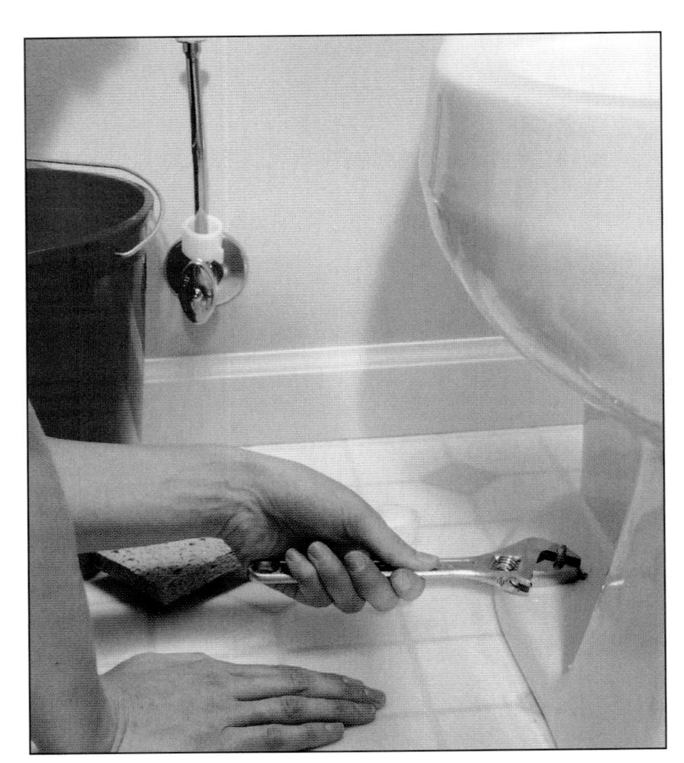

1 TURN OFF THE WATER SUPPLY, FLUSH THE TOILET, AND PLUNGE ONCE TO REMOVE EXCESS WATER IN THE TRAP. Remove the tank, then the closet bowl bolts. Disconnect the nuts with an adjustable wrench. Once the nuts are removed, rock the toilet back and forth until the base separates from the wax ring.

2 LIFT OFF THE BASE. Position yourself so you can lift the toilet straight up. If it's too heavy, enlist a helper. Place the bowl on its side on an old towel or rug to absorb any water remaining in the bowl.

Remove the old flange on the drain horn before you install the new wax ring.

3 REMOVE THE OLD WAX RING. Stuff a rag in the opening to prevent noxious gases from entering the house. Use a putty knife to scrape the old wax ring from the closet flange and the toilet base. Wipe clean any excess wax. Remove the old flange from the base of the bowl.

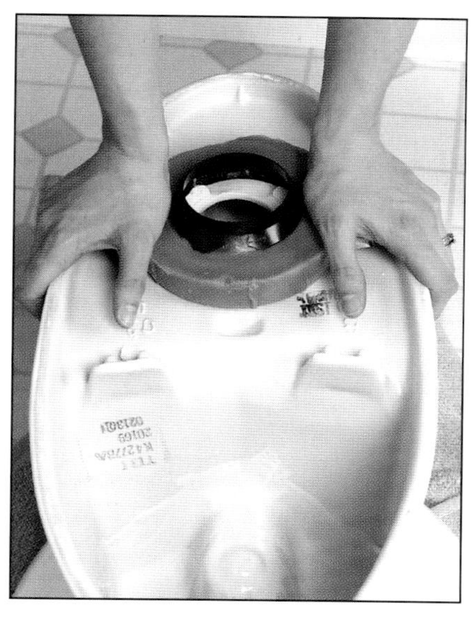

4 REPLACE FLANGE BOLTS AND INSTALL A NEW WAX RING. Turn the toilet upside down on an old towel to protect the surface. Place the new wax ring and flange on the drain horn of the toilet. Remove the rag from the drain opening. Lift the toilet and set it down slowly over the flange bolts. Press down on the base to compress the wax ring.

CLOSER LOOK

Use a penny as a shim.

LEVEL THAT ROCKING TOILET.

No matter how carefully you tighten the bolts, the toilet may still rock. This could be caused by an uneven floor or a closet flange that was not properly installed. Use plastic wedge-shaped shims to level the toilet. (Wood shims will rot away.) You also can try this penny-pinching trick: Level the toilet with pennies. They're a practical alternative to store-bought plastic shims.

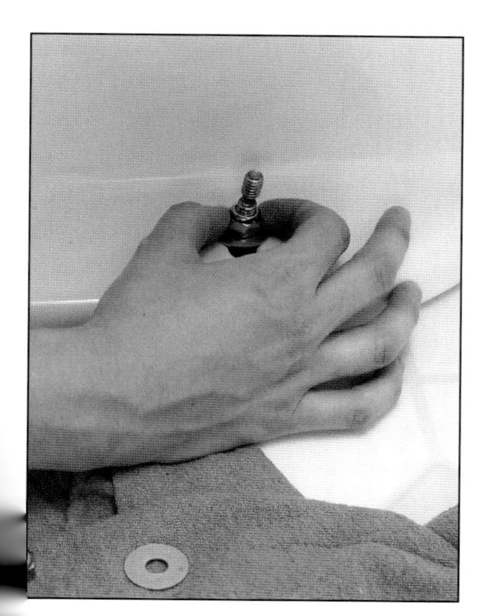

5 LEVEL THE TOILET AND TIGHTEN THE CLOSET NUTS. Level the toilet and shim if necessary. (See "Level That Rocking Toilet," above right.) Tighten with an adjustable wrench, alternating from one side of the toilet to the other to keep the bowl level. Don't overtighten; you can break the base.

6 CAULK AROUND BASE. Fill the bolt caps with plumber's putty and put them on the bolts. Replace the tank and reconnect the supply line. Turn on the supply valve and check for leaks. Wait several days before applying silicone caulk or a mixture of plaster and white vinegar to seal the base.

BUYER'S GUIDE

NO-WAX CLOSET RING.

You can avoid sticky hands when you replace the closet ring on your toilet if you use a "no-wax" model. Instead of the traditional wax ring that molds itself to the drain to prevent leaks and the release of sewer gases, this model creates a compression seal with a heavy-duty rubber O-ring.

ADJUSTING THE TANK HANDLE AND WATER LEVEL

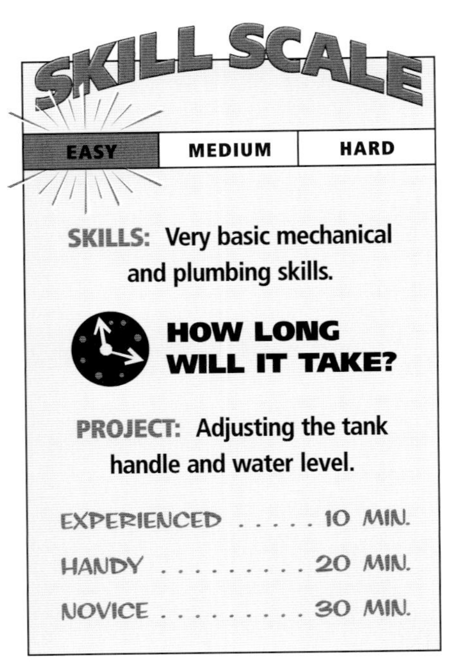

SKILL SCALE
EASY
MEDIUM
HARD
SKILLS: Very basic mechanical and plumbing skills.
HOW LONG WILL IT TAKE?
PROJECT: Adjusting the tank handle and water level.
EXPERIENCED . . . . . 10 MIN.
HANDY . . . . 20 MIN.
NOVICE . . . . 30 MIN.

TOOLS: Adjustable wrench, screwdriver

MATERIALS: Bucket, sponge

GOOD IDEA

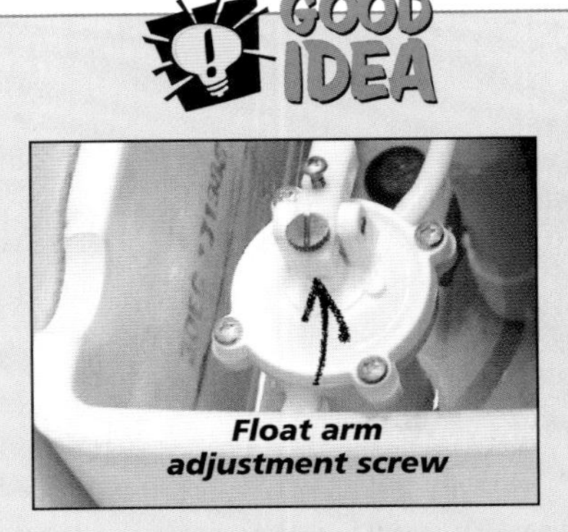

ADJUST THE FLOAT ARM.

Before you resort to more complicated measures, try adjusting the float arm first. Turn the screw on top to lift or lower the float.

The sound of water bubbling from a fountain may be relaxing. But if it's coming from your toilet, it's just adding to the water bill. Quick fixes include shortening a chain, bending a wire, or adjusting a float clip. Each solution is easy and requires little time.

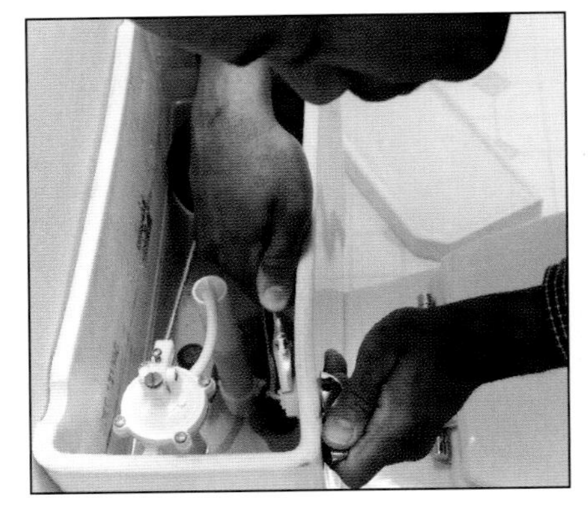

1 ADJUST THE TANK HANDLE. If the handle has too much play or binds, use an adjustable wrench to tighten the nut inside the toilet. Unlike other nuts and bolts, the threads on a tank handle are left-handed, so you'll tighten clockwise.

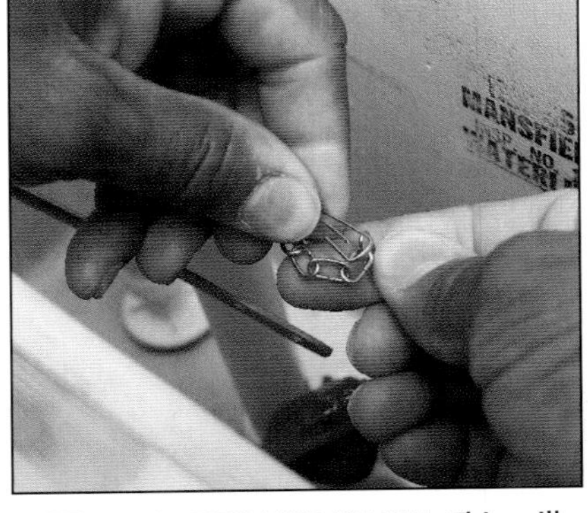

2 SHORTEN THE CHAIN. This will help if you have to hold the handle down to flush all the water from the toilet tank.

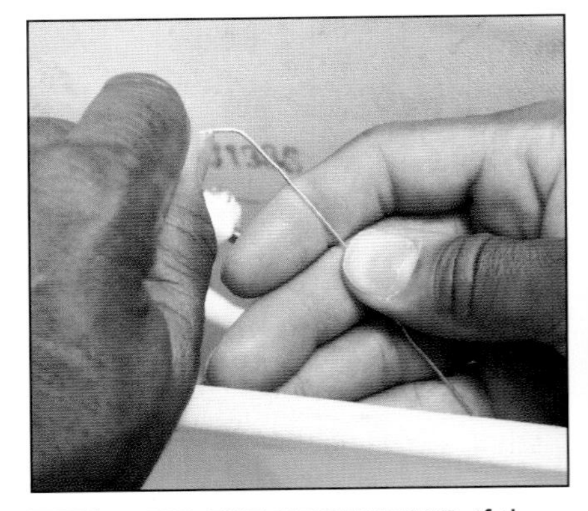

3 BEND THE UPPER WIRE. If the tank doesn't have a chain, it has a wire you can bend.

4 BEND THE FLOAT ARM. To adjust the water level in the tank, bend it up for a higher water level or down to lower the water level.

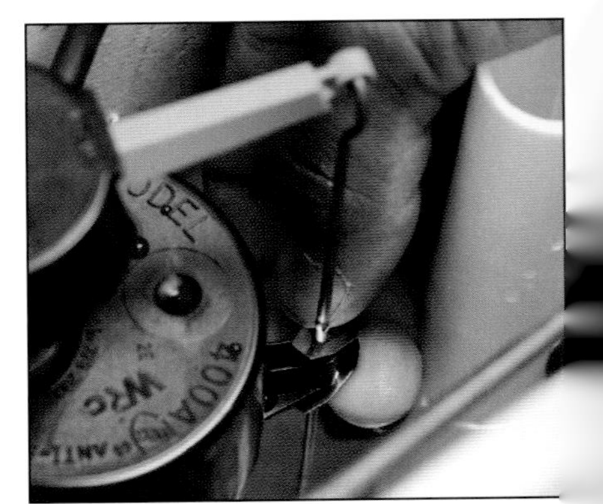

5 SLIDE THE FLOAT CYLINDER. Squeeze the float clip to release the float cylinder, which can be raised or lowered to adjust the water level in float style toilets.

CURING A SWEATING TANK

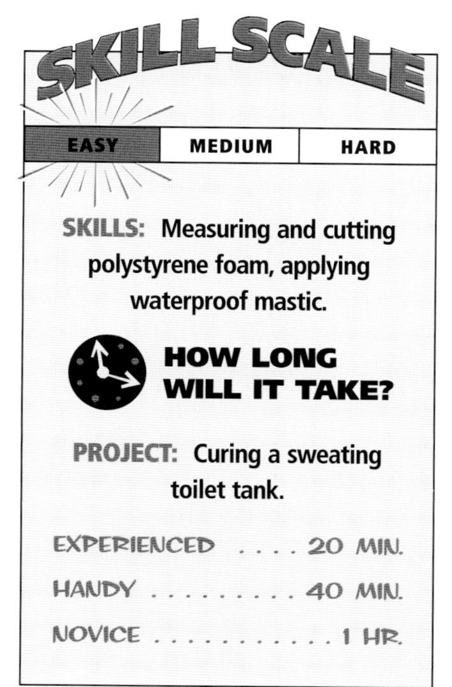

SKILL SCALE
EASY
MEDIUM
HARD
SKILLS: Measuring and cutting polystyrene foam, applying waterproof mastic.
HOW LONG WILL IT TAKE?
PROJECT: Curing a sweating toilet tank.
EXPERIENCED .... 20 MIN.
HANDY .... 40 MIN.
NOVICE .... 1 HR.

✓ STUFF YOU'LL NEED

TOOLS: Pencil, straightedge, scissors, utility knife

MATERIALS: Paper, polystyrene foam, waterproof mastic

Humid weather can cause condensation to form on the outside of the tank. You will have to buy a kit to insulate the tank and stop it from sweating, but you can do the job yourself. Remember, you will not be able to use the toilet overnight while the mastic is drying.

PREPARE THE TOILET

Turn off the water to the toilet at the supply valve. Flush the toilet and remove as much water as possible from the tank with a sponge and bucket. Completely dry the inside with an old towel or rag.

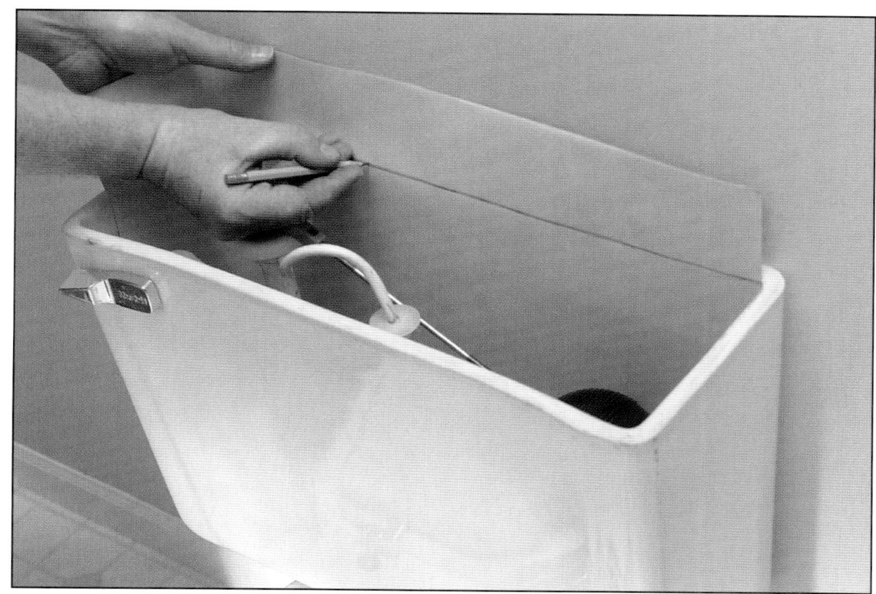

1 MAKE THE TEMPLATE. Make a pattern out of a sheet of paper placed over the rear wall of the tank. Mark the edges with a pencil and cut the lines with a straightedge.

2 CUT THE FOAM. Lay the pattern you created on top of a sheet of polystyrene foam. Fasten it with tape or pins so it won't move when you cut it. Use a utility knife to cut along the pattern. You now have insulation panels for the tank.

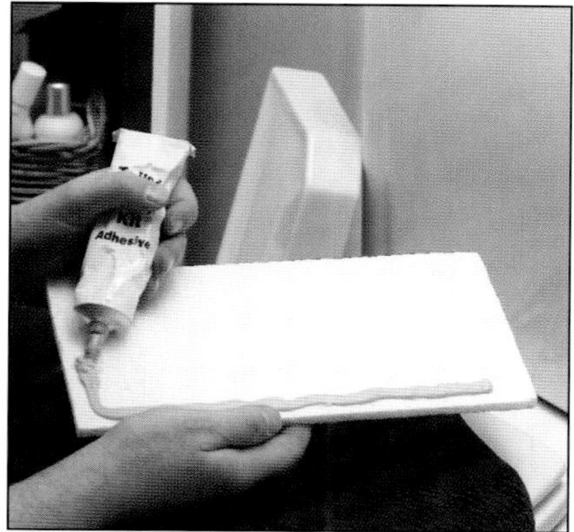

3 APPLY WATERPROOF MASTIC. Press the panel into place against the rear wall. Repeat the process for the front and all sides of the tank. Let the mastic dry overnight before refilling the tank.

REPLACING A TOILET FILL VALVE

SKILL SCALE
EASY
MEDIUM
HARD
SKILLS: Connecting and disconnecting fittings.
HOW LONG WILL IT TAKE?
PROJECT: Replacing a toilet fill valve.
EXPERIENCED . . . . 20 MIN.
HANDY . . . . 40 MIN.
NOVICE . . . . 1 HR.

Old toilet fill valves can develop leaks. If water continues to run after you have made the adjustments on page 46, you may need to remove and replace the fill valve. You may also have to replace the supply line so the new valve will fit.

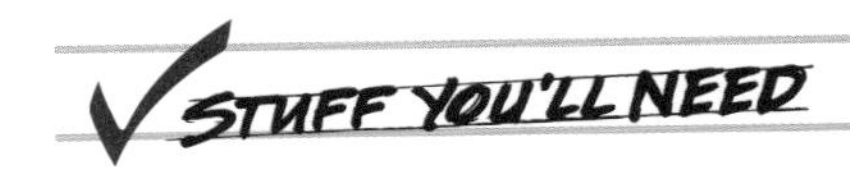

TOOLS: Adjustable wrench

MATERIALS: Fill valve, fill valve gasket

Know the locations of the main shutoff valve or the supply valves for each fixture in your home.

TOILET FILL VALVE COMPONENTS

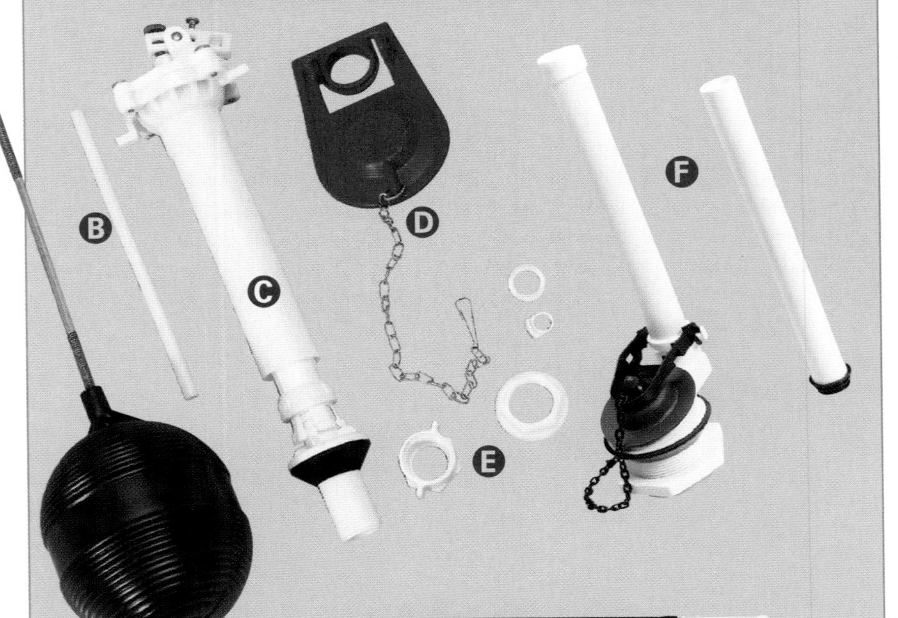

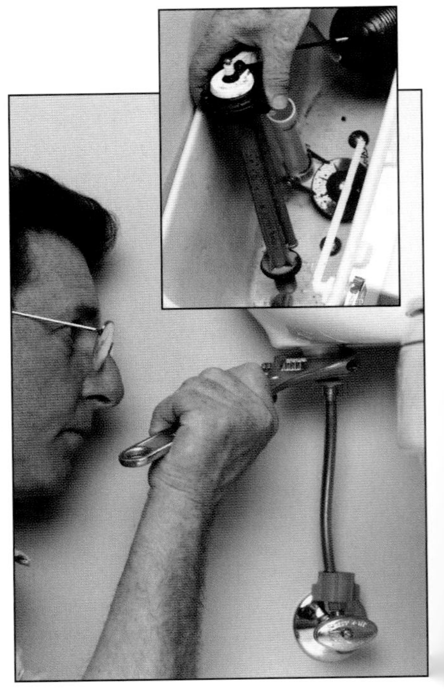

1 SHUT OFF THE WATER AND DRAIN THE TANK. Disconnect the fill valve nut and remove the old fill valve. (See inset.)

◀ **FILL VALVE COMPONENT NAMES**

- Ⓐ Float with float arm
- Ⓑ Refill tube
- Ⓒ Fill valve assembly
- Ⓓ Flapper
- Ⓔ Plastic washers and nuts
- Ⓕ Overflow tubes
- Ⓖ Fluidmaster™ valve assembly

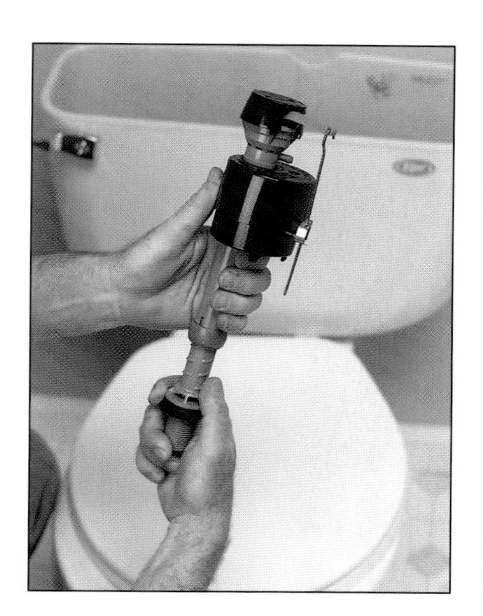

2 ADJUST THE HEIGHT OF THE VALVE. Adjust the height so the marking on the top of the valve is at least 1 inch above the overflow tube.

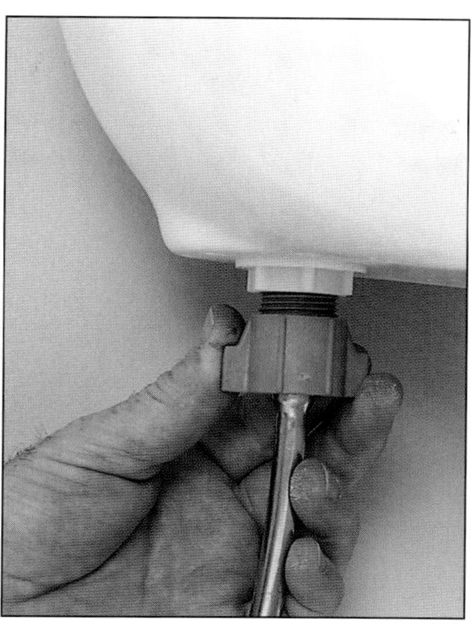

3 POSITION THE VALVE IN THE TANK. Push down on the valve shank and tighten the locknut one-half turn beyond hand-tight. Connect the supply.

4 ATTACH REFILL TUBE AND ANGLE ADAPTOR TO THE OVERFLOW. Trim the tube if necessary so there aren't any kinks.

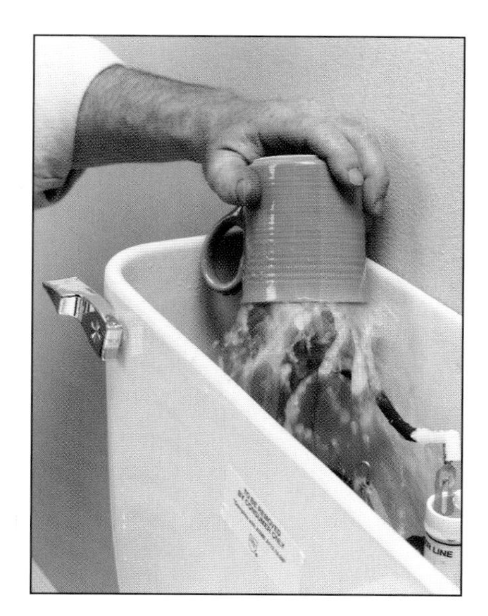

5 FLUSH THE SYSTEM. Remove the top valve. Hold a cup over the uncapped valve and turn on the water supply to flush the system of rust and debris. Turn off the water.

6 REPLACE THE TOP VALVE. Replace the top valve by engaging the lugs and rotating one-eighth turn clockwise. Make sure it's firmly locked into position.

7 ADJUST THE WATER LEVEL. Adjust the water level by squeezing the adjustment clip and moving the float cup up or down.

THREE VALVE TYPES—ONE WAY TO REPLACE.

There are three basic types of fill valves: the plunger valve, the diaphragm, and the float cup. You replace all three the same way. It may not be necessary to replace the fill valve with the exact same type that you remove, but it may be easier for you because the installation process is the opposite of the removal sequence.

REPLACING A FLAPPER

SKILL SCALE

EASY	MEDIUM	HARD

SKILLS: Removing the flapper and adjusting the chain.

HOW LONG WILL IT TAKE?

PROJECT: Replacing a toilet flapper.

EXPERIENCED . . . . . 5 MIN.
HANDY . . . . 10 MIN.
NOVICE . . . . 15 MIN.

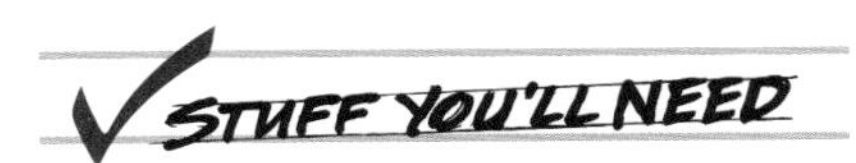

STUFF YOU'LL NEED

TOOLS: None

MATERIALS: Bucket, sponge, scrub pad, flapper or tank ball

If you have hard water, you will probably need to change the toilet flapper occasionally. Minerals in hard water build up around the base of the flapper and the opening of the toilet. The sediment will eventually destroy the flapper, causing the toilet to leak. No tools are needed for this project.

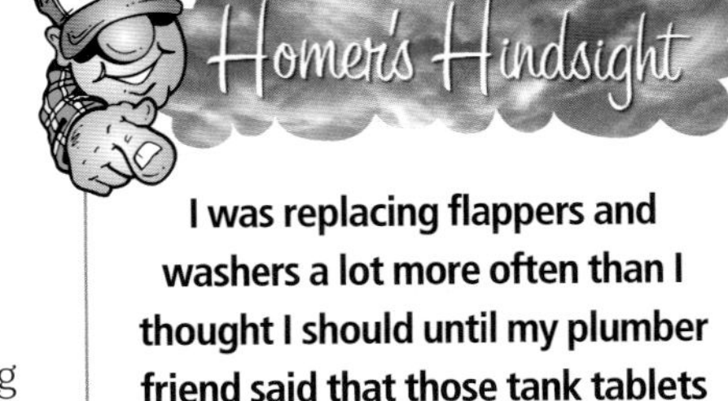

Homer's Hindsight

I was replacing flappers and washers a lot more often than I thought I should until my plumber friend said that those tank tablets I was putting in the water contained chlorine and were eating away at the rubber. I don't use them anymore.

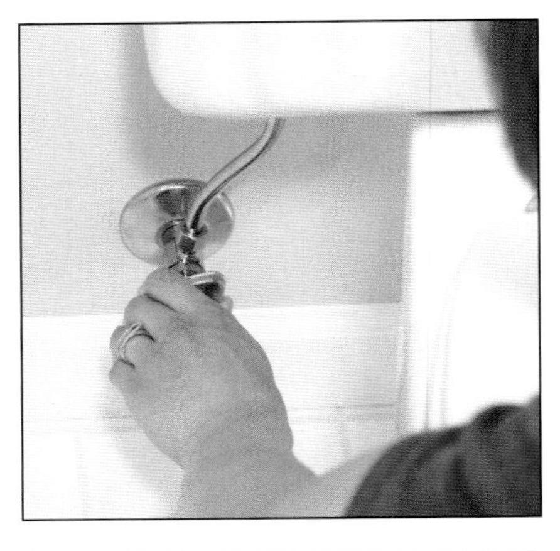

1 TURN OFF THE WATER SUPPLY SHUTOFF VALVE. If there isn't a shutoff valve, turn off the water at the main valve; this is also a good time to install a shutoff valve for future use. (See pages 140–141.)

2 FLUSH THE TOILET. Dry the inside walls and base of the tank with a sponge. Have a bucket handy to wring out the sponge.

WORK SMARTER

FLAPPER TIPS.

Have about a half-inch of slack in the chain that connects the flush lever to the flapper. Most replacements have side tabs or a ring that slides over the overflow tube so they can work with any system. Make sure the flapper can move up and down freely.

3 REMOVE THE OLD FLAPPER. Pull the flapper from the pivot arm. For ball-style toilets, grip a loop of lift wire and unscrew the old tank ball. Clean the surface area of the opening with a scrub pad to remove the sediment.

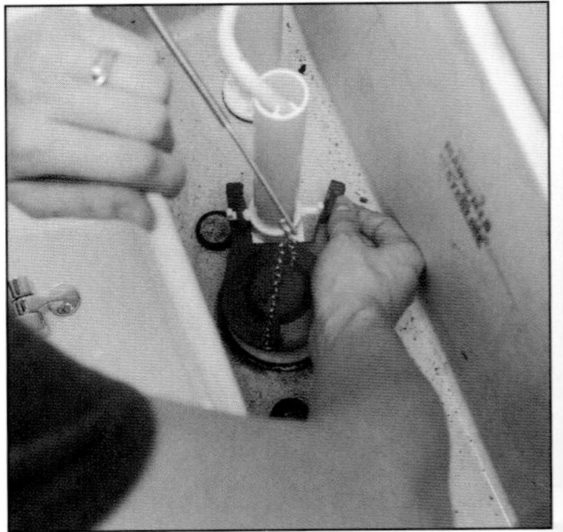

4 INSTALL A NEW FLAPPER. Line up the flap or ball with the valve seat by straightening the lift wire or adjusting the guide arm. This will provide a sufficient seat and keep the tank from leaking.

REPLACING A TOILET SEAT

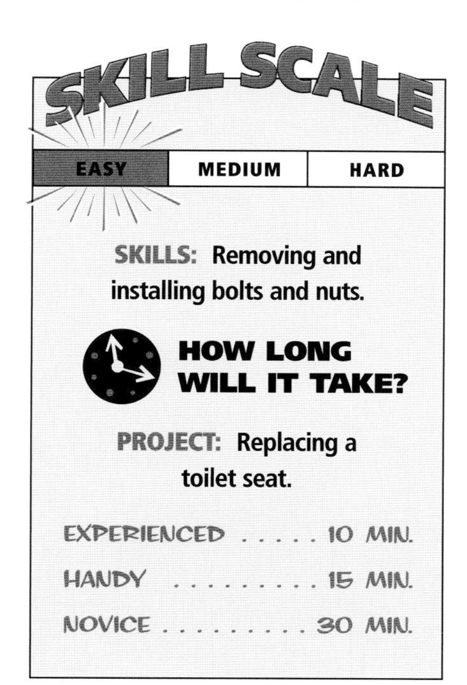

SKILLS: Removing and installing bolts and nuts.

HOW LONG WILL IT TAKE?

PROJECT: Replacing a toilet seat.

EXPERIENCED . . . . . 10 MIN.
HANDY . . . . 15 MIN.
NOVICE . . . . 30 MIN.

Style and comfort are two of the many reasons for changing the toilet seat. The seat or cover can become cracked or damaged. Appearance is the usual reason for changing the seat and cover. With an adjustable wrench and screwdriver, this is very easy because you don't have to shut off the water.

STUFF YOU'LL NEED

TOOLS: Screwdriver, adjustable wrench

MATERIALS: Replacement toilet seat, scrub pad

KNOW YOUR TOILET SEAT!

People often have to make extra trips to the store when replacing toilet seats because they don't realize there are two basic styles—round, which is 16 inches front to back, and elongated, which is 18 inches front to back. Measure before you go or bring the old seat to the store.

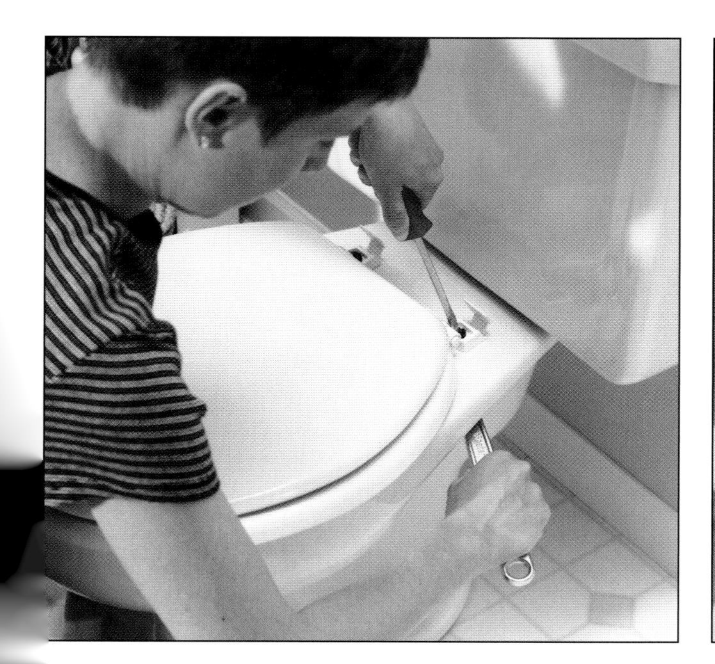

1 REMOVE THE OLD TOILET SEAT. Pry open the bolt caps. Use a screwdriver and adjustable wrench to remove the anchor bolts. Turn the bolt with the screwdriver while holding the nut steady. Lift off the old seat. Clean around the mounting holes with a scrub pad to remove any sediment or debris.

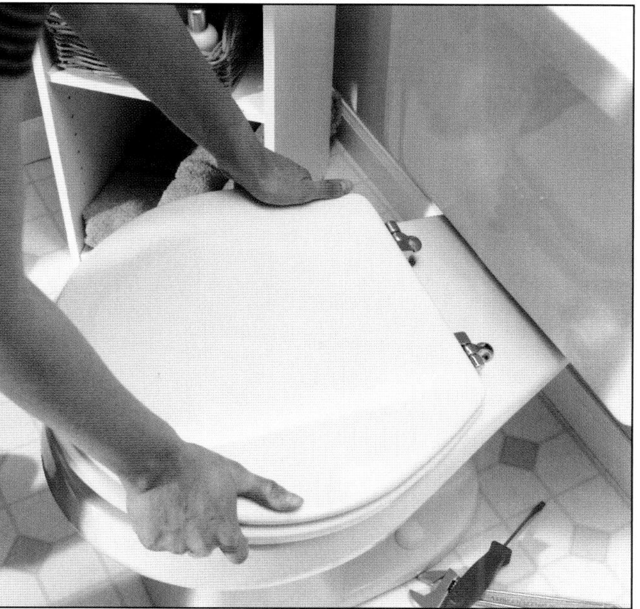

2 INSTALL THE NEW SEAT. To install a seat with metal hinges, align the holes of the toilet with the hinge bolts of the seat. Drop the bolts through the holes in the bowl. Fasten the nuts hand-tight and adjust the seat so it's centered properly. Tighten (but don't overtighten) with an adjustable wrench. (If you're installing the same type of seat you removed, reverse the procedure in step 1.)

MAINTAINING A WATER HEATER

REPAIR AND MAINTENANCE

EASY	MEDIUM	HARD

SKILLS: Attaching wire to a terminal. Operating plumbing fittings.

HOW LONG WILL IT TAKE?

PROJECT: Maintaining a water heater.

EXPERIENCED	20 MIN.
HANDY	20 MIN.
NOVICE	20 MIN.

TOOLS: Garden hose, screwdriver, bucket

MATERIALS: None

BE WARY OF EXPOSED WIRES! Always use care when working around exposed wires. If possible, turn off the electrical supply at either the switch box or the circuit breaker.

Today's water heaters are designed to last 10 years. Regular maintenance can extend the life of the water heater by 15 years or longer.

Standard water heaters are designed to be easily maintained. Access panels allow you to remove and replace worn parts. If you need to replace any parts, make sure you replace them with parts that match the make and model of the existing water heater. A plate attached to the outside of the water heater usually has product information.

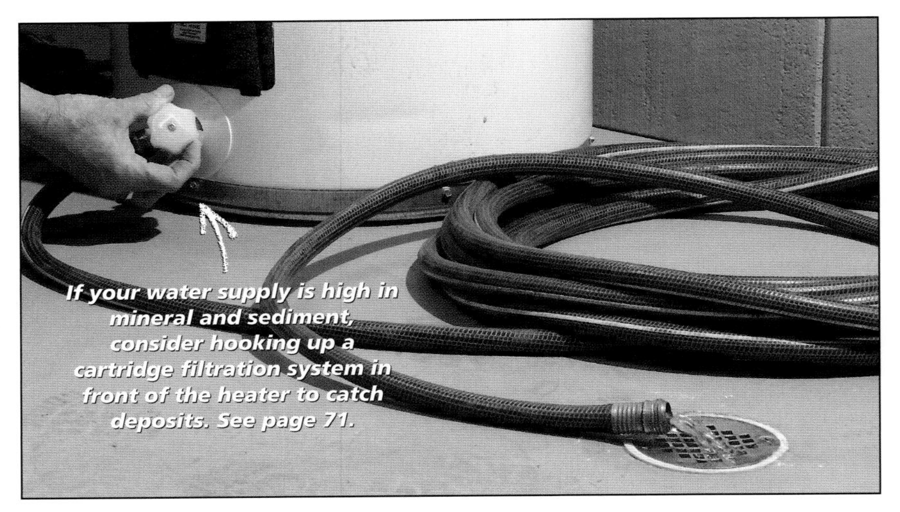

If your water supply is high in mineral and sediment, consider hooking up a cartridge filtration system in front of the heater to catch deposits. *See page 71.*

1 TURN OFF THE ELECTRICITY OR SHUT OFF THE GAS AND ATTACH A GARDEN HOSE TO THE DRAIN VALVE AT THE BASE OF THE HEATER. Find the water heater drain near the base of the water heater. Connect a standard garden hose to the outlet. Place the other end of the hose near a floor drain or in a large bucket. Leave the water supply on. Open the drain spigot to flush the water heater system until the water runs clear, usually less than 3 or 4 gallons.

WATER HEATER MAINTENANCE TIPS.

Water heaters are workhorses unless you don't take care of them; then they'll reward you with a cold shower!

1 Flush the system at least once a year. Sediment from the water supply builds up in the tank and makes it heat less efficiently. Sediment is more of an issue in gas heaters because the burner is at the base of the unit and has to heat through the gunk to get to the water.

2 Check the elements in an electric water heater. There are two of them and the top one is usually the first to fail.

3 Don't test the pressure-relief valve; once it's been opened, it will need to be replaced.

4 Check the flue on a gas water heater at least once a year to make sure the ducts are properly aligned and the tape is secure.

ADJUSTING THE WATER TEMPERATURE

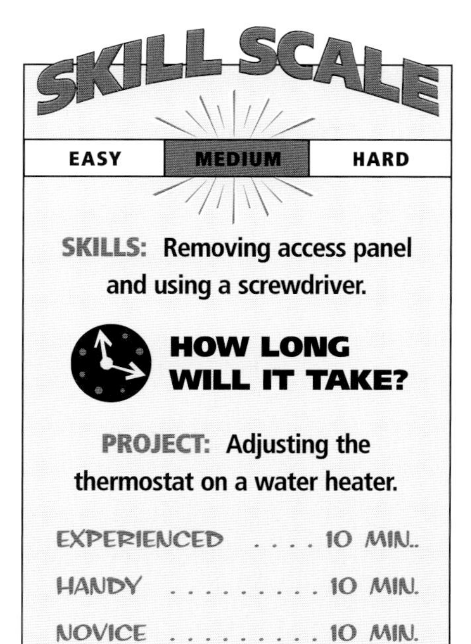

SKILLS: Removing access panel and using a screwdriver.

HOW LONG WILL IT TAKE?

PROJECT: Adjusting the thermostat on a water heater.

EXPERIENCED .... 10 MIN.

HANDY .... 10 MIN.

NOVICE .... 10 MIN.

STUFF YOU'LL NEED

TOOLS: Screwdriver

MATERIALS: None

Adjust the temperature of the water heater yourself. Wait one hour and test the temperature of the water at a remote faucet. If you need to make another adjustment, do it in 10-degree increments. Be aware that increasing the water temperature may shorten the life of the unit; lower settings can lengthen the life of the water heater. You will also see an increase on your utility bill if you increase the water temperature. You should follow the manufacturer's suggested temperature setting.

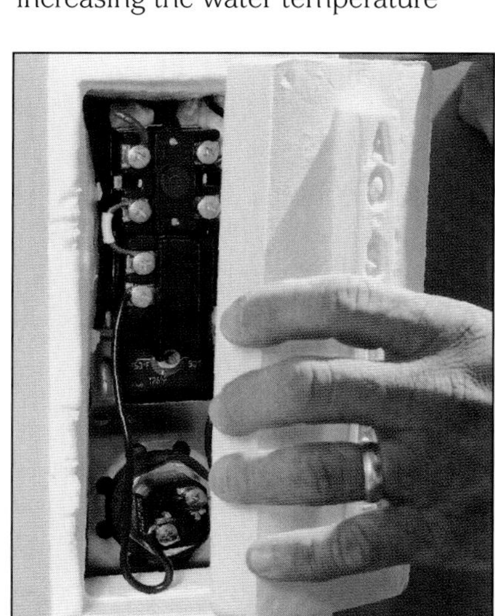

1 TURN OFF THE ELECTRICITY AND REMOVE THE ACCESS PANEL. Pull back any insulation to expose the thermostat.

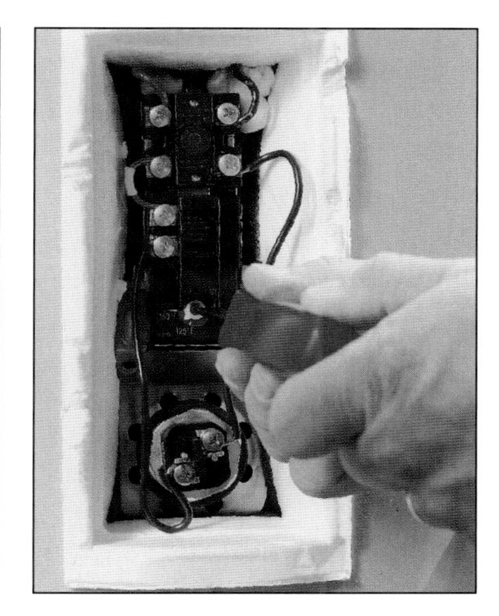

2 ADJUST THE SETTING. Use a screwdriver to change the thermostat in 10-degree increments. Close up the panel, restore power, and check the water temperature after one hour. Repeat until it's where you want it.

PREVENT SCALDING!

Don't set the thermostat too high. Water can scald a person quickly. Water that is 65.5 degrees C scalds in just half a second. Water that is set at 60 degrees C scalds in just one second. It will take water at 49 degrees C four minutes to scald a person.

ADJUSTING THE THERMOSTAT ON A GAS WATER HEATER.

It's easy to adjust the temperature on a gas water heater. The thermostat is on the outside of the tank and can be easily adjusted with the control knob. Change the setting in 10-degree increments. Wait an hour and test the temperature of the water at a remote faucet. Readjust another 10 degrees if you want. Don't set the temperature too high; that will shorten the life of your water heater. You'll also see an increase on your energy bill.

REPAIR AND MAINTENANCE

TROUBLESHOOTING A GAS WATER HEATER

NEVER IGNORE GAS ODORS!

If you think you smell natural gas odor, call the gas company immediately. Don't turn on any light switches! Open all the windows and leave the house.

SKILL SCALE
EASY
MEDIUM
HARD
SKILLS: Ability to light a gas pilot.
HOW LONG WILL IT TAKE?
PROJECT: Relighting gas water heater pilot lights.
EXPERIENCED . . . . . 10 MIN.
HANDY . . . . 15 MIN.
NOVICE . . . . 30 MIN.

✓ STUFF YOU'LL NEED

MATERIALS: Long wooden fireplace matches or a barbecue lighter

Cold water pouring out of the hot shower tap is an annoying surprise, to say the least. You may need to flush the sediment out of the system. (See "Maintaining a Water Heater," page 52.) Or you may just need to relight the pilot as you would on a gas stove. There will be detailed instructions for relighting your water heater on a plate located on the side of the tank next to the pilot and thermostat control knob. Long wooden matches or a barbecue lighter will easily reach the pilot through the access opening. Don't use a propane torch; it will melt the thermocouple. If the pilot does not stay relit, replace the thermocouple. (See pages 60–61.)

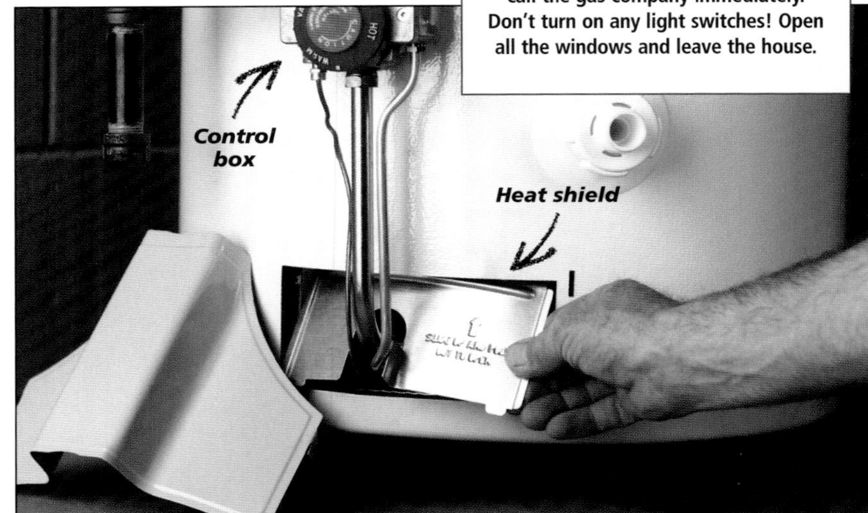

1 REMOVE THE COVER PANEL. Lift off the access panel at the bottom of the tank. This is where the gas line enters the water heater tank and where the pilot is located. Remove the heat shield and look inside to see if there is a flame. If not, shut off the gas at the supply valve. Disperse excess gas by blowing gently into the opening before you relight the pilot.

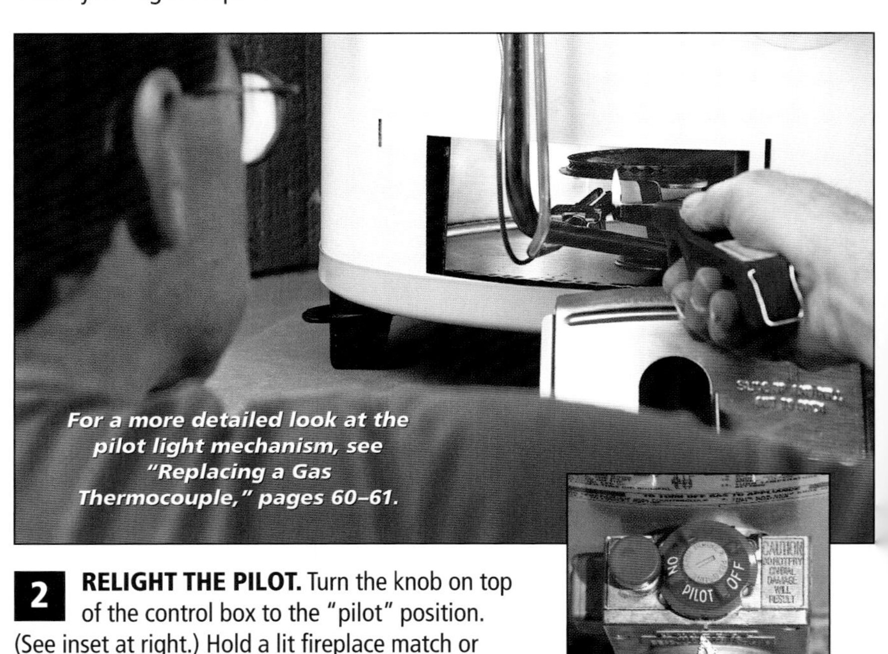

For a more detailed look at the pilot light mechanism, see "Replacing a Gas Thermocouple," pages 60–61.

2 RELIGHT THE PILOT. Turn the knob on top of the control box to the "pilot" position. (See inset at right.) Hold a lit fireplace match or barbecue lighter to the pilot access opening below. Press and hold down the "reset" button on the control box for 60 seconds after the pilot ignites, then turn the control knob to the "on" position. The main burner should ignite. If it doesn't, turn the control knob to the "off" position and call the gas company.

TROUBLESHOOTING AN ELECTRIC WATER HEATER

If your hot water is discolored, try flushing the tank. Sediment can cause discoloration.

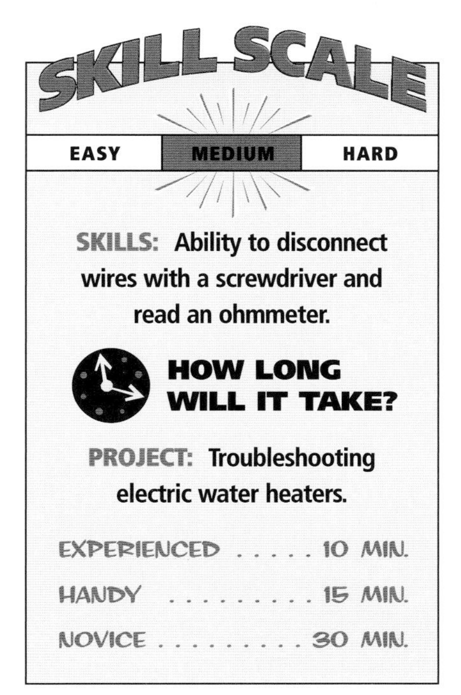

✓ STUFF YOU'LL NEED

TOOLS: Screwdriver, continuity tester

MATERIALS: None

WATER HEATER OVERHEATING!

Turn on all hot water faucets in the house to relieve pressure buildup. Turn off the circuit breaker. If you suspect a faulty thermostat, see pages 58–59 for how to replace it.

Before you call the plumber, troubleshoot your electric water heater. You can do it with a screwdriver and a continuity tester.

Water heaters heat the water in an insulated cylinder with a heating element—smaller units will have one element; larger ones may have two. The temperature within the tank is controlled by one or two thermostats attached to the outside.

UPPER OR LOWER?

A simple way to troubleshoot is to turn on a faucet in the house and let the water run for a while. If you only get a large amount of lukewarm water, chances are the upper element or thermostat is not working. (See pages 56–57 for replacing the element and pages 58–59 for replacing the thermostat.) A small volume of hot water at the faucet means the lower thermostat or element is not working. (See pages 56–57.)

1 REMOVE THE ACCESS PANEL. Turn off the water heater circuit breaker at the main electrical panel. Remove the access panel from the water heater. Remember that large-capacity water heaters may have two access panels.

2 DISCONNECT WIRES. Use a screwdriver to disconnect the wires to the water heater element terminals.

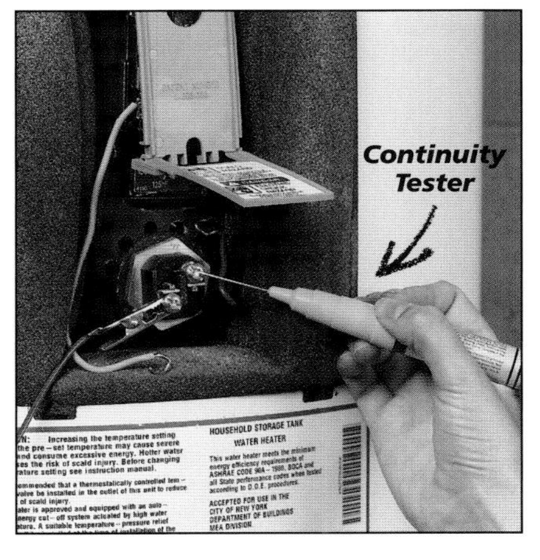

3 TEST THE ELEMENT FOR CONTINUITY. A continuity tester tells you if a circuit is complete. A heating element is like a circuit. If the light comes on, the element is OK. If not, it needs to be replaced.

REPLACING AN ELECTRIC WATER HEATER ELEMENT

EASY | MEDIUM | HARD

SKILLS: Attaching wire to a terminal. Removing fittings with a wrench or pliers.

HOW LONG WILL IT TAKE?

PROJECT: Replacing an electric water heater element.

EXPERIENCED .... 30 MIN.
HANDY .... 50 MIN.
NOVICE .... 1.5 HRS.

STUFF YOU'LL NEED

TOOLS: Voltage tester, element wrench or water-pump pliers, silicone grease, garden hose

MATERIALS: Heater element

TRIP SAVER

SAVE YOURSELF SOME TIME!

Most plumbers replace the thermostat when they replace a heating element on a water heater. (See "Replacing a Water Heater Thermostat," pages 58–59.)

Larger water heaters have an upper and lower heating element. Separate thermostats control each one. If your hot water doesn't get hot enough, the upper element is usually the problem. If you run out of hot water too quickly, replace the lower heating element.

SCREW-IN OR FLANGE?

Depending on the heater, the elements will either screw in or attach with a flange.

DON'T BLOW IT!

When working on an electric water heater, the first step is to turn off the power. Don't restore power until the job is done. Also, make sure the heater is full of water before you restore power; otherwise you will destroy the element.

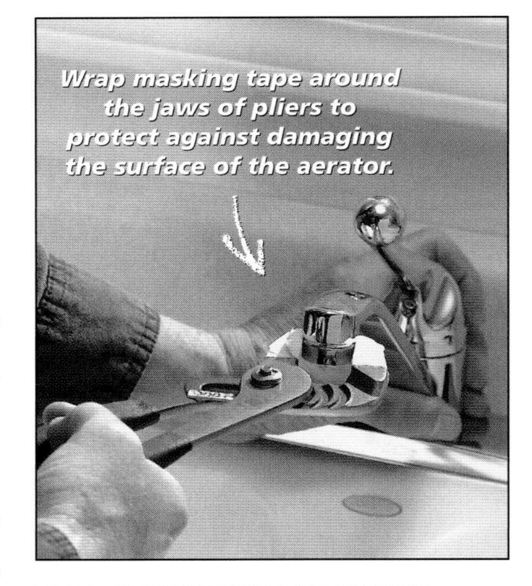

Wrap masking tape around the jaws of pliers to protect against damaging the surface of the aerator.

1 REMOVE THE AERATORS. Replacing a water heater element loosens sediment in the tank, which can go through the pipes and clog faucet aerators. Remove the aerators with water-pump pliers. Protect the aerator surface by wrapping tape around the jaws of the pliers.

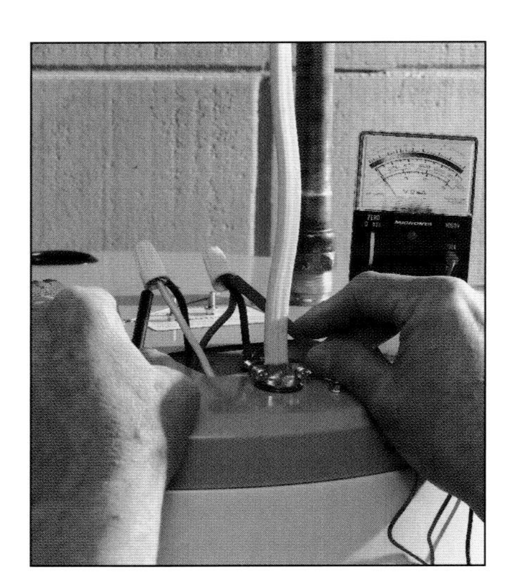

2 TEST FOR VOLTAGE. Turn off the power to the water heater at the circuit breaker in the main electrical panel. Use a voltage meter to make sure the power is off. Shut off the water flow to the water heater at the supply valves.

3 DRAIN THE TANK. Connect a garden hose to the water heater drain. Place the end of the hose near a floor drain. Open the drain valve to drain the water in the tank. Close the valve and remove the hose when it's empty.

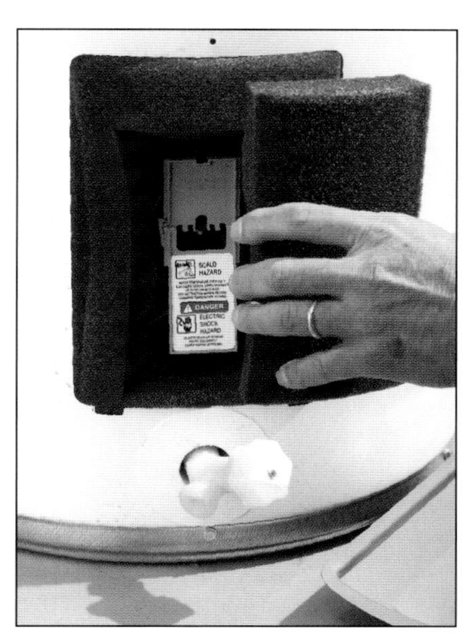

4 REMOVE THE ACCESS PANEL. Remove the access panel to access the heating elements.

5 DISCONNECT THE WIRES. Loosen the terminals connecting the wires to the heating element and remove the wires.

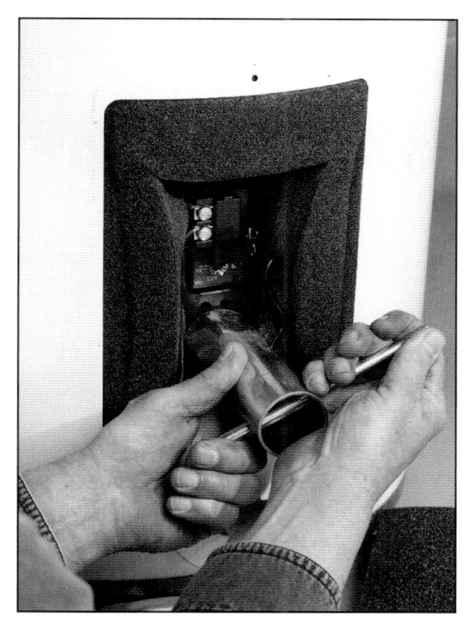

6 REMOVE THE HEATING ELEMENT. Removal is easy with an element wrench, or you can use water-pump pliers or an adjustable wrench. Turn the heating element counterclockwise to remove it.

The tank must be filled with water before the electricity is turned on — otherwise the heating element will melt.

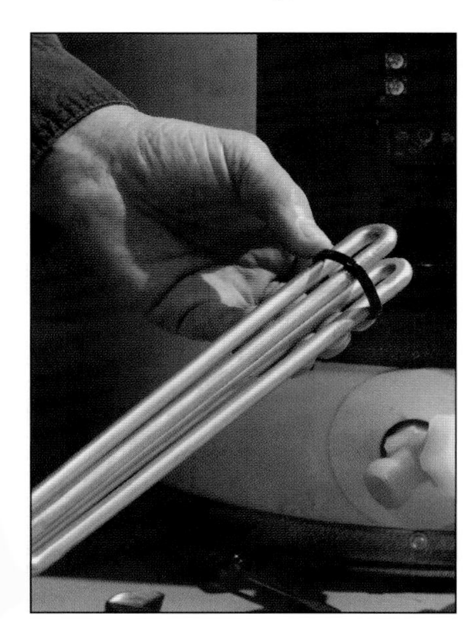

7 INSTALL A NEW GASKET. Coat both sides of the new element gasket with heat-resistant silicone grease and slide it over the heating element to the base.

Slip gasket over base threads. Seat firmly at the bottom of the base.

8 SCREW THE NEW ELEMENT INTO THE TANK. Screw the new element into the tank with the element wrench, then hook up the wires in the same order you removed them.

9 RESTORE THE SYSTEM. Turn on the water supply and open all faucets in the house. When there is a steady flow through the hot side, turn the faucets off. Reconnect the wires to the element terminals. Replace the access panel. Restore power to the water heater by turning on the circuit breaker. Press the "reset" button to activate the unit.

REPLACING A WATER HEATER THERMOSTAT

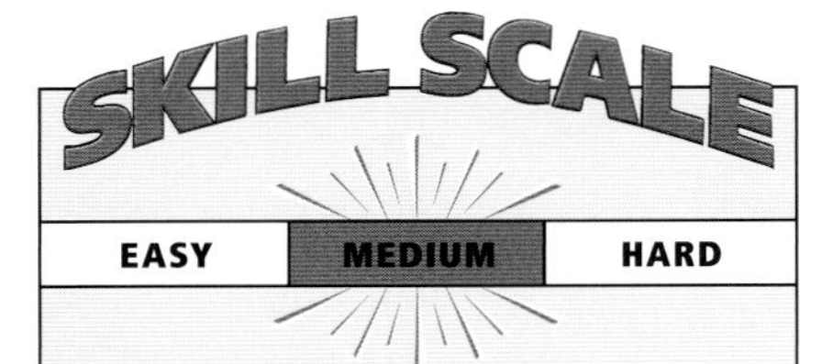

SKILL SCALE

EASY | **MEDIUM** | HARD

SKILLS: Using a screwdriver and attaching wire to a terminal.

HOW LONG WILL IT TAKE?

PROJECT: Replacing an electric water heater thermostat.

EXPERIENCED .... 30 MIN.
HANDY .... 45 MIN.
NOVICE .... 1 HR.

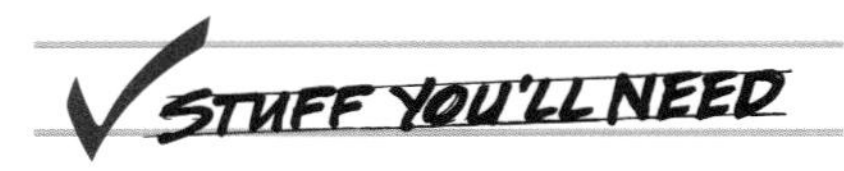

STUFF YOU'LL NEED

TOOLS: Screwdriver

MATERIALS: Thermostat, masking tape, marker

GOOD IDEA

EVERYBODY NEEDS A BLANKET.

Older water heaters can gain a new lease on life when you fit them with an insulating blanket of fiberglass. Kits that slip over the entire heater are simple to install and keep heat from escaping into the basement or garage.

A faulty thermostat or high-temperature cutoff can allow the water in the tank to heat well past recommended safe temperature ranges. Water heated above 49 degrees C can cause serious burns.

The same thing can happen if the high-temperature cutoff is defective. The thermostat and high-temperature cutoff are often linked into one unit. If they are separate, follow a basic rule of water heater maintenance: Replace both elements while you're at it. The parts are inexpensive and you'll save time and an extra trip to the home center.

1 TURN OFF THE POWER.
Disconnect the power to the water heater by turning off the circuit breaker at the main panel. If you haven't mapped your breaker box, shut off all the breakers one-by-one until you find the heater switch. Since you've gone to all the trouble to find the right breaker, mark it so you don't have to search next time.

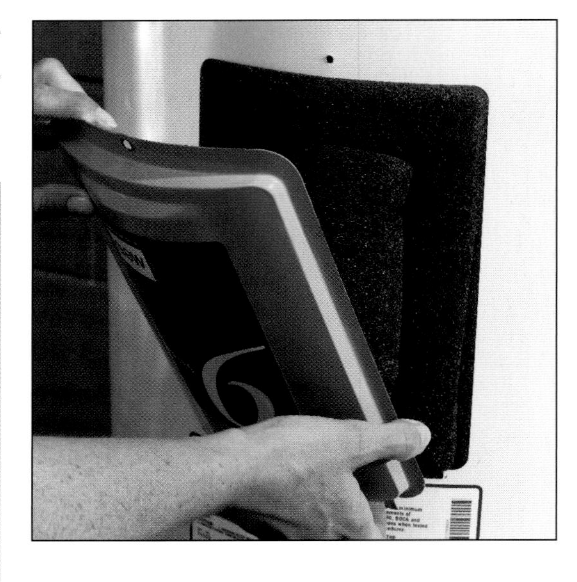

2 REMOVE THE ACCESS PANEL.
Use a screwdriver to remove the screws from the access panel. Lift off the panel.

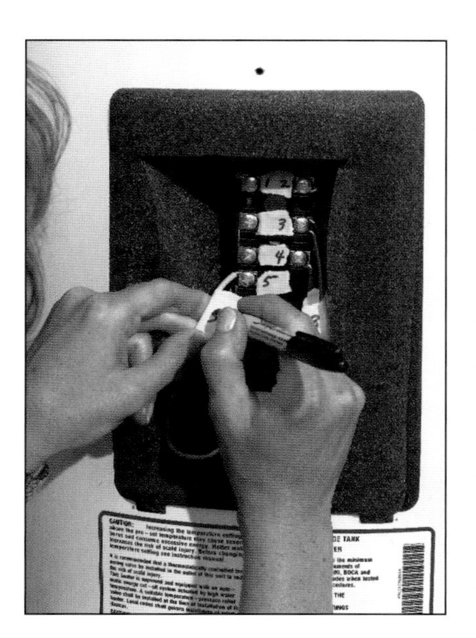

3 LABEL THE WIRES AND TERMINALS. Before removing the thermostat, number the wires and terminals. This will help you reconnect each wire to the proper terminal.

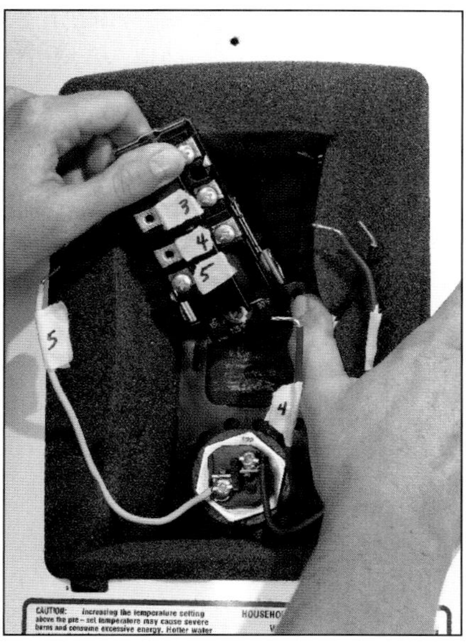

4 LIFT OUT THE THERMOSTAT. Disconnect the old thermostat from the mounting clip and lift it out. Replace it with a new thermostat of the same model.

DOUBLE-CHECK FOR SAFETY!

Be careful when working with electric water heaters. Most household appliances use 120 volts, which can give you quite a jolt. Electric water heaters require 240 volts. This is enough to do serious harm and could even be fatal. Always turn off the power at the circuit breaker in the main panel or remove the fuse. Use a voltage meter or a neon tester to make sure the power is off.

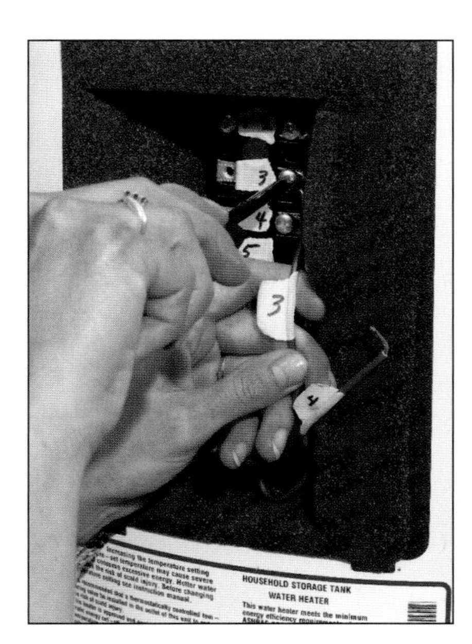

5 CONNECT THE WIRES. Install the new thermostat to the mounting clip. Reconnect the wires by the numbers you wrote on them in step 3.

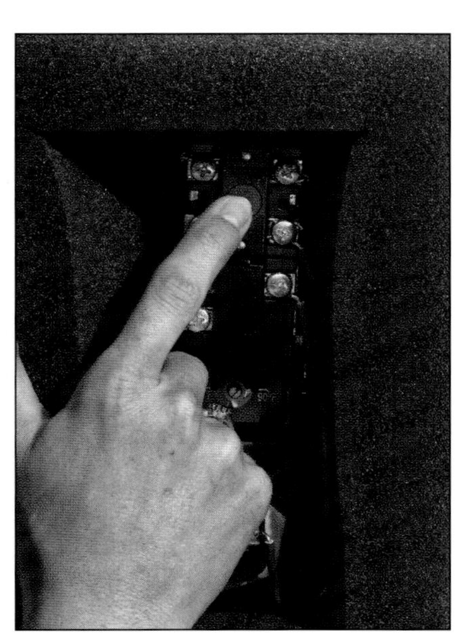

6 RESTORE THE SYSTEM. Set the thermostat to the manufacturer's recommended temperature. Make sure the tank is full. Restore power to the water heater at the circuit breaker. Press the red "reset" button on the thermostat.

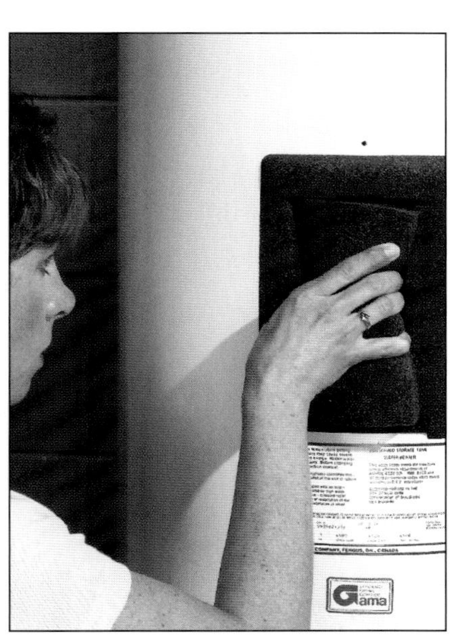

7 REPLACE THE ACCESS PANEL. Replace insulation and reinstall the access panel.

REPLACING A GAS THERMOCOUPLE

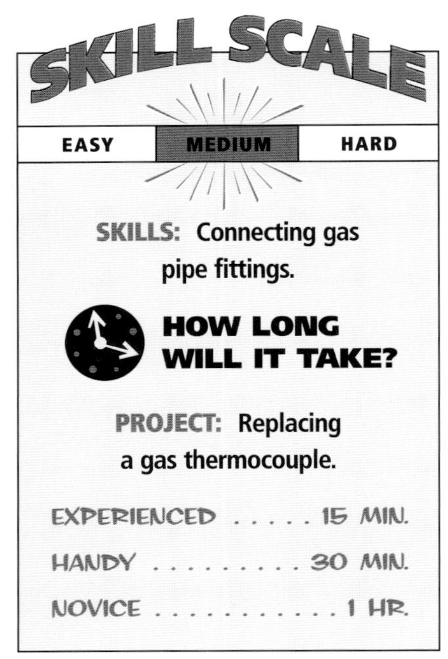

SKILL SCALE
EASY
MEDIUM
HARD
SKILLS: Connecting gas pipe fittings.
HOW LONG WILL IT TAKE?
PROJECT: Replacing a gas thermocouple.
EXPERIENCED ..... 15 MIN.
HANDY .... 30 MIN.
NOVICE .... 1 HR.

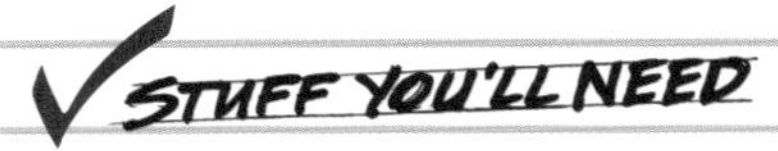

STUFF YOU'LL NEED

TOOLS: Screwdriver, adjustable wrench, fireplace matches or barbecue lighter, stiff wire or paper clip

MATERIALS: Thermocouple

The water heater isn't working if cold water comes from a hot water tap. Remove the access panel and check to see if there's a flame at the pilot light. If the pilot light will not stay lit, the thermocouple may be the problem.

The thermocouple is a thin copper line with a tip that rests in the flame of the pilot light. It runs from the burner to the control box on the outside of the water heater. The thermocouple shuts off the gas when the pilot light goes out. A worn thermocouple has to be replaced.

While replacing the thermocouple, you should also clean the burner and tubes. Unscrew the burner from the tube. Use a stiff wire or paper clip to clean out debris clogging the burner holes. Shake out any debris that has fallen into the chamber. You may want to use a vacuum cleaner to remove any debris remaining in the tubing and the burner chamber.

SMELLING NATURAL GAS?

Natural gas does not have an odor. Utility companies add an odorant to the gas so you can tell if you have a leak. If the smell persists, open all windows, do not turn on any electrical switches or appliances. Leave the house immediately and call the utility company from a neighbors.

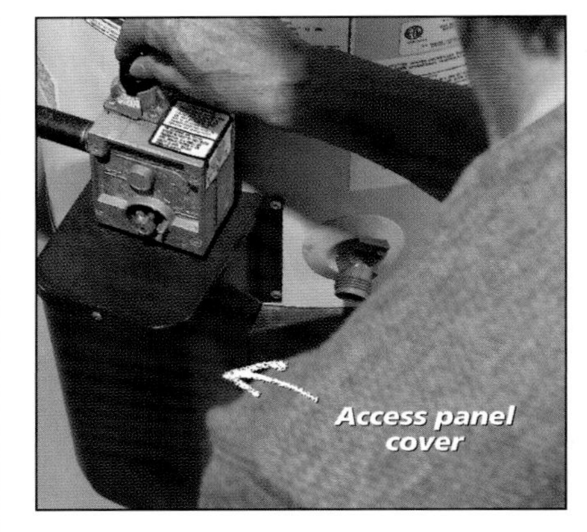

1 TURN OFF THE GAS. Locate the gas control knob on the top of the control box. Turn it to the "off" position, then wait five minutes for gas to dissipate. Remove the access panel cover.

2 DISCONNECT THE BURNER. Remove the access cover. Use an adjustable wrench to remove the connectors for all three tubes connecting the burner to the control box.

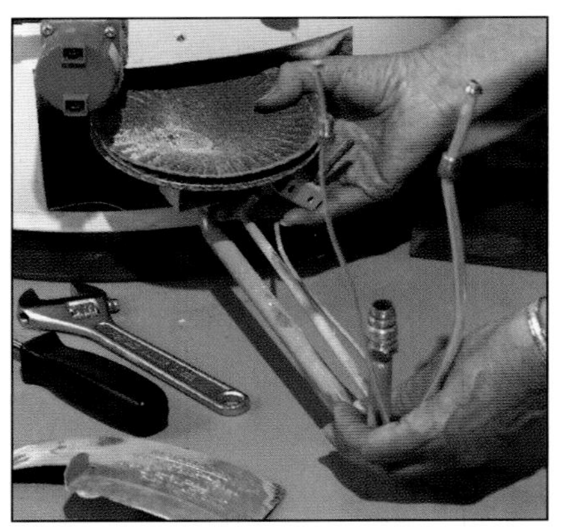

3 REMOVE THE BURNER. Remove the burner access panel and slide the burner out.

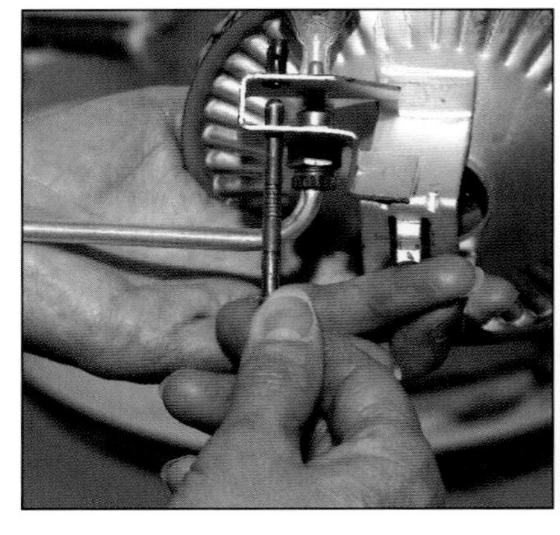

4 REMOVE THE OLD THERMOCOUPLE. Unsnap the old thermocouple from its mounting bracket and replace it with one of the same model.

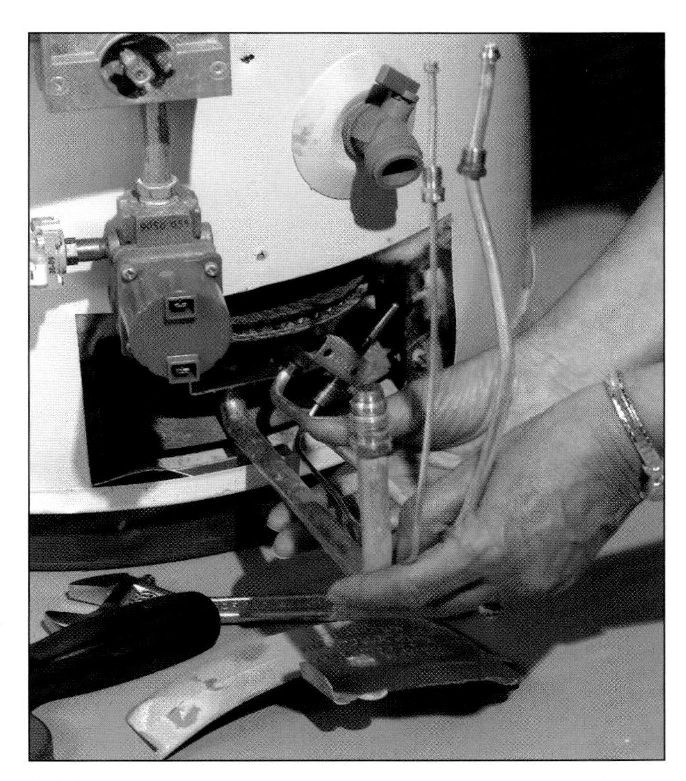

5 **INSERT THE BURNER AND THERMOCOUPLE.** Slide the burner unit back into the water heater. The flat end of the burner tube will slip into a slot in the water heater bracket. Carefully bend any excess thermocouple tubing into a coil that is at least 2 inches wide. If there are any kinks in the line, you'll have to replace it—again.

6 **RECONNECT THE LINES.** Align the tubes and reconnect them to the control box using an adjustable wrench. Test all the newly connected fittings for leaks. (See step 8, page 170.) Relight the pilot (see "Closer Look," below).

CLOSER LOOK

RELIGHTING THE PILOT.

To relight the pilot, turn the control knob to the "pilot" position—this will allow you to depress the reset button. Light a long fireplace match or use a barbecue lighter. Depress the button; hold it down while relighting the pilot for 35 seconds. Release the button and the pilot should stay lit. Turn the control knob to the "on" position and the burners will ignite. Check that the tip of the thermocouple is just inside the flame. **(See "Troubleshooting a Gas Water Heater," page 54.)**

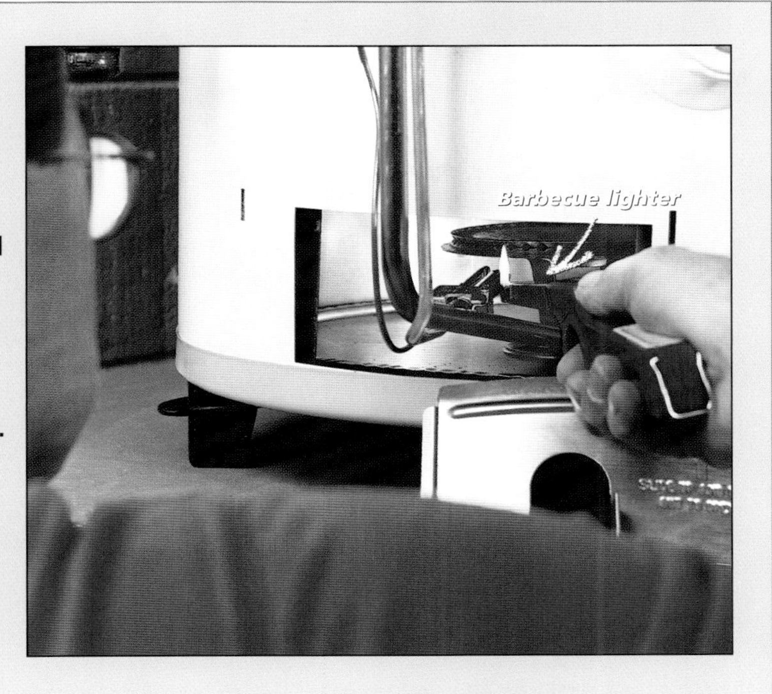

FIXING A SLOW-FILLING CLOTHES WASHER

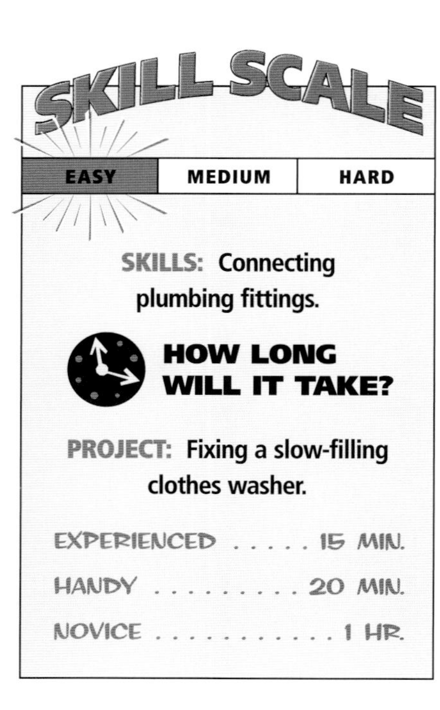

SKILL SCALE
EASY
MEDIUM
HARD
SKILLS: Connecting plumbing fittings.
HOW LONG WILL IT TAKE?
PROJECT: Fixing a slow-filling clothes washer.
EXPERIENCED . . . . . 15 MIN.
HANDY . . . . 20 MIN.
NOVICE . . . . 1 HR.

Does it take too long for the washer to fill? This may be a sign that the inlet screens are filled with sediment.

Hard water results in sediment buildup, caused by excess quantities of calcium carbonate and magnesium from soil and limestone.

Inlet screens prevent the sediment from entering the components of the washer and causing damage to the internal parts. Clogged parts can be expensive to replace and time-consuming to repair. Cleaning the screens is an easy project for a beginning do-it-yourselfer. While you're at it, replace the hoses with the "nonburst" braided type for a longer lasting repair.

STUFF YOU'LL NEED

TOOLS: Water-pump pliers, screwdriver or tweezers, small wire brush or old toothbrush, small bowl

MATERIALS: Inlet screens, white vinegar or cleaning solution

CHECK THE HOSES!
Hoses become brittle after several years. Instead of waiting for them to bubble, crack, and fail, buy a new set. Look for "nonburst"-type hoses. A reinforced cover prevents the hose from bursting. They're a little more expensive than the regular hose; but if a hose bursts, you're not going to appreciate the money you saved.

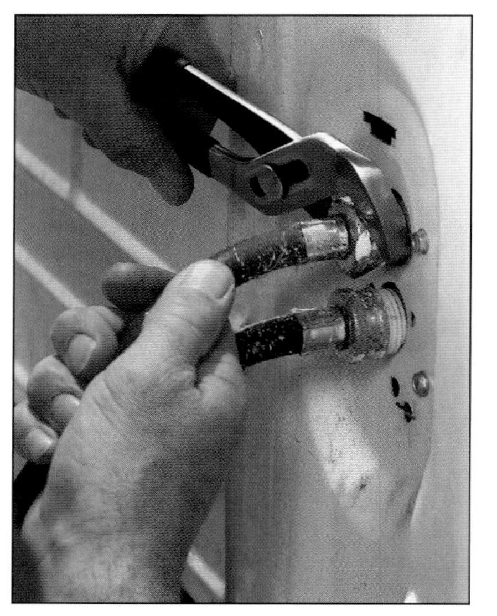

1 DISCONNECT THE FILL HOSES. Unplug the washer and pull it away from the wall so you'll have enough room to work. Turn off the hot and cold water to the unit at the shutoff valves. Remove the hoses from the back of the washer using water-pump pliers. Drain any water left in the lines into either a bucket or the floor drain. If water continues to drip out of the lines, you may need to repair the shutoff valves; sediment may build up on the faucet washers because of lack of use.

2 PRY OUT THE INLET SCREENS. Use tweezers, a screwdriver, or a pick to remove the inlet screens. These screens prevent debris from entering the washer. Work carefully—if you damage a screen, you will need to replace it. You can reuse the screens after cleaning them. You can find replacements at your local home center.

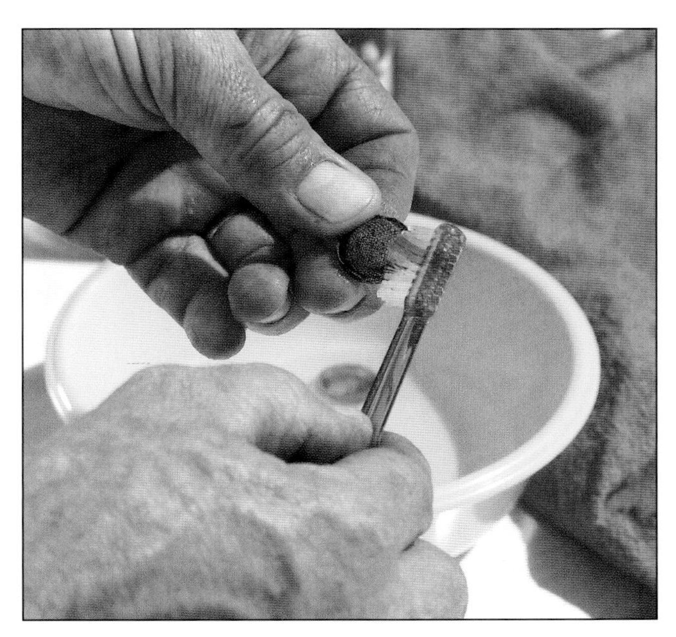

3 **CLEAN THE INLET SCREENS.** Soak the screens overnight in a white vinegar solution or purchase a cleaning solution for removing hard water deposits. This will loosen the mineral deposits that tend to build up on the screens. Use an old toothbrush or a small wire brush to clean any remaining sediment. Wash under running water and inspect to see if more soaking and cleaning are necessary.

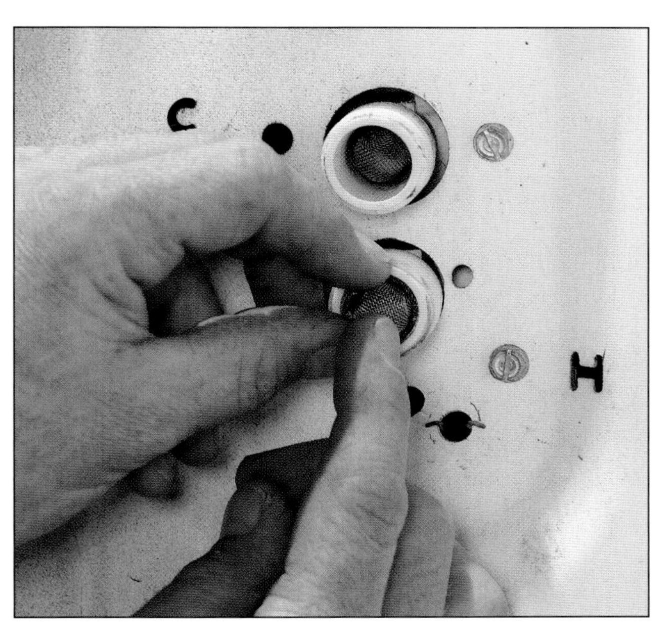

4 **INSERT THE INLET SCREENS INTO THE WASHER.** Push the inlet screens into the inlet openings with the dome-shaped side facing out. Be careful not to damage the screens when you insert them. You may need to use tweezers or a screwdriver to seat the screens properly.

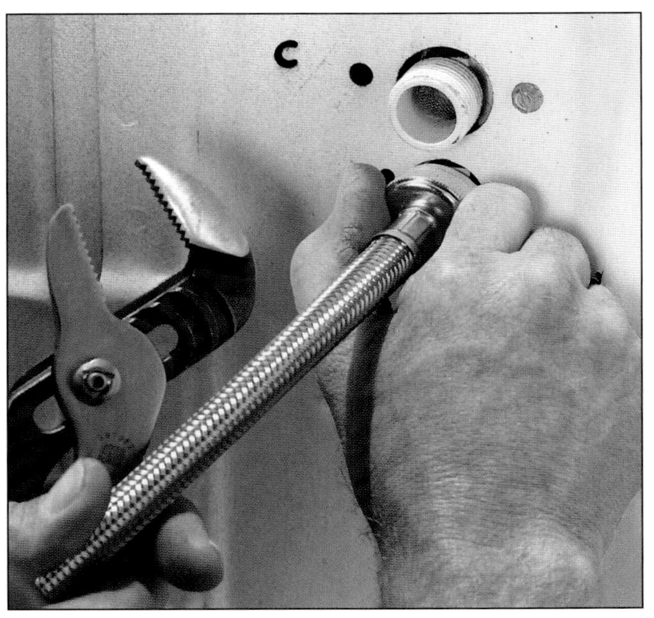

5 **CONNECT THE NEW HOSE LINES TO THE WASHER.** Attach nonburst hot and cold water lines to the inlet openings. Hand-tighten the hoses. Tighten a quarter turn with water-pump pliers. Push the washer back in place and check to make sure it is level. Add shims or adjust the feet under the washer until it's level. Turn on the hot and cold water at the supply valves and check for leaks.

FIX ALL THE PROBLEMS WHILE YOU'RE AT IT.

While fixing my slow-filling washer, I noticed that the valve faucets had a slow leak. I cranked down on the valve until the leak stopped. Feeling satisfied, I cleaned out the inlet screens and reconnected the washer. I opened the valves, inspected the washer, and patted myself on the back for a job well done. Later, when I came down to wash a load of laundry, I noticed a wet spot on the floor. Moving the washer out of the way, I found that the leaks were from the valve seats in the faucets being overtightened. If you notice the faucet valves are leaking, replace the washers and valve seats. It's always easier to do the job right the first time!

PROTECTING PIPES FROM FREEZING

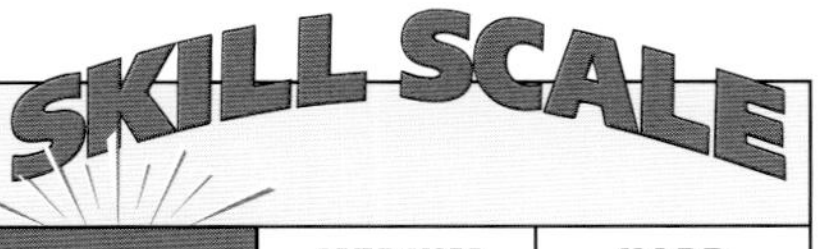

EASY	MEDIUM	HARD

SKILLS: Cutting and fitting various types of insulation around pipes.

HOW LONG WILL IT TAKE?

PROJECT: Protecting pipes from freezing.

EXPERIENCED . . . . VARIABLE
HANDY . . . . .VARIABLE
NOVICE . . . . VARIABLE

STUFF YOU'LL NEED

TOOLS: Carpenter's level, hair dryer or heat lamp, watering can, towel, bucket

MATERIALS: Polystyrene foam, pipe insulation tubes, hot water

NEVER PLACE NEW HEAT TAPE OVER OLD!

Remove old heat tape before wrapping the pipe with new tape. Placing new heat tape over old creates a potential fire hazard.

Freezing climates pose a threat to both outdoor and indoor plumbing. Water expands when it freezes, fracturing pipes and valves. It's important to bury pipe below the frost line outdoors—call your local public works department for frost depth in your area. You should protect piping indoors that is exposed to exterior walls. Drain unused lines that may be exposed to freezing. If you're not using them, you don't need water in them.

NO NEED TO FREEZE

You can prevent pipes from freezing by not running supply lines in exposed areas or against exterior walls, wrapping pipes with sleeve-type foam insulation, or protecting them with an insulation wrap, such as heat tape. Be careful when using heat tape, however. It can deteriorate over time and pose a fire hazard. You should inspect it occasionally and replace it if it is worn.

PIPE INSULATION

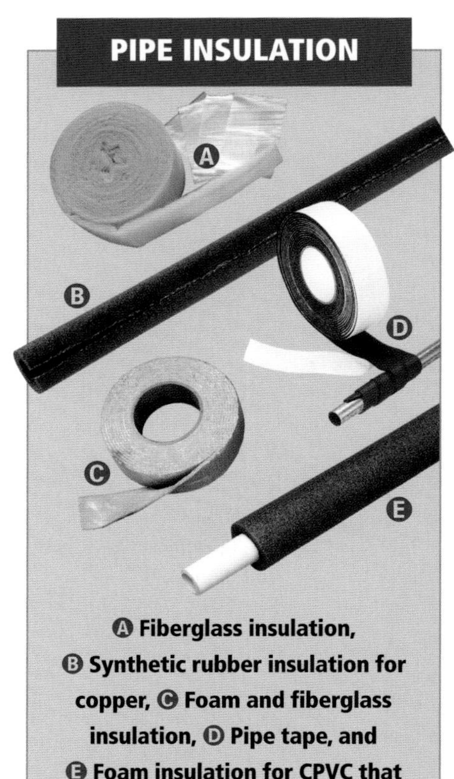

Ⓐ Fiberglass insulation, Ⓑ Synthetic rubber insulation for copper, Ⓒ Foam and fiberglass insulation, Ⓓ Pipe tape, and Ⓔ Foam insulation for CPVC that won't soften or damage pipe.

1 PROTECT SHALLOW BURIED PIPES. Outdoor piping should be buried below the frost line. Call your local public works department and ask for the frost depth in your area. If it's too deep or you can't dig, you can still protect pipes from damage during the winter months. Cover shallow pipes with a sheet of polystyrene foam to insulate them. For extra protection, drain a shallow line before the winter, if possible.

2 SLOPE PIPES AWAY FROM THE HOUSE. Outdoor pipes that are not buried can be a problem in areas that freeze. Slope the pipes away from the house and install an air gap in the line. You also should install a valve to drain the line during severe cold weather to prevent freezing and ruptures.

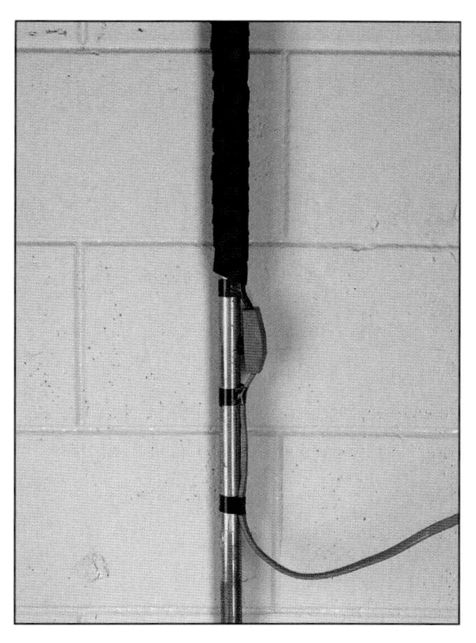

3 INSULATE PIPES ALONG EXPOSED EXTERIOR WALLS. Pipes installed against exposed exterior walls can freeze during winter months. Insulate them with a foam jacket. This protects the pipes from freezing and is an excellent way to conserve energy. Hot water pipes tend to lose heat rapidly, so wrapping them will help maintain water temperature.

4 TRY HEAT TAPE. Heat tape or cable wraps around or attaches to pipes and plugs into a standard outlet. Use only UL-approved materials and follow the manufacturer's instructions carefully.

Some types of insulation will soften and damage CPVC. Check manufacturer's instructions and use only foam-type insulation on plastic pipe.

5 LEAVE THE FAUCET OPEN. We can't always plan or build for temperature extremes. Should your area experience unusually cold temperatures, use this stop-gap method for preventing pipes from freezing. Leave faucets that are connected to piping on outside walls open so water trickles from them. This is not energy efficient, but it will get you through a day or two of extremely cold weather. Also, leaving the cabinet doors open will help keep pipes warm.

REPAIR AND MAINTENANCE

CLOSER LOOK

THAWING FROZEN PIPES.

If you forget to do something to prevent your pipes from freezing and one pipe is frozen, there's still a solution. Turn off the water from the main shutoff valve. Inspect the pipe for damage. Look for ruptures along the pipe that may have been caused by expansion of the freezing water. If you find a rupture, fix it. **(See "Emergencies—Quick Fixes for Leaking or Burst Pipes," pages 68-69.)** **If the pipe does not appear to be fractured, drain out any water in the line by opening a downstream faucet. Use A, a hair dryer or heat lamp to thaw the pipe. Or B, place a bucket under the pipe, wrap the pipe with an old towel, and pour hot water over the towel to thaw out the pipe. Once the pipe is thawed, have someone turn on the water supply while you inspect the pipe for leaks and damage.**

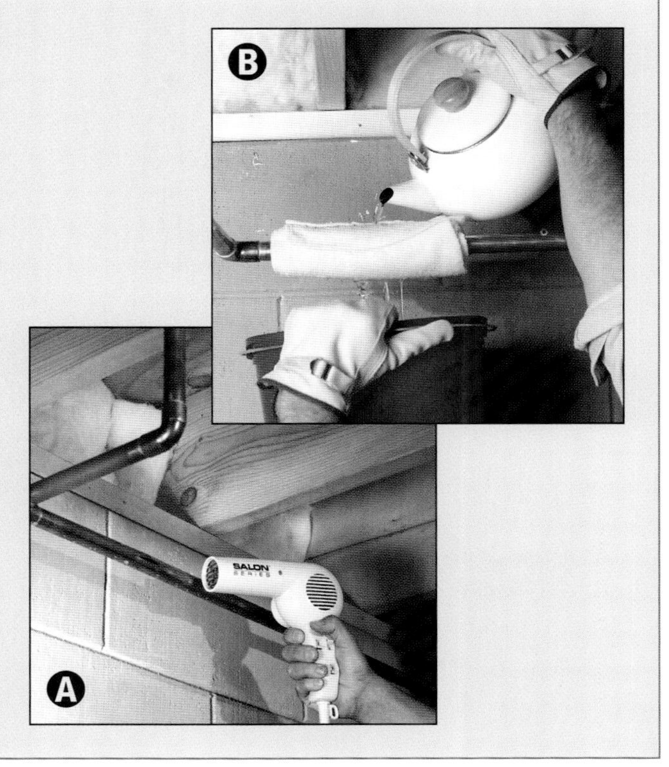

FIXING NOISY PIPES

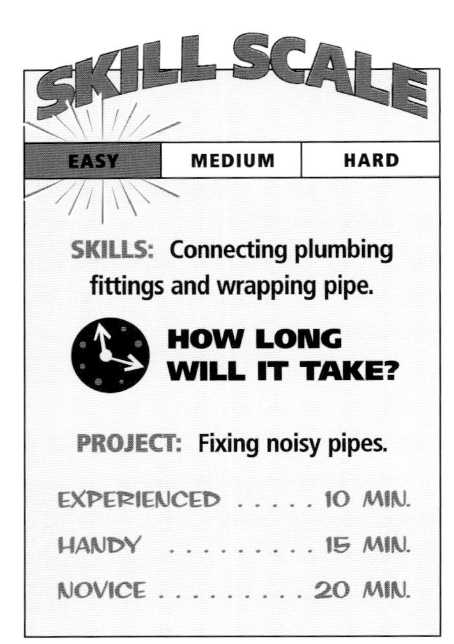

SKILL SCALE
EASY
MEDIUM
HARD
SKILLS: Connecting plumbing fittings and wrapping pipe.
HOW LONG WILL IT TAKE?
PROJECT: Fixing noisy pipes.
EXPERIENCED ..... 10 MIN.
HANDY .... 15 MIN.
NOVICE .... 20 MIN.

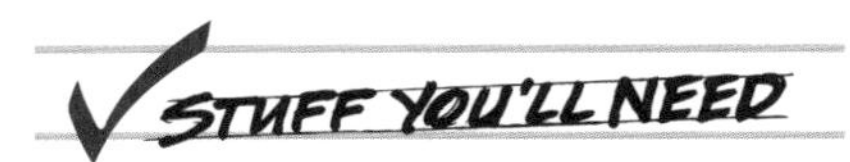

STUFF YOU'LL NEED

TOOLS: Water-pump pliers, utility knife

MATERIALS: Water hammer shock absorbers, pipe foam insulation tubes, vibration clamps

If your car starts making strange sounds, your first thought is to get it repaired. The same should apply to your house. Banging or vibrating pipes are more than an annoyance; they may eventually damage the piping.

Noise that starts when you open your faucet may be a sign that the seat washer is defective, causing vibration or rattling. Replacing the washer may solve the problem.

Most local codes require a pressure of 80 pounds per square inch (psi) entering the house. This is too high for home piping. Install a pressure-reducing valve downstream of the water meter to lower the water pressure and to keep pipes and fittings from bursting.

Clothes washers—with their stop and start operation for supplying water to the drum—often cause pipes to bang. Quiet them with water hammer shock absorbers that you can buy at your local home center.

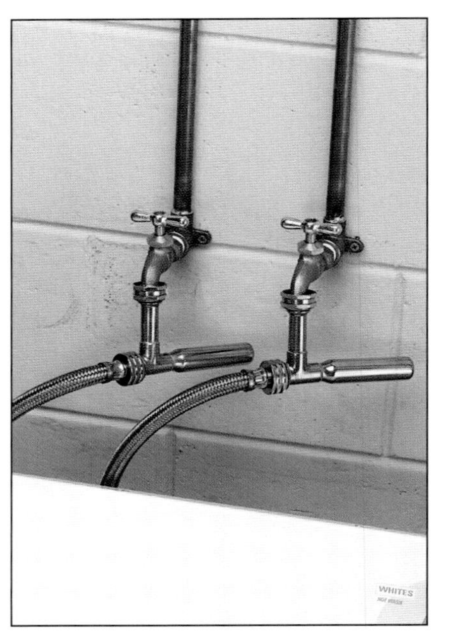

1 INSTALL WATER HAMMER SHOCK ABSORBERS ON NOISY PIPES. Turn off the hot and cold water supply valves. Disconnect the supply hoses from the valves using water-pump pliers. Connect water hammer shock absorbers to the valves and attach the supply hoses to the shock absorbers. Water hammer shock absorbers come in many styles with different installation requirements. They can be attached anywhere you have noisy pipes.

2 APPLY INSULATION TO PREVENT NOISE FROM VIBRATING PIPES. Have a helper turn on the supply to the section of line you suspect is causing the noise while you find the source of the banging. Once you find it, wrap foam insulation around the pipe at the point of impact. There are a variety of clamps and hangers available that will stop pipes from vibrating. (See "Clamps and Hangers" at left.)

CLAMPS AND HANGERS

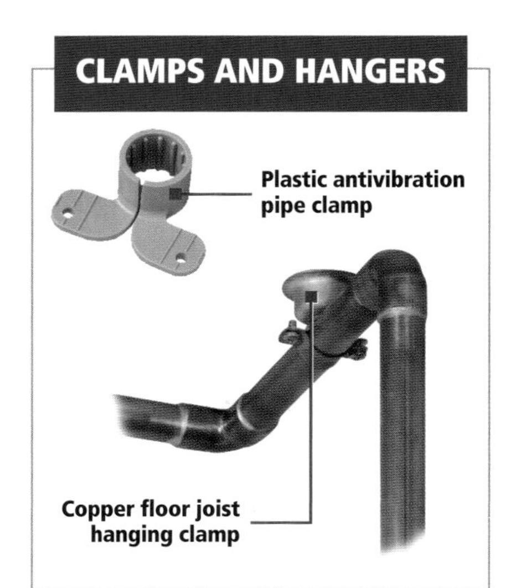

FIXING SWEATING PIPES

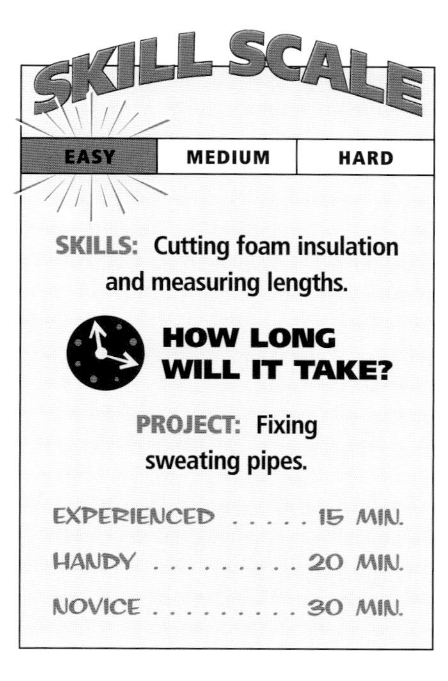

SKILL SCALE
EASY
MEDIUM
HARD
SKILLS: Cutting foam insulation and measuring lengths.
HOW LONG WILL IT TAKE?
PROJECT: Fixing sweating pipes.
EXPERIENCED ..... 15 MIN.
HANDY .... 20 MIN.
NOVICE .... 30 MIN.

Pipes sweat in areas of high humidity. Most basements tend to be damp, so you may have sweating pipes there.

The combination of the humidity in the area and water running through the pipes will cause condensation to form on the surface of the pipe. Operating a dehumidifier in the basement may solve the problem. Covering the pipes with foam tubing insulators will also prevent the buildup of condensation on the surface of the pipes. Tube insulators are available in most home centers and come in standard diameters to fit over indoor water lines.

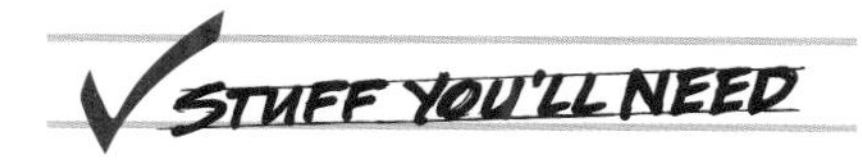

✓ STUFF YOU'LL NEED

TOOLS: Tape measure, serrated kitchen knife

MATERIALS: Pipe foam insulators

Install a vibration clamp here.

1 MEASURE THE PIPE. Use a measuring tape to measure the length and diameter of the pipe you wish to insulate. Go to your local home center and purchase pipe insulators to fit the diameter of the pipe. These usually come in packages of a standard length. Use a serrated knife, such as a kitchen knife, to cut the insulation to length.

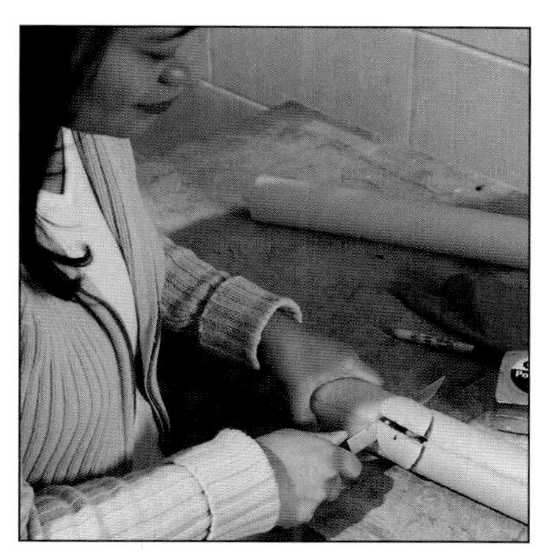

2 CUTTING CORNERS. To make 90-degree bends in the pipe, cut a 45-degree notch in the tubing. This will allow you to fit the tubing around the elbow.

3 WRAP INSULATION AROUND THE PIPE. Separate the insulation along the seam. Slip it over the pipe. Press the seam back together to seal.

EMERGENCIES—QUICK FIXES FOR LEAKING OR BURST PIPES

EASY	MEDIUM	HARD

SKILLS: Connecting plumbing fittings and applying epoxy.

HOW LONG WILL IT TAKE?

PROJECT: Quick fixes for leaking or burst pipes.

EXPERIENCED .... VARIABLE
HANDY .... VARIABLE
NOVICE .... VARIABLE

Finding a leak is half the battle. You might be amazed at how far water can travel before reappearing to ruin a wall or ceiling. Water can travel across joists and other surfaces. You may have to cut into walls and ceilings to find the source.

Most leaks occur at pipe fittings. Fractures result from corrosion, dents, or freezing.

Temporary fixes include epoxy and clamps. Permanent solutions involve cutting the pipe and installing dresser couplings. Never install a temporary fix behind a wall, because it will fail, and when it does, you'll be in the same trouble all over again.

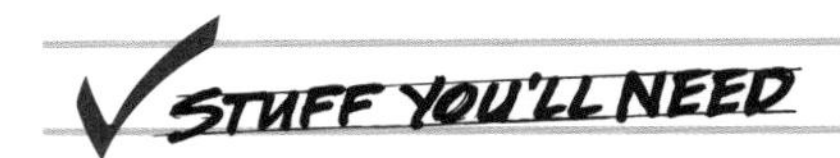

TOOLS: Tape measure, screwdriver or nut driver, water-pump pliers, flat metal file, scissors

MATERIALS: Plumber's epoxy putty; 1/16-inch-thick neoprene rubber or bicycle tube patch; sleeve clamp, hose clamp, or dresser coupling

BUYER'S GUIDE

PATCHING THE LEAK.

Quick fixes can be made for small leaks and wet surfaces with Ⓐ plumber's epoxy putty. More serious leaks require more drastic measures, but solutions abound. Try 1/16-inch-thick neoprene rubber or a bicycle tube patch with Ⓑ hose clamps, Ⓒ sleeve clamps, or Ⓓ dresser couplings.

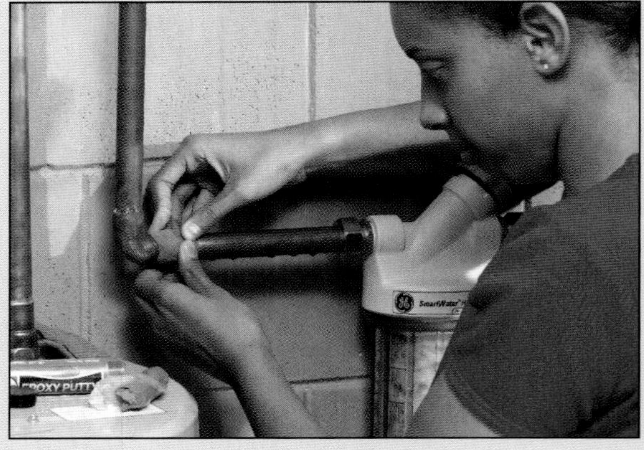

A QUICK FIX FOR A LEAKING PIPE.

Plumber's epoxy is a good quick fix for a small leak at a pipe joint. Turn off the water supply upstream of the leak. Tear off two pieces of the claylike ribbon. Knead enough plumber's epoxy putty to cover the surface around the leak. Apply the putty according to the manufacturer's instructions. Turn on the water and inspect for leaks. Remember, this is only a temporary fix, so make sure you make a permanent repair to the line.

1 **PREPARE THE SURFACE FOR PATCHING.** Clean the area around the rupture. Use a flat metal file on any sharp edges that may cut through the patch. Measure the length of the area you will need to patch and the diameter of the pipe. Buy the appropriate patching supplies at your local home center.

2 **CUT THE PATCH TO THE RIGHT LENGTH.** Use scissors to cut out the size of the neoprene rubber patch you will need to repair the leak. For a pinhole leak, cut a 1-inch square. For a fracture, cut a patch that is 1 inch wider and 1 inch longer than the split in the pipe.

REMEMBER: QUICK FIXES ARE ONLY TEMPORARY!

It's easy to convince yourself that a quick fix on a leaking pipe is the final solution. It's not; quick fixes are temporary. If you don't take time to go back and properly repair the leak, you'll forget about it. Before you know it, the fix will have failed or developed a new leak. Don't be tempted to apply another temporary measure. You'll be spending more time applying a new bandage than it would have taken to go back and do the job right the first time. Don't procrastinate!

3 **TIGHTEN THE CLAMPS TO SEAL THE LEAK.** Wrap the pipe with the neoprene rubber. Secure the piece with hose clamps. This type of patch will work for most small fractures. You may need to cut out the broken section of a larger fracture. Install either a dresser coupling or a sleeve clamp to fix this type of leak.

REPLACING A SUMP PUMP

SKILL SCALE

EASY | **MEDIUM** | HARD

SKILLS: Connecting plumbing fittings.

HOW LONG WILL IT TAKE?

PROJECT: Replacing a sump pump.

EXPERIENCED . . . . 20 MIN.
HANDY . . . . 40 MIN.
NOVICE . . . . 1 HR.

Sump pumps remove excess water from basements through a discharge line that exits the house. You should test the system once a year by pouring a couple gallons of water into the sump pit to activate the pump. Sump pumps are either submersible or pedestal style activated by a float. This project shows you how to replace a pedestal-style sump pump with a submersible pump.

✓ STUFF YOU'LL NEED

TOOLS: Water-pump pliers, carpenter's level, tubing cutter

MATERIALS: Sump pump, discharge pipe if necessary, plastic shims

BUYER'S GUIDE

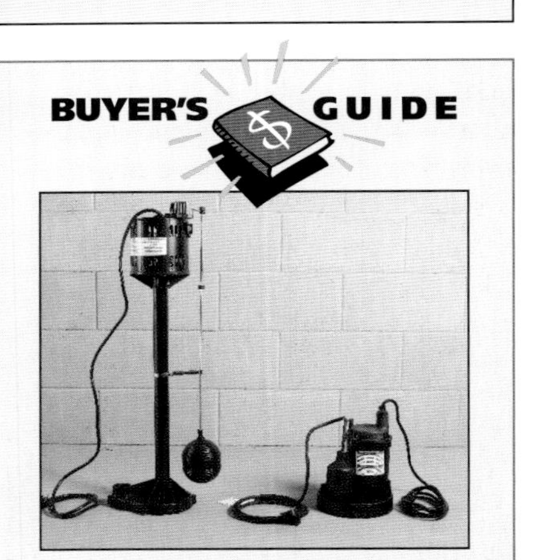

THE RIGHT SUMP PUMP. A float activates a pedestal sump pump. Submersible pumps must be under water to operate.

GOOD IDEA

GUIDE WATER AWAY. If water continues to come back into the system, it may be because the exterior drain is too close to the house. Make sure discharge water is guided away from the foundation.

1 DISCONNECT THE DISCHARGE UNION. Unplug the sump pump cord from the grounded receptacle. Use two pairs of water-pump pliers to disconnect the union on the discharge line entering the sump pit. Remove the old sump pump and any debris in the pit.

2 INSTALL THE NEW SUBMERSIBLE PUMP. Make sure the new pump sits level in the pit. Use plastic shims; wooden shims will rot. Reuse the old piece of discharge line from the old pump or cut a new piece. Connect the outlet pipe to the main discharge line with the union. Plug the sump pump cord into a grounded outlet. Never plug a sump pump into an ungrounded outlet because of potential electric shock. Test the new pump by filling the pit with water until the pump is activated.

CHANGING A FILTER CARTRIDGE

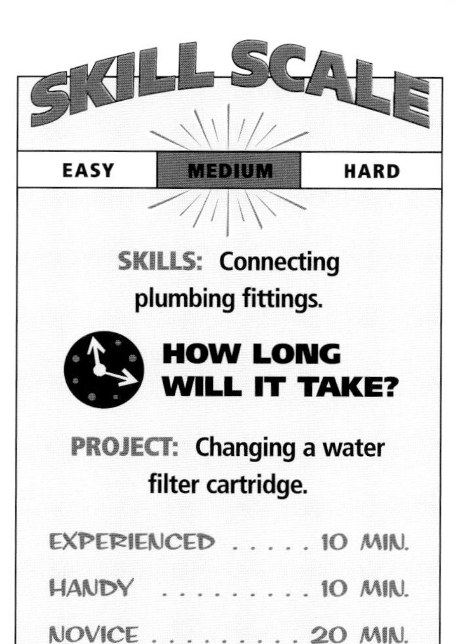

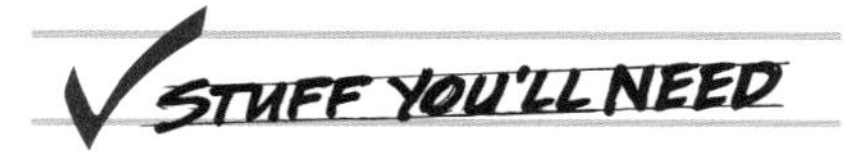

✓ STUFF YOU'LL NEED

TOOLS: Filter or strap wrench, bucket, sponge

MATERIALS: Filter cartridge, O-ring, silicone grease

KEEP SOME SPARES ON HAND.

Save on trips to the store by buying extra filters for your cartridge. You'll be able to change the filters on schedule, keeping the water free of sediments, and you'll save time spent running unnecessary errands.

Concern about minerals and sediment in the water supply has increased the installation of sediment filtration systems. The in-line cartridge filtration system is a popular style. These systems require filter replacement every 300 to 3500 liters. Follow the manufacturer's instructions for setting up a cartridge replacement schedule. Only use solutions recommended by the manufacturer for cleaning the inside of the cartridge housing.

WATER FILTRATION

Water softening and filtration systems range from very simple and basic to quite complex, but they do a great job of treating problem water. Consider installing a system if your water supply is high in mineral content.

1 TURN OFF THE FILTER VALVE. If there is no valve, turn off the water at the main shutoff valve. Place a bucket under the cartridge. Release the water pressure in the line by opening a downstream valve. Some styles will have a pressure-relief valve you press to release pressure.

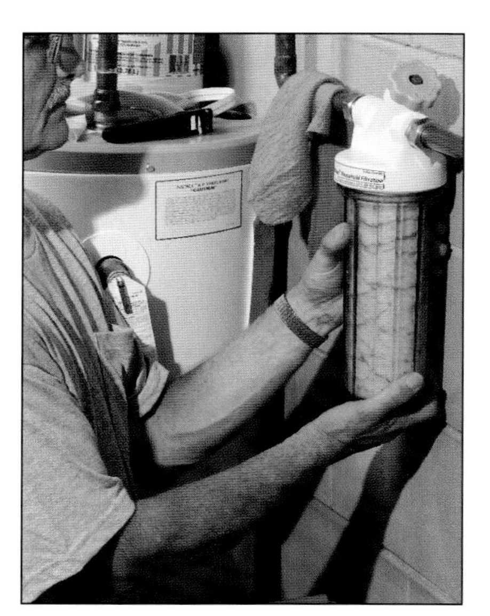

2 REMOVE THE CARTRIDGE HOUSING. Grasp the housing body with both hands and unscrew it counterclockwise. Use a water filter wrench or strap wrench to remove the housing if it is too tight to remove by hand. Remove the old filter cartridge.

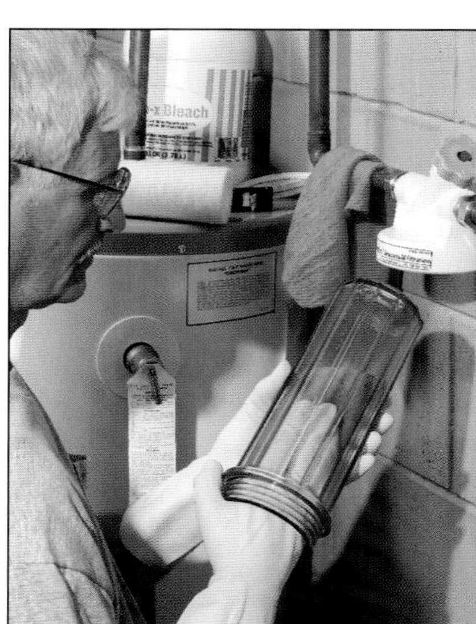

3 CLEAN THE HOUSING. Follow the manufacturer's recommendations for cleaning the housing. If the housing has an O-ring, replace it with a new one. Coat the O-ring lightly with silicone grease. Place the new cartridge in the housing and connect it. Tighten by hand.

UNCLOGGING THE CLOGS

You can clear a drain through the elbow with a hand spinner but for a stubborn clog, remove the elbow and go straight into the drain line.

GET A FEEL FOR THE CLOGS!

When you hit resistance with your auger, you've found the clog. Figuring out what's causing it is a matter of feel. Solid resistance may mean an object like a sponge or a tangle of hair. Use the auger to either retrieve the clog or push it into the main stack. If the auger meets continuous resistance but still advances slowly, it's probably a buildup of grease or soap. Bore through the clog several times, then flush the system with hot water to push what's left into the main drain.

Stopped or slow-moving drains are seldom the result of collapsed or defective pipes. Blockage caused by the accumulation of solid waste in the lines such as small objects, hair, or clumps of soap and grease are usually the culprits.

ISOLATING THE PROBLEM

If one fixture seems to be draining slowly or not at all, first check to see if other fixtures are having the same problem. If only one fixture is affected, the drain line to the fixture is likely to be the problem. Two or more affected fixtures means that the clog may be in the branch line, the main line, or the vent stack. If more than one fixture is affected and they are all on the second level of the house, the blockage may be high in the main line or in a vent stack. Isolating the affected area will help you to decide how to clean out the line.

DRAIN LINES ARE FRAGILE

Drain lines are more fragile than you might expect, especially because of all the fixtures attached to them. Be careful when using chemicals and augering—some chemicals can weaken the walls of the drain lines, and augers can shatter porcelain fixtures. Try using a plunger first. If that doesn't work, move on to snakes and augers, but work carefully and slowly.

THE RISKS OF USING CHEMICAL CLEANERS

Never use a drain cleaner on a clogged drain or in a toilet. It will probably make matters worse. Use cleaners only on sluggish drains. Be skeptical of cleaners that claim not to harm pipes and gaskets.

IF YOU DO USE CHEMICAL CLEANERS

1. Pour the drain cleaner carefully so it doesn't come in contact with the skin.
2. Never plunge a drain with drain cleaner in it. When used on a blockage, the chemicals can build up at the blockage and damage gaskets or the walls of the pipe because of the heat created by the chemical reaction.
3. If the cleaner doesn't work and you have to remove the trap to clear the line, be very careful—the cleaner is caustic.
4. Before using chemical drain cleaners, read the instructions and the safety directions. Wear heavy clothing that you don't mind throwing out, safety gloves rated for caustic solutions, and eye protection.
5. Never pour acid in standing water.

HOUSEHOLD DRAIN-CLEANING TOOLS

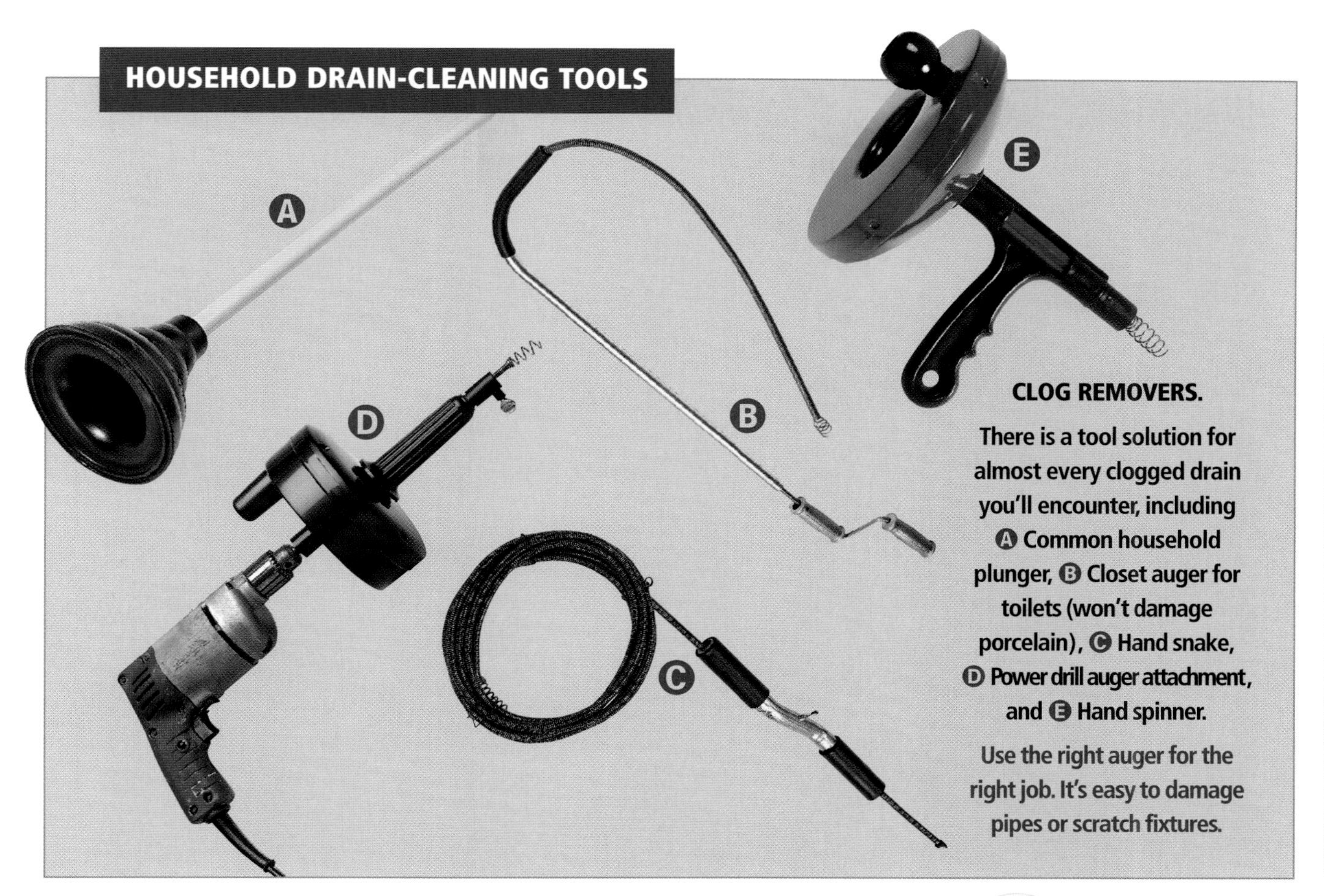

CLOG REMOVERS.

There is a tool solution for almost every clogged drain you'll encounter, including Ⓐ Common household plunger, Ⓑ Closet auger for toilets (won't damage porcelain), Ⓒ Hand snake, Ⓓ Power drill auger attachment, and Ⓔ Hand spinner.

Use the right auger for the right job. It's easy to damage pipes or scratch fixtures.

PREVENTION IS GOOD.

The only fixture or appliance that is designed to handle solid waste is the toilet. Use preventive measures to avoid clogs, and try to keep solid matter out of drain lines.

1. Don't put dental floss, sanitary napkins, or paper towels down the toilet.
2. Hair and food clogs can be very difficult to clear. Clean tubs and sinks regularly.
3. Don't pour grease down the drain unless you have a grease trap.
4. Disposers grind food into a paste that can collect in the drain line and eventually form a clog. Keep oily or dense matter out of the disposer and flush the lines thoroughly with cold water after each use.
5. A capful of bleach in the drain lines once a week will help keep them odor free.

SAFETY FIRST. Waste lines are sources of biohazards and other potential health risks. Avoid contact with any solids or fluids from drain, waste, or vent lines.

1. Always tell a plumber if you've used a chemical drain cleaner in lines that will be worked on.
2. Wear safety goggles to protect your eyes from splashes.
3. Wear rubber gloves, a long-sleeve shirt, and heavy pants to prevent contact with skin.
4. Clean tools and launder clothes immediately after use.
5. Wash hands and exposed skin with antibacterial soap.
6. If you want to tackle the big clogs, rent a power auger from your local rental house. Read the instructions that come with the unit carefully and get a lesson from the salespeople before you attack the drain.
7. If you have a severe clog in a main or service line, you might want to consider hiring a pro to clear it out.

UNCLOGGING A LAVATORY SINK DRAIN

EASY | MEDIUM | HARD

SKILLS: Connecting plumbing fittings and using a plunger or sink auger.

HOW LONG WILL IT TAKE?

PROJECT: Unclogging a lavatory sink drain.

EXPERIENCED . . . . VARIABLE
HANDY . . . . VARIABLE
NOVICE . . . . VARIABLE

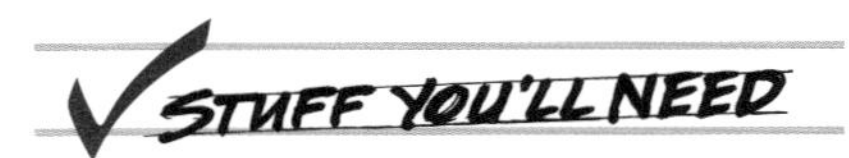

STUFF YOU'LL NEED

TOOLS: Flange plunger, bucket, water-pump pliers, sink auger

MATERIALS: Rag

SAFETY ALERT!

DRAIN CLEANERS CAN BE DANGEROUS.

Commercial drain cleaners may help sluggish drains, but they are extremely caustic and won't get rid of stubborn clogs. Don't use drain cleaners on stubborn clogs, and don't remove a trap that's full of drain cleaner because you can get severely burned.

Most clogs in lavatories are caused by hair getting caught in the pop-up mechanism. Often just removing the stopper will reveal the source of the problem. If you can't lift out the stopper, pull the pivot rod back. The pivot rod is located under the sink. It connects the pop-up stopper to the operating lever on top of the sink. Removing the rod and the stopper makes plunging easier.

USING A PLUNGER

- Fill the sink with about 4 inches of water. Stuff a rag in the overflow so you don't lose water pressure while plunging.
- Vigorously plunge the drain. When water drains freely, the blockage is dislodged. If the water drains slowly, the blockage is still in place. Add more water and plunge again.

USING A SINK AUGER

- Place a bucket under the trap. Use water-pump pliers to remove the slip nuts connecting the trap.
- Remove the trap, then drain any water in the line into the bucket.
- Insert the end of the auger into the drain and turn the crank to advance the auger toward the blockage. Pull out or dislodge the blockage.
- Reconnect the trap.

STILL CLOGGED?

Unable to dislodge or remove the blockage with the sink auger? It's time to call a plumber.

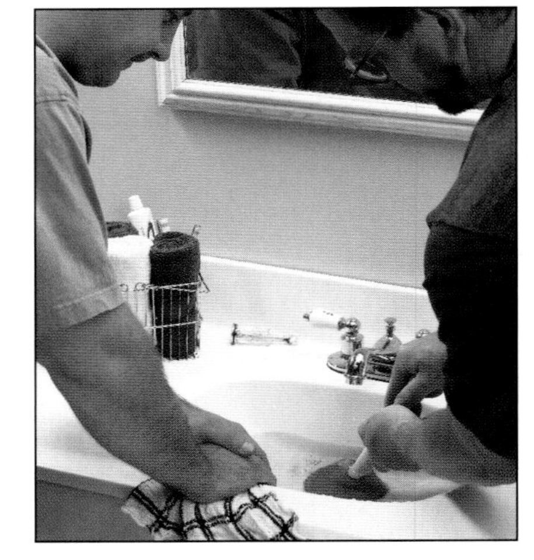

PLUNGE FIRST. Plungers are the safest and most economical tools for getting rid of clogs. They won't damage the fixture or the drain, and they're cheap.

SINK-AUGER STUBBORN CLOGS. Stubborn blockages may require using a sink or hand auger to pull out or dislodge the blockage.

WORK SMARTER

SHARE THE BURDEN.

Have a helper hold a rag blocking the overflow drain while you plunge away. It not only makes the job easier, you have someone to talk to.

If you're on your own, duct tape over the opening will work just fine.

UNCLOGGING A SINK WITH A DISPOSER

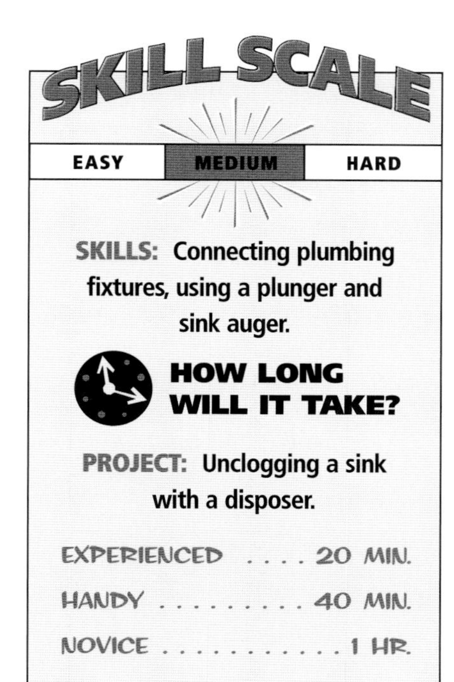

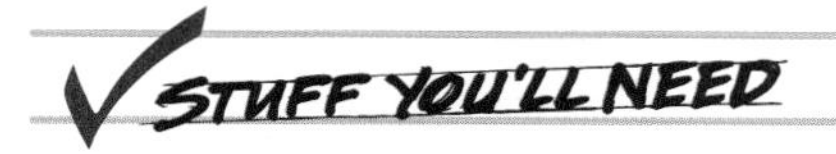

STUFF YOU'LL NEED

TOOLS: Flange plunger, bucket, water-pump pliers, sink auger

MATERIALS: Rag

The steps for unclogging a kitchen sink basin that has a disposer are the same as unclogging anything else; the only issues are whether the sink has a single or double basin and whether or not there is a dishwasher hooked up to the system. You'll need to find some means of clamping the hose so you don't back flush waste into the dishwasher.

SINGLE-BASIN SINKS

Fill the sink so there is 4 inches of standing water in the basin. Using a plunger with a fold-out lip, plunge forcefully up and down a dozen times. If the blockage does not dislodge, turn on the disposer and plunge vigorously again.

Clogs that are more stubborn will need to be augered. Place a bucket under the sink drain trap. Disconnect the trap slip nuts with water-pump pliers. Remove the trap. Loosen the setscrew and push about 6 inches of the cable into the drain—too much cable will kink. Tighten the setscrew and turn the crank clockwise while pushing with medium pressure. A sudden lack of resistance will mean that you have pushed past the blockage. You will need to either pull out the blockage by turning the crank counterclockwise, or push through, then flush out the drain. If the drain does not run freely, repeat the steps.

DOUBLE-BASIN SINKS

Fill both basins with 4 inches of water. Have a helper hold a rag or a closed strainer over the opening of the disposer drain. Use a plunger to vigorously plunge the drain in the other sink. A dozen times should be enough for most clogs. If the drain is still clogged, switch positions with your helper and plunge the drain in the disposer basin.

You may have to use a sink auger to clear stubborn clogs. Augering a blockage in a double-basin sink is done the same way as for a single-basin sink.

HELPING HANDS. Double-basin kitchen sinks require an extra set of hands to clear the drain. Have someone block one drain with a rag while you plunge the other. That way, you can attack the clog with more force than if you try to plunge one drain while blocking the other by yourself.

UNCLOGGING JAMMED DISPOSERS

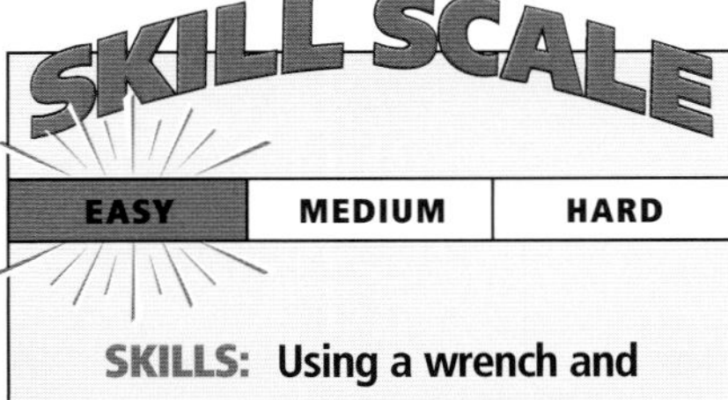

SKILLS: Using a wrench and operating a flashlight.

HOW LONG WILL IT TAKE?

PROJECT: Unclogging a disposer.

EXPERIENCED ..... 15 MIN.
HANDY .... 30 MIN.
NOVICE .... 45 MIN.

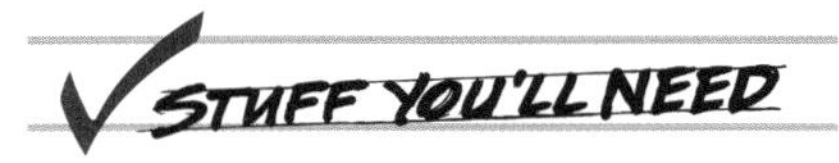

STUFF YOU'LL NEED

TOOLS: Flashlight, ¼-inch allen wrench, a disposer wrench or an adjustable wrench

MATERIALS: None

WORK SAFELY!
Before reaching into a disposer, disconnect the power. Unplug the unit from the outlet. If there is no outlet, turn off the circuit breaker. Never stick your hand into an operating disposer.

Turn off the power to the jammed disposer and unplug it. Look inside the opening with a flashlight to see what is jamming it. Remove the waste and restore power. If it's still jammed, try the following:

- Check the bottom of the unit; find the reset button and press it. Turn on the power. The disposer should run freely. If the disposer is still stuck, turn the power off.
- Insert a broom handle into the drain opening and try to free the impellers.
- Next, insert a ¼-inch allen wrench into the hex socket on the bottom. The socket is connected to the impellers that crunch up the waste. Using the hex key turn the impellers in both directions to free them up.

To keep a garbage disposer smelling fresh and clean, grind up some lemons followed by a few ice cubes.

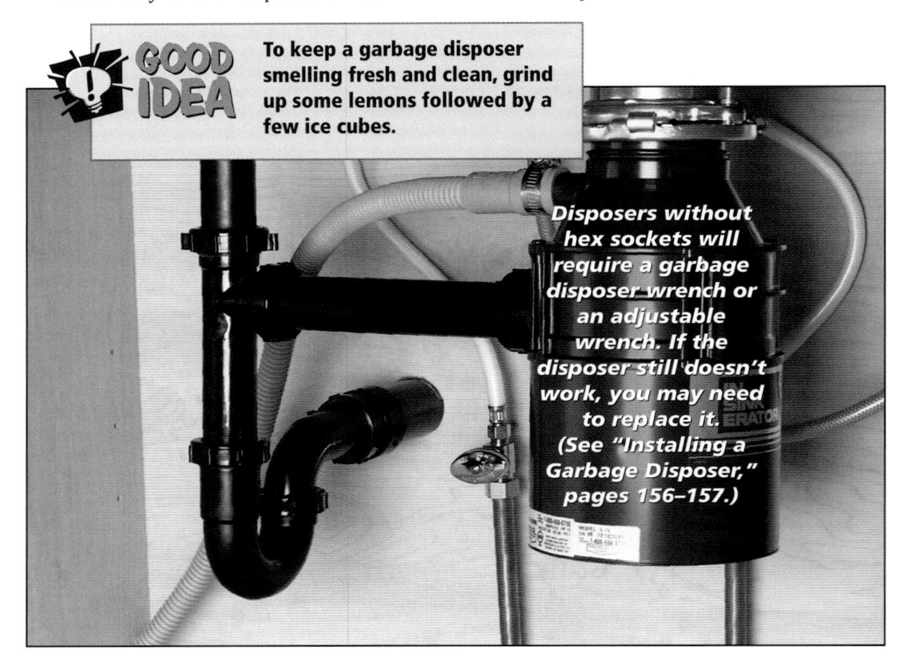

Disposers without hex sockets will require a garbage disposer wrench or an adjustable wrench. If the disposer still doesn't work, you may need to replace it. (See "Installing a Garbage Disposer," pages 156–157.)

WORK SMARTER

SOLID DISPOSER INFORMATION.

1. **Make sure your sink is strong enough to support the unit. A disposer will shake and rattle in a thin stainless-steel sink.**
2. **Garbage disposers use a lot of water, which can cause problems with septic systems. If you're willing to clean the tank more frequently, you shouldn't have any problems.**
3. **Stringy vegetables, such as artichoke pieces and corn husks, can become wrapped around the water seal and shorten its life.**
4. **There are two types of units on the market—continuous feed, which operates without having to close the unit's lid; and batch feed, which requires the lid in place to operate. Continuous feed is the most commonly used and most often recommended for the home kitchen.**
5. **Models with a self-reversing feature jam less often.**

UNCLOGGING A BATHTUB POP-UP DRAIN

SKILL SCALE
EASY
MEDIUM
HARD
SKILLS: Operating a hand auger and removing plumbing fittings.
HOW LONG WILL IT TAKE?
PROJECT: Unclogging a bathtub pop-up drain.
EXPERIENCED .... VARIABLE
HANDY .... VARIABLE
NOVICE .... VARIABLE

A slow-draining bathtub can be a quick-fix project. Bathtub drains are usually clogged by soap scum and hair that collects in the trap. Stuff a rag into the overflow and fill the tub with 4 inches of water, if it doesn't already have standing water in it. While plunging, the vacuum created by the standing water helps to dislodge the blockage.

Place a flange plunger over the drain and plunge vigorously at least a dozen times. If the blockage is not dislodged, continue plunging vigorously until the line drains freely.

More serious clogs may have to be flushed or augered. Flushing the drain with a blow bag can be messy, especially if you don't have a good seal. The best way to clear more difficult blockages is to use a hand auger. The following steps describe how to use a hand auger to clean out the drain.

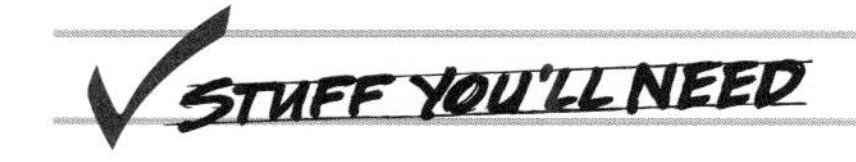

TOOLS: Flange plunger, screwdriver, hand auger

MATERIALS: Rag

Clogs of hair are often found wrapped around the pop-up drain assembly.

1 REMOVE THE POP-UP DRAIN ASSEMBLY. Flip the drain lever up and pull out the drain plunger. Remove the screws from the overflow cover plate with a screwdriver, then lift out the linkage.

2 AUGER THE DRAIN. Insert the end of the hand auger into the overflow drain. Turn the handle clockwise, feeding out the cable until you meet resistance. Slowly withdraw the auger to dislodge or remove the blockage. Repeat until the tub drains normally. Insert the drain linkage back into the overflow. Install the cover. Flip the lever up and insert the stopper. Don't use a power-driven auger in a bathtub drain; the assembly is too fragile and can be damaged easily.

UNCLOGGING A SHOWER DRAIN

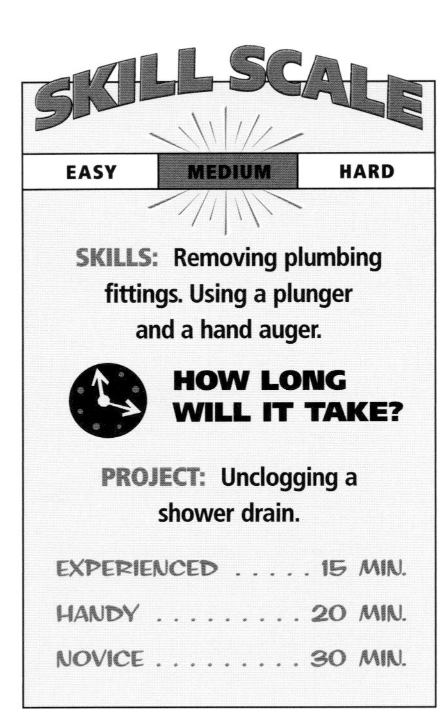

Shower stalls work hard, especially the ones in utility areas. We use them to wash the dog and we stick plants in the shower to wash the leaves and freshen them up. A shower is the perfect place to scrub down miniblind window treatments. When the weather is too cold to hose things down outside, we use the shower stall to clean dirty boots. It's no wonder that the drain will occasionally clog.

Plunge the shower drain vigorously with a flange plunger to unclog it. For stubborn clogs, use a hand or power-driven auger. Try both before calling a plumber or a drain cleaning service to unclog the drain.

STUFF YOU'LL NEED

TOOLS: Screwdriver, flange plunger, hand auger

MATERIALS: None

Don't be surprised if your plumber isn't interested in going after a tough clog. He may suggest a drain cleaning service to handle the job instead. Plumbers have specialties too.

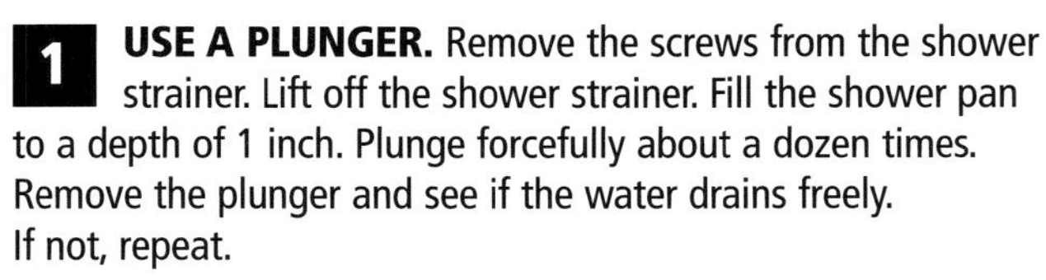

1 USE A PLUNGER. Remove the screws from the shower strainer. Lift off the shower strainer. Fill the shower pan to a depth of 1 inch. Plunge forcefully about a dozen times. Remove the plunger and see if the water drains freely. If not, repeat.

2 HAND AUGER OR POWER AUGER THE DRAIN. Tougher clogs can use a tougher approach. Feed a hand auger into the drain until there is resistance. Turn the auger handle in a clockwise direction and slowly withdraw the auger. Repeat until the shower drains normally. If the blockage refuses to clear, call a plumber or a drain service.

UNCLOGGING A TOILET

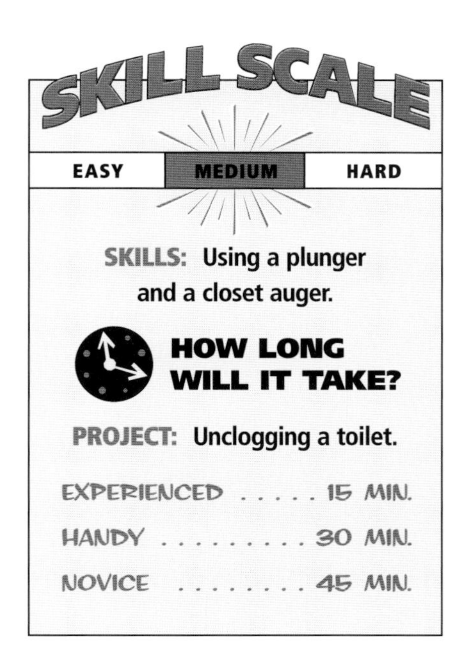

SKILLS: Using a plunger and a closet auger.

HOW LONG WILL IT TAKE?

PROJECT: Unclogging a toilet.

EXPERIENCED . . . . . 15 MIN.
HANDY . . . . 30 MIN.
NOVICE . . . . 45 MIN.

✓ STUFF YOU'LL NEED

TOOLS: Flange plunger, closet auger, bucket

MATERIALS: Rags

HOW TO TAME A CLOG.

This recipe for a clog buster really works. Add 3 tablespoons of dishwashing soap to the bowl. The soap will lubricate the interior of the drain and help to loosen the clog when you plunge. Wait a few minutes for the soap to do its work, then plunge vigorously.

Toilets have built-in traps. Objects stuck in the trap cause slow draining or clogged toilets. They usually can be dislodged by plunging.

Don't be tempted to use a coat hanger to unclog a toilet. You will scratch the bowl. If the bowl is scratched, try removing the black marks with a heavy-duty powder-type cleanser with bleach.

Never use chemical drain cleaners in a toilet.

If there is no water in the bowl, pour some water into it. Water helps seal the plunger flange, creating a vacuum and allowing you to apply pressure to dislodge the blockage. **Ⓐ Plunge forcefully about a dozen times.** Remove the plunger to allow the toilet to drain. If it doesn't drain, or drains slowly, repeat plunging.

IT'S AUGER TIME

You may need an auger to remove stubborn objects, such as small toys. Never use a hand auger on a toilet. The force of the auger when turning the crank may shatter the porcelain bowl. A closet auger is designed specifically to be used on toilets. The closet auger has a long handle with a crank. A bend in the handle covered with a protective sleeve prevents scratching of the porcelain. **Ⓑ Turn the crank clockwise and push.** The auger can shove the blockage forward into the drain system. If the auger catches on the object, continue turning the crank as you pull out the cable until you can retrieve the object.

REMOVE THE TOILET

When plunging or augering doesn't work, your only option will be to remove the toilet (see "Removing an Old Toilet," page 150), and try to fish out the object from the other end.

A *Common household plunger*

B *Closet auger*

UNCLOGGING DRAINS AND WASTE LINES

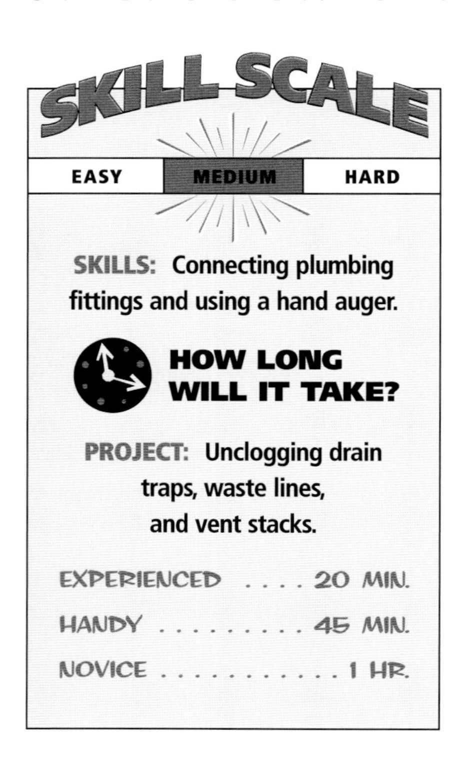

SKILL SCALE
EASY
MEDIUM
HARD
SKILLS: Connecting plumbing fittings and using a hand auger.
HOW LONG WILL IT TAKE?
PROJECT: Unclogging drain traps, waste lines, and vent stacks.
EXPERIENCED .... 20 MIN.
HANDY .... 45 MIN.
NOVICE .... 1 HR.

Small objects, such as toys, can be difficult to remove from a drain line. No matter how cleverly you try to snag them, they seem to defy removal. The only recourse may be to push the blockage further down the drain line system to a cleanout or to flush the line with water when the blockage has been jarred loose.

Remove the drain trap under the sink and use a hand auger to try to snag the blockage. If you can't pull it out, try pushing it into a larger waste line. Then open up the line cleanout and try snagging it with the hand auger.

If you don't have accessible cleanouts in your home, push the blockage out through the roof stack vent. Be careful when climbing onto the roof, and don't try it in bad weather.

TOOLS: Water-pump pliers, bucket, hand auger

MATERIALS: Rags for cleanup

Pushing an object farther down the system can cause a clog where it's difficult to auger. You might have to call a plumber or drain cleaning specialist to clear out the main drain line to the house.

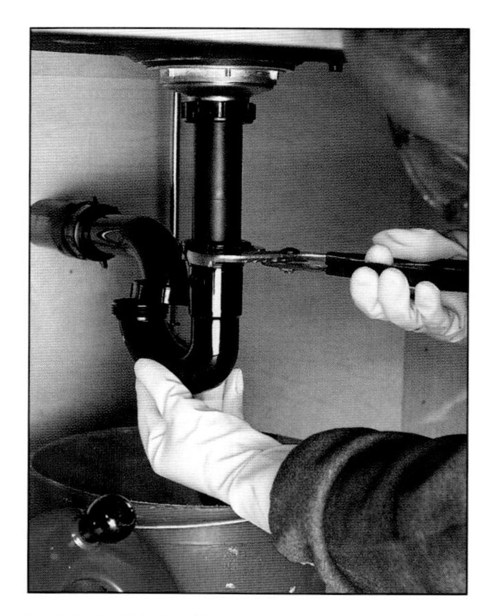

1 DISCONNECT THE DRAIN TRAP. Place a bucket under the drain trap to catch wastewater. Loosen and slide back the slip nut couplings with water-pump pliers. Remove the trap. Clean out any debris stuck in the trap. Look for cracks in the pipe or sediment buildup in the trap; either means you'll have to replace it (see "Replacing a Sink Trap," page 37).

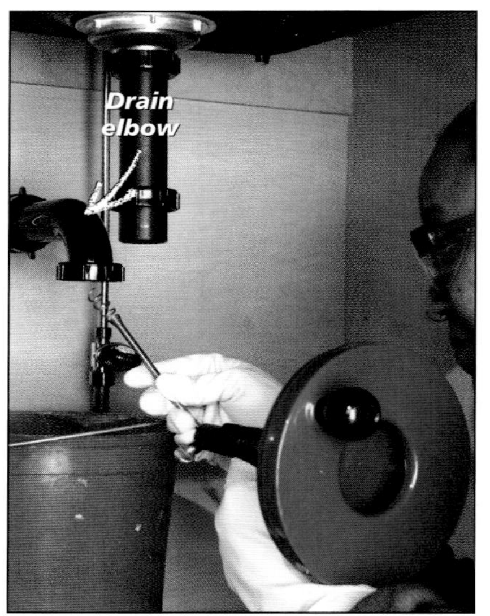

2 INSERT THE AUGER. Feed 6 inches of the hand auger cable into the drain pipe by turning the auger handle clockwise—a greater length may crimp the cable. Extend the auger until you find resistance. Give yourself a little more room to work by removing the drain elbow.

3 TIGHTEN THE AUGER LOCKNUT on the side of the auger. Maintain the 6 inches of exposed cable. Rotate the handle clockwise and continue to push. Repeat until the auger will no longer advance. If you are able to snag the blockage, rotate the auger handle clockwise and pull the cable out slowly. If nothing is snagged, replace the trap and try augering the main drain line.

BANG THE DRAIN, SLOWLY!

If you are having trouble locating the exact position of a clog in the main drain line, here's a tip to speed up the search. The street side of the drain line to the point of the clog will be empty and the house side will be full of waste and water. Tap along the pipe from the street side with a broomstick or a piece of scrap lumber. (Don't use a hammer or piece of metal.) A ringing or hollow sound means a clear pipe. When you hear a thud, you've found the clog. Open up the nearest cleanout and put the auger to work!

You can learn from the pros at a local rental center.

MAYBE IT'S TIME FOR A HEAVY-DUTY SOLUTION. Sometimes it's easier to get to a blockage in the main line through the vent system. If that's the case, you will need to rent a commercial auger to remove the blockage. This might be the time to bring in a pro; but if you want to do it yourself, get some lessons from the rental people, take every precaution, and work carefully on the roof.

CLOSER LOOK

CLEARING A DRAIN THROUGH A CLEANOUT.

Wearing protective clothing, gloves, and eye protection, place a bucket under the cleanout. Carefully remove the end cap. Use water-pump pliers to loosen the end and slowly remove it by hand. Be careful! Wastewater trapped in the line will be under pressure, so it may gush or spray out of the end. Insert the hand auger cable through the opening, leaving 6 inches of cable exposed. Advance until you meet resistance. Tighten the locknut. Rotate the handle clockwise and continue pushing. Try snagging the blockage and slowly pulling it out. If you are unable to clear the drain, try augering through the roof vent.

DON'T FORGET ACCESS PANELS. While remodeling my house, I decided to finish the basement. I took my time when I built the walls and paneled. It looked great. Not too long after that, one of my kids accidentally flushed a toy down the toilet. No problem! I got out my trusty closet snake. Unable to snag the toy, I decided to push it farther down the drain into a larger line. This way I could snag it and remove it through a cleanout. Having successfully pushed it through, I grabbed a hand auger and proceeded downstairs to the cleanout tap. When I got there, I stood there dumfounded, looking at a beautifully paneled wall with no access panel to where the cleanout was. I got my tools and cut one. It's a good idea to build access panels to all cleanouts.

CHAPTER

4 THE SUPPLY SYSTEM

MULTISTORY CONSTRUCTION SUPPLY SYSTEM

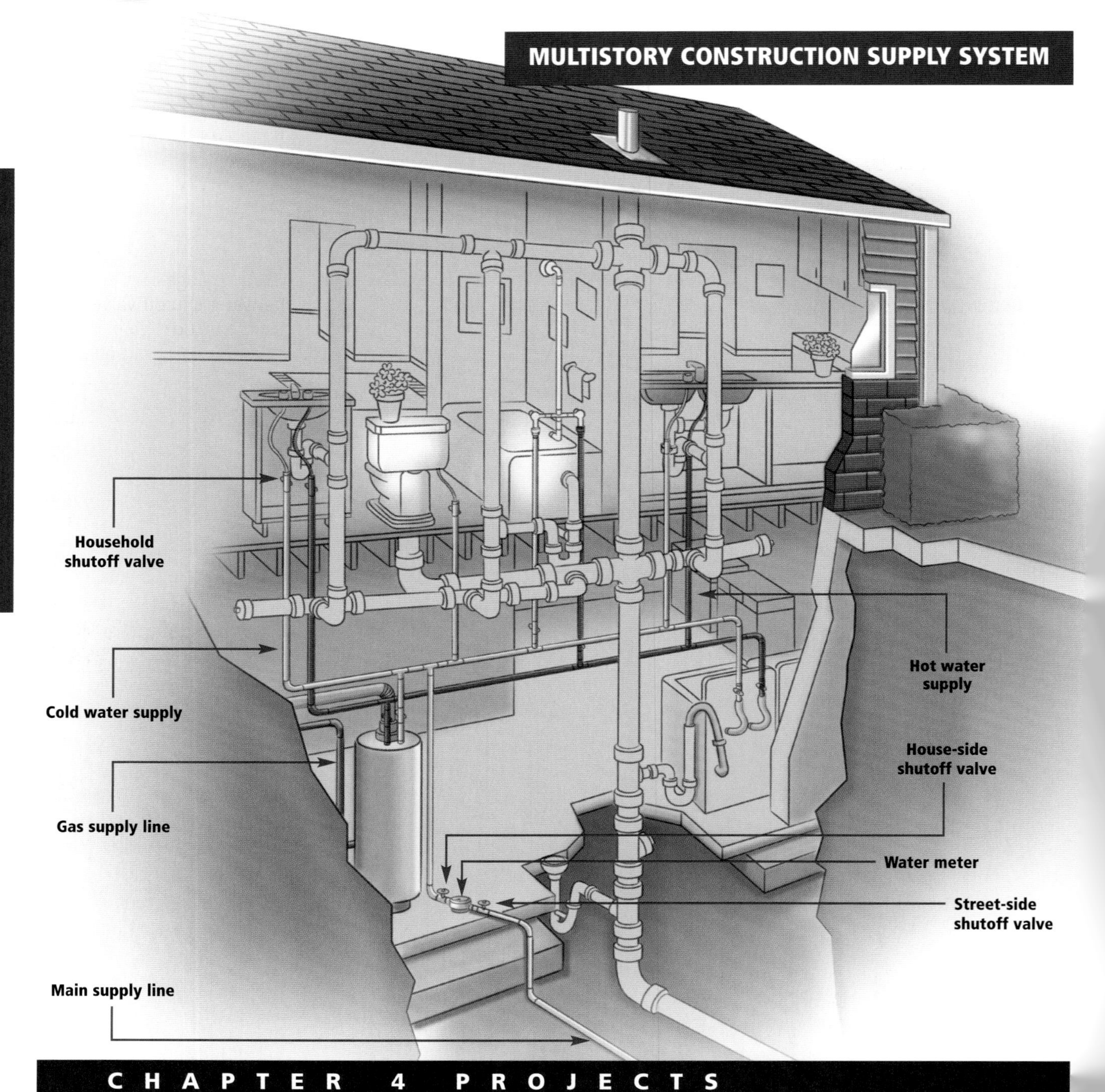

CHAPTER 4 PROJECTS

The quality of the water that enters your house from a municipal water system or from a private well is your first concern. If your source is a public system, the responsibility for the safety of the drinking water is with the provider. Communities must abide by state and federal regulations concerning the potability and safety of drinking water. Standards for private well water are regulated by state and federal governments as well, but the owner of the well is responsible for water purity and maintenance of the well and its pumping system.

THROUGH THE SERVICE LINE

Water from the municipal water company enters your property from the main water line in the street through a stop or "buffalo" box with a shutoff valve (which you may or may not have access to) into a service line that runs to your home. Repair and maintenance of the service line—once it passes the stop-box valve—is the homeowner's responsibility.

TO THE WATER METER

The service line enters the house through a water meter owned and maintained (sometimes for a fee) by the municipality that measures the amount of water you use. Homes on a grade or slab have a utility area where the service line enters and where the water meter is located. Homes with basements usually have the water meter in the basement where the service line enters the house. The meter has a valve on each side. The valve on the street side is controlled by the provider. The valve on the home side is controlled by the homeowner and is called the main shutoff valve.

THE MAIN SHUTOFF VALVE

The main shutoff valve is near the meter. Know its location and how to shut it off in case of an emergency. It's a good idea to turn off the water at the main and drain the system before attempting major repairs or maintenance work. Local shutoff valves should also be installed near every fixture and appliance so you don't have to shut off the water to the entire house to fix the toilet.

BRANCHING OUT

The sizes of the lines and branch piping in the house will depend on the demands of the appliance. The main supply line is usually ¾- or 1-inch pipe. A ¾-inch pipe is the standard size for most water heaters—usually the first connection to the water line as it enters the house. Internal branch lines to appliances and fixtures are usually made of ½-inch copper, CPVC (chlorinated polyvinyl chloride) or, in some areas of the country, PEX (cross-linked polyethylene). See "Plastic vs. Metal: Plumbing Pipe Explained," page 189.

Branch lines from the main supply line are run in the subfloors and in the walls to the fixtures they serve. Branch lines should not be run along outside walls where freezing is common. Freezing can cause the pipes to rupture or burst.

THE RIGHT PRESSURE

Nothing in your supply system operates properly without the correct water pressure. Correct pressure is usually 40 to 55 PSI although pressure as low as 20 PSI, can be acceptable. High water pressures, especially greater than 85 PSI, can cause damage. Test the water pressure at the service line with a gauge, which you can buy or rent. Install the gauge on the home side of the water shutoff valve. If water pressure is too high or too low, contact your municipal supplier.

SLAB CONSTRUCTION

Homes built on a concrete slab have drain and supply lines buried beneath the slab. The slab has to be broken through or tunneled under to get to the source of the problem.

SLAB CONSTRUCTION SUPPLY SYSTEM

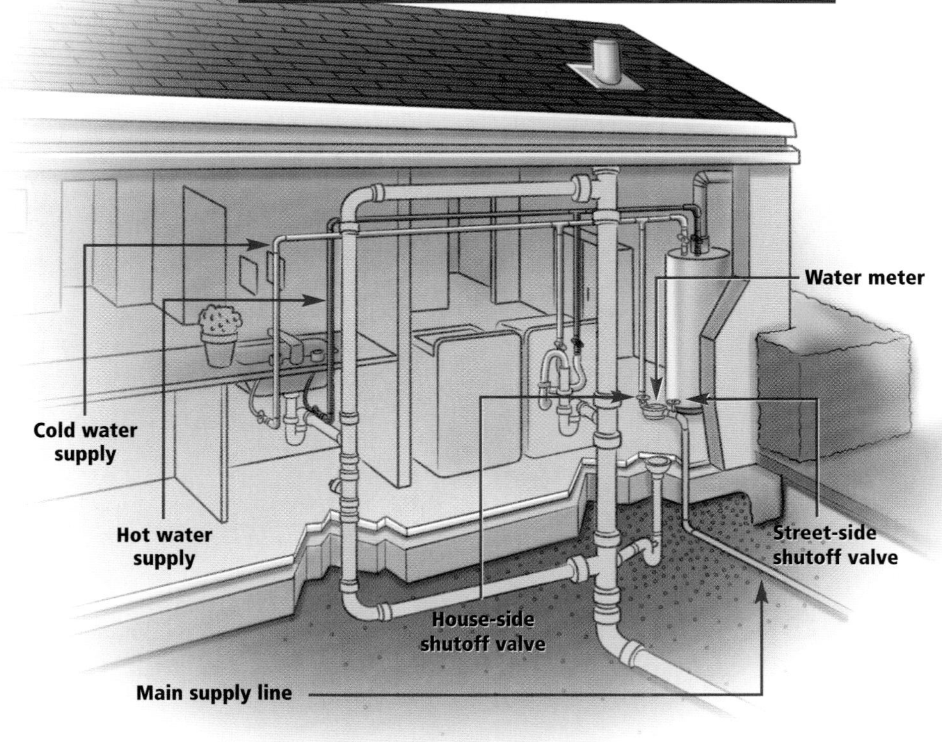

Water meter
Cold water supply
Hot water supply
Street-side shutoff valve
House-side shutoff valve
Main supply line

RIGID SUPPLY PIPES

If your home is older, you'll discover that many different plumbing materials were probably used—some of which meet code and some of which don't. Lead was common for supply lines 20 years ago, but much of it has been replaced with copper. You will find galvanized steel and cast iron. You will likely be able to replace the older materials with modern equivalents as long as they meet code requirements.

PIPE REPLACEMENT GUIDE

Here's what you'll need to know when replacing pipe:

- All pipe is designated by a grade, which will tell you how it can be used in a system to meet code requirements.
- Rigid copper pipe is sized by its interior diameter (ID).
- Soft copper pipe is sized by its outside diameter (OD).
- Schedule 40 PVC (for drain lines) is ID. Other PVC is OD.
- Galvanized steel is ID. (If you're measuring the length of galvanized steel, include the depth of the socket it will screw into.)

If you're unsure of the proper size or grade for a replacement piece, take it to the store so you can buy an exact duplicate or find a viable alternative that meets code. For more information on plumbing pipes, see page 94.

COPPER, CPVC, AND PEX

Copper and **CPVC** (chlorinated polyvinyl chloride) are the most common choices for supply piping. **PEX** (cross-linked polyethylene), an easy-to-connect, flexible plastic pipe, is common in warmer areas of the country where frost is not an issue. (Check local codes.) Galvanized pipe still meets code, but is difficult for the home plumber to use and not as durable. Each withstands the pressure of water being driven into the supply system from the service.

Copper pipes used for supply lines in the home are sold in three grades—M, L, and K—based on the thickness of the pipe wall. M is the thinnest and adequate for most home uses. L is required by code in commercial installations, and K (the thickest) is used for underground water service.

CPVC, with its resistance to high temperatures (up to 82 degrees C) and high water pressures, makes it an ideal pipe for water supply lines. It's easier to work with and less expensive than copper.

Check with local authorities to find out if there are restrictions on using CPVC as supply lines.

UP TO CODE

When you're researching code issues, go to your home center and see what's on the shelves. If there are long rows of one item and not much of another for the same application, you've learned something about common use in your area. You've still got to check local codes to confirm that all materials meet standards, but you've got a place to start.

TESTING FOR LEAKS.

Never cover plumbing pipes or fixtures before testing for leaks. You have two options for testing—air pressure and water.

Testing with water. Isolate the new lines, then turn on the water to both hot and cold supplies. Inspect for leaks at each junction.

Testing with air pressure. Isolate the line being tested from existing supply lines. Install a threaded adapter for a pressure gauge and an air compressor fitting. Force air into the pipes with an air compressor. Local building codes will define the test pressure and time the line needs to maintain the pressure.

You may need to have an inspector on site, depending on the size of the project. Set up an appointment if necessary.

GETTING THE LEAD OUT.

Lead is acknowledged as a health hazard, but it's still present as supply pipes in many older homes. Lead pipes may not be a problem if they have been properly installed and are well maintained. Some communities add a trace of phosphate, which coats the inside of the pipes and prevents lead from leaching into the system. If you have any concerns about potential health issues and lead pipes in your home, have the system tested. The results will tell you if pipes need to be replaced with copper or CPVC. In the meantime, run water through the pipes for a few minutes in the morning to get the lead out.

SUPPLY PIPES AND FITTINGS

Galvanized 90° elbow

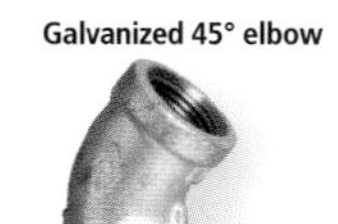
Galvanized 45° elbow

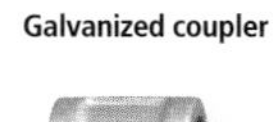

Galvanized coupler

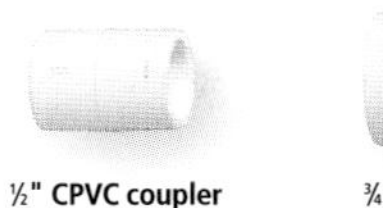
½" CPVC coupler

¾" CPVC coupler

Galvanized cap

Galvanized reducer

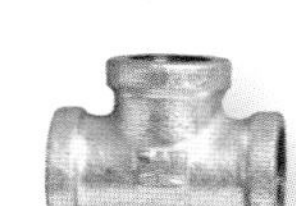

Galvanized tee

½" CPVC 45° elbow

½" CPVC 90° elbow

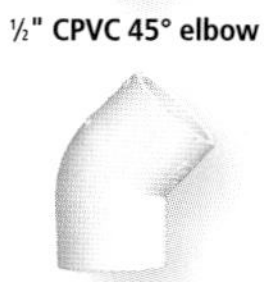
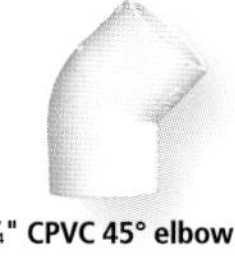
¾" CPVC 45° elbow

¾" CPVC 90° elbow

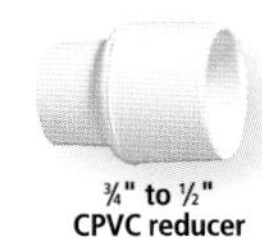
¾" to ½" CPVC reducer

½" CPVC cap

½" CPVC threaded drop elbow

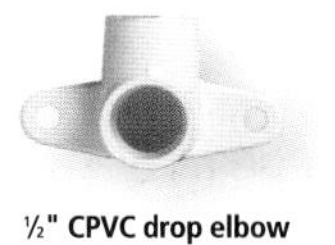
½" CPVC drop elbow

CPVC to brass union

½" CPVC street elbow

½" CPVC tee

¾" CPVC tee

½" CPVC strap

¾" CPVC strap

CPVC compression fitting

½" 45° elbow

½" 45° street elbow

½" 90° elbow

½" 90° street elbow

½" brass drop elbow

½" cap

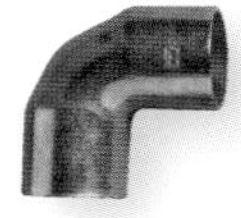
¾" 90° elbow

Supply stop compression

¾" to ½" reducer

½" coupler

½" tee

½" to ¼" reducer

¾" 45° street elbow

¾" 90° street elbow

¾" 45° elbow

Screw-type copper valve stops

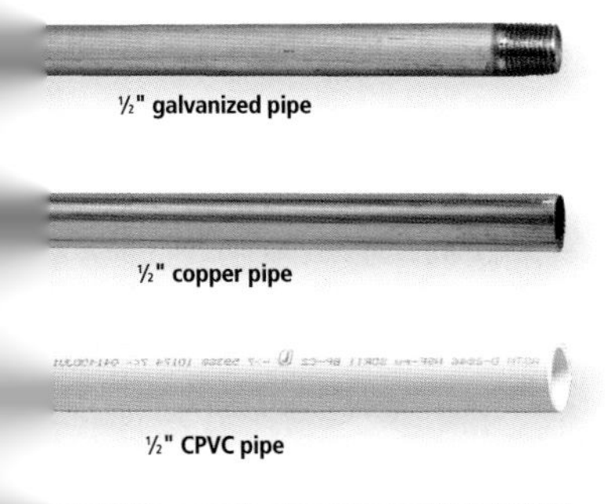
½" galvanized pipe

½" copper pipe

½" CPVC pipe

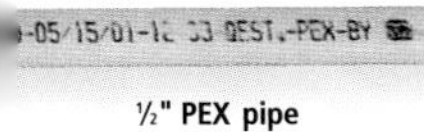
½" PEX pipe

½" flexible copper tube

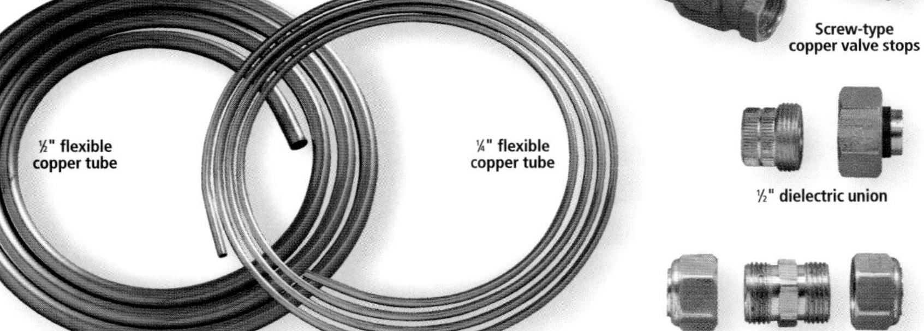
¼" flexible copper tube

½" dielectric union

Compression fitting

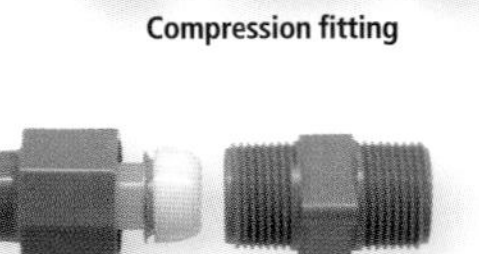
PEX compression fitting

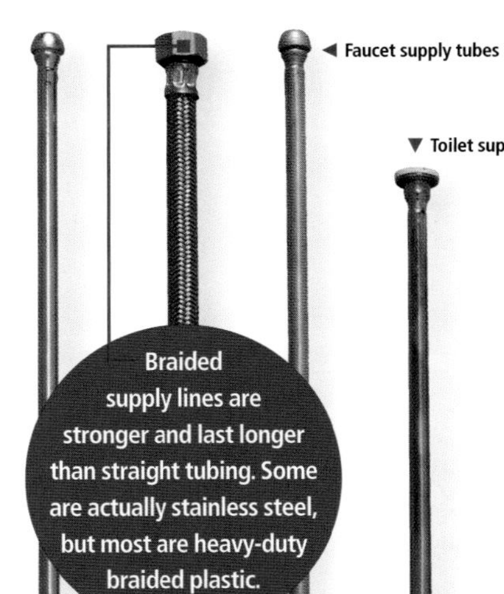
◀ Faucet supply tubes

Braided supply lines are stronger and last longer than straight tubing. Some are actually stainless steel, but most are heavy-duty braided plastic.

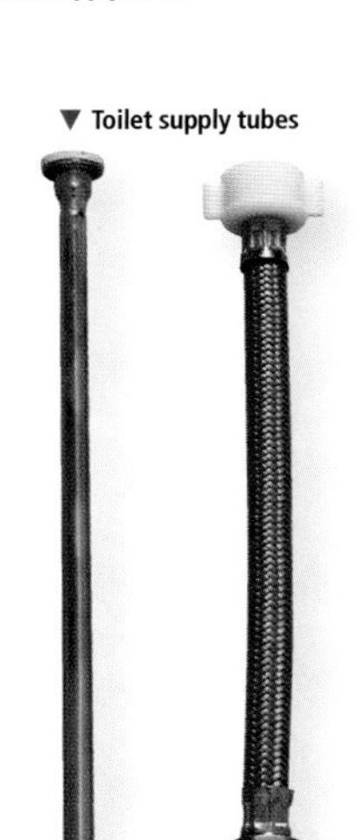
▼ Toilet supply tubes

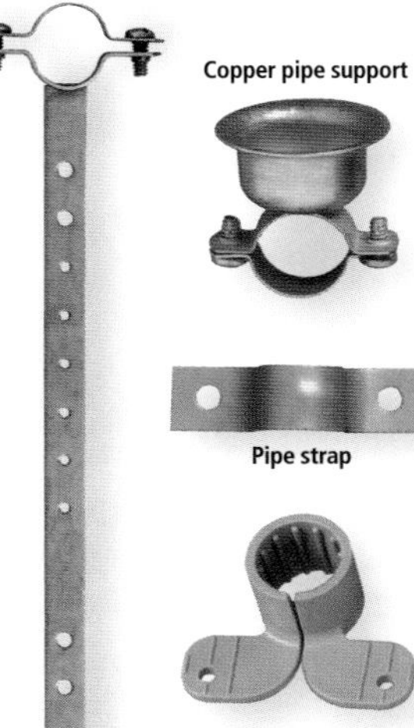
Milford hanger

Copper pipe support

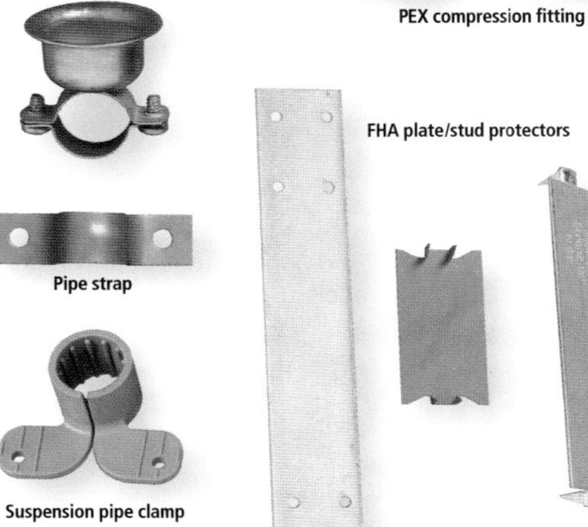
Pipe strap

Suspension pipe clamp

FHA plate/stud protectors

SOLDERING COPPER PIPES

EASY | **MEDIUM** | HARD

SKILLS: Operating a propane torch. Connecting plumbing fittings.

HOW LONG WILL IT TAKE?

PROJECT: Soldering copper pipes.

EXPERIENCED . . . . . 10 MIN.
HANDY . . . . 20 MIN.
NOVICE . . . . 30 MIN.

TOOL TIP

THE RIGHT COMBINATION.
Wire brushes work well for cleaning fittings, but the 4-in-1 cleaning tool is a solderer's friend. It's a combination deburrer and cleaner for preparing ½- and ¾-inch copper pipe.

WORK SMARTER

KEEP YOUR COPPER DRY.

Whether you're adding fittings to a system or starting from scratch, water in the line will keep the pipe from getting hot enough to make a secure solder joint.

1 Drain the line and open a faucet down the run to release steam.

2 Stuff bread into the pipe to absorb moisture; it dissolves when you run water later.

Soldering copper pipe fittings isn't difficult, but you'll need to practice before you're perfect. Once you've mastered the skill, you'll see why copper plumbing is appreciated for its professional look. Copper is a durable, clean, and functional connecting system.

BE A PATIENT PLUMBER

The most important thing with any fitting is that it doesn't leak. Your first few attempts may not be works of art, but soon you'll be soldering like a pro.

NO-FREEZE SOLUTION

In colder climates, run supply lines at a slight slope so they can be easily drained. Add bleeder caps at the low points so you can get rid of excess water.

STUFF YOU'LL NEED

TOOLS: Propane torch, spark lighter, emery cloth, round wire brush or 4-in-1 tool, flux brush, fire extinguisher, water, bucket

MATERIALS: Copper pipe, copper fitting, lead-free soldering paste (flux), lead-free solder, fire barrier, rag, bread

Don't ever apply heat directly to the solder. It will just melt away and you'll never make a sealed joint.

1 PREPARE THE INSIDE OF THE FITTING. Ream the inside of each fitting with a round wire brush. Clean, grease-free connections ensure a good seal.

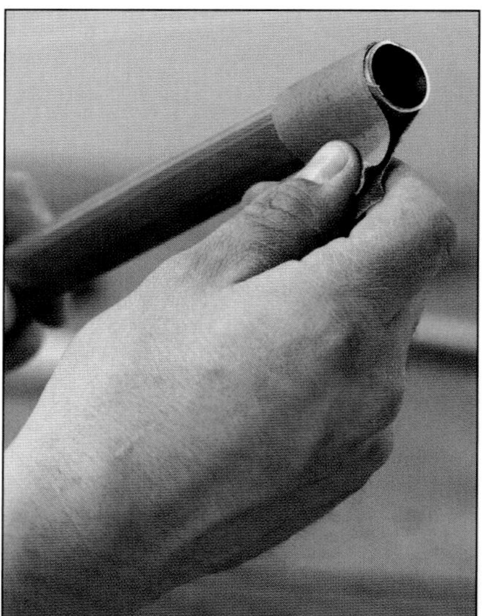

2 CLEAN THE OUTSIDE OF THE PIPE. Use emery cloth or steel wool to clean the outside of the pipe. Use a deburring tool or the handle of a pair of pliers to deburr the inside of the pipe. Be careful—the edges may be sharp!

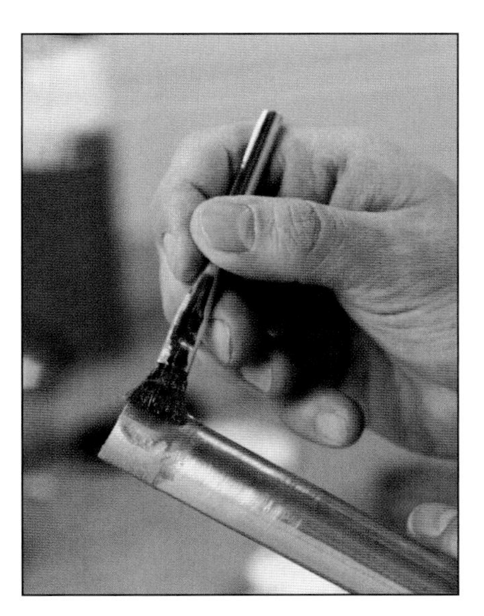

3 **APPLY FLUX TO THE PIPE.** Apply a thin layer of lead-free soldering paste (flux) to the end of the pipe using a flux brush. The paste should cover about 1 inch of pipe. Insert the pipe into the fitting. Make sure the pipe is tight against the bottom of the fitting. Twist the fitting slightly to spread the flux.

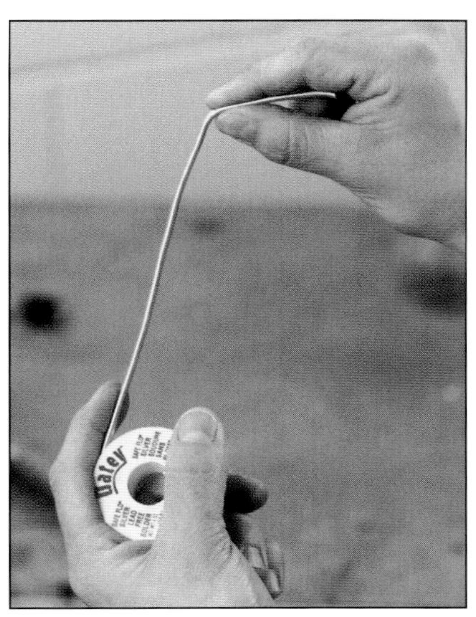

4 **UNWIND THE SOLDER WIRE.** You will need 8 to 10 inches of the wire extended from the spool. Bend the first 2 inches to a 90-degree angle.

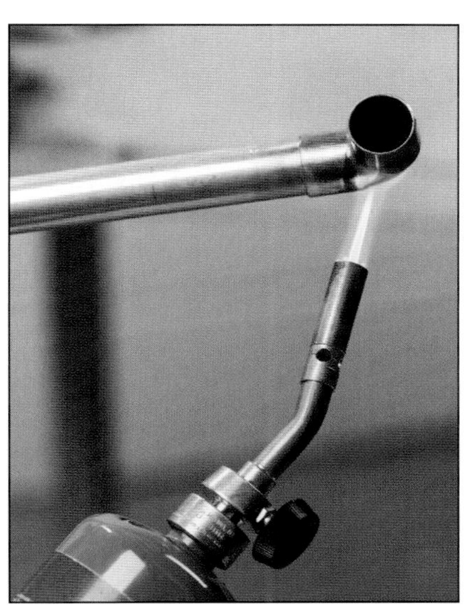

5 **HEAT THE FITTING.** Light the propane torch. Hold the tip of the flame against the middle of the fitting for 4 to 5 seconds or until the soldering paste begins to sizzle. Heat the opposite side of the fitting to ensure the heat is evenly distributed.

6 **TOUCH THE SOLDER TO THE PIPE.** If the solder melts, the pipe is ready to solder.

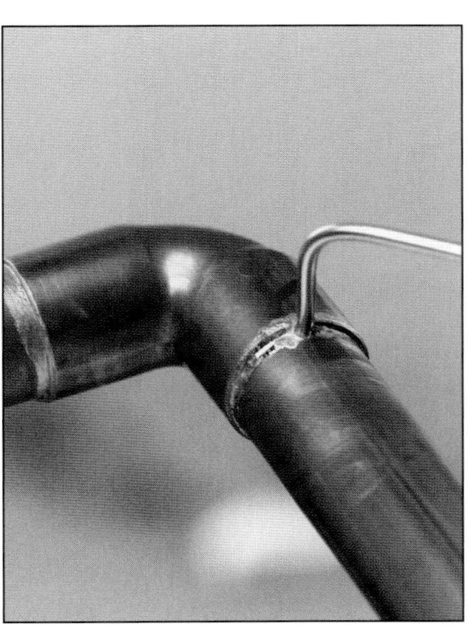

7 **REMOVE THE FLAME FROM THE FITTING.** Quickly insert ½-inch to ¾-inch of solder into the joint. Capillary action will draw the liquid solder into the joint. A properly soldered joint should show a thin bead of solder around the fitting.

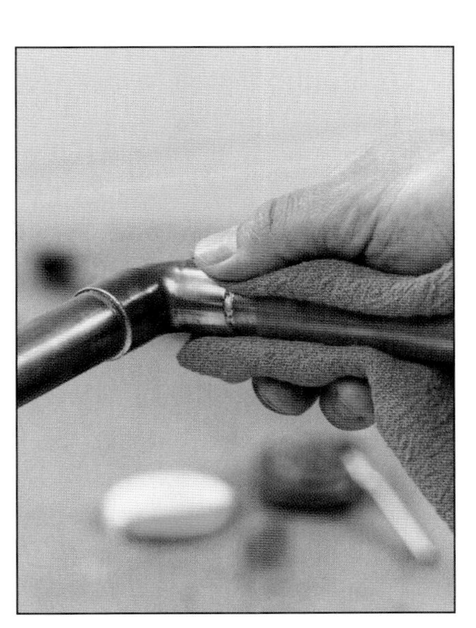

8 **CLEAN THE FITTING.** Wipe away the excess solder with a rag. The pipe will be hot, so be careful while handling it. After the pipe and fitting have cooled, turn on the water and check for leaks. If the joint leaks, take it apart and resolder it.

USING A PROPANE TORCH

Light the propane torch by opening the valve and striking a Spark Lighter (below right) or match next to the nozzle until the gas ignites. Adjust the torch valve until the blue portion of the flame is 1 to 2 inches long.

Always read the manufacturer's operating instructions before using a propane torch. Make sure the area where you will solder is well-ventilated—propane torches can produce carbon monoxide. Inspect the area for potential fire hazards and isolate them by moving them or placing a fireproof barrier in front of them.

1. Inspect the seals on both the cylinder and nozzle for damage before you connect them. If either is damaged, do not use the damaged part. Follow the manufacturer's recommendation for disposal of the damaged piece.
2. When you're finished with the torch, turn the valve to the off position. Hand-tighten the seal to shut off the flow of gas. Do not overtighten the valve; you can damage the seals.

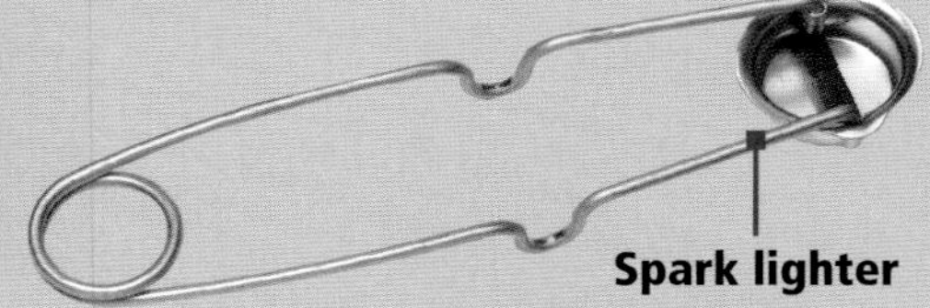

TOOL TIP

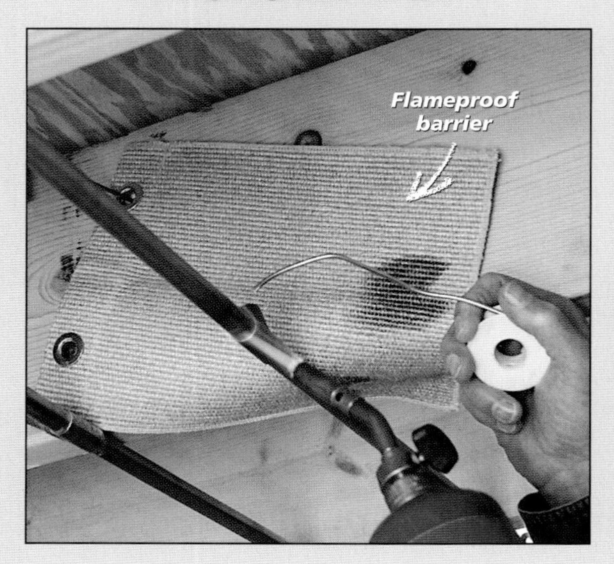

SOLDERING IN PLACE SAFELY.

Soldering isn't a difficult job when you're assembling a system on your workbench, but a lot of soldering projects require you to do the work against a wall, inside an access panel, or in the ceiling or floor joists. Fire is a constant danger whenever live flame meets combustible materials, so you've got to take some common-sense precautions. Purchase a flameproof barrier to hang on the surface behind the soldering position. Made of fire-retardant material, these barriers usually come with grommets to assist in hanging. Follow the manufacturer's instructions. Keep a fire extinguisher and a bucket of water near you when you work—just in case.

CONNECTING CPVC PIPE

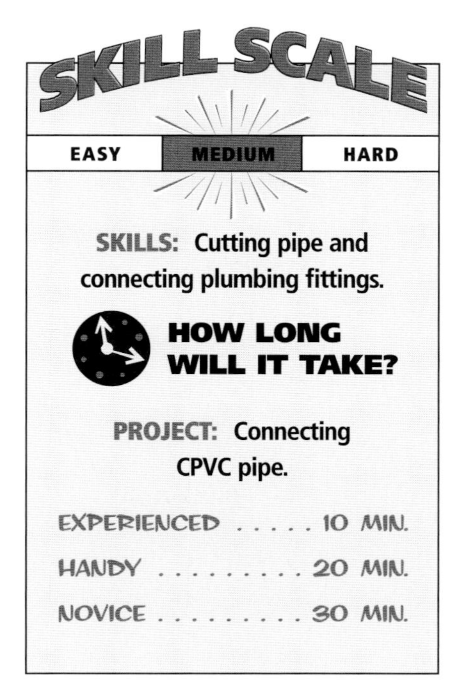

SKILL SCALE
EASY
MEDIUM
HARD
SKILLS: Cutting pipe and connecting plumbing fittings.
HOW LONG WILL IT TAKE?
PROJECT: Connecting CPVC pipe.
EXPERIENCED . . . . . 10 MIN.
HANDY . . . . 20 MIN.
NOVICE . . . . 30 MIN.

CPVC is less expensive than copper but just as durable, and it withstands high temperatures and pressure in the supply system. It cuts easily with a tubing cutter or hacksaw, connections are easy, and assembly is quick.

ASSEMBLE QUICKLY

Once you glue a fitting in place, it can't be removed—you have to cut it apart and start over. You can't twist fittings apart, and fine-tuning adjustments is impossible. Cut sections to length and test-fit the complete run before cementing the lines in place.

DO NOT use sparking or open-flame equipment (power drills or torches) near areas where solvent vapors may exist—the vapors are potentially explosive.

TOOLS: Pipe cutter, hacksaw, or miter saw and box; deburring tool

MATERIALS: Cleaner, primer, cement, rags

WORK SMARTER

TIPS FOR CONNECTING RIGID PLASTIC PIPE.

1 Use the correct primers and cements for the pipe you're installing. CPVC and ABS are not interchangeable without transition fittings.

2 Cure time depends on the cement used, the size and tolerance of the pipe and fitting, and the air temperature. You will weaken the bond by trying to speed or retard the cure.

3 Keep the lid on cements and primers when not in use.

4 Stir or shake cement before using.

5 Use a ¼-inch dauber on small-diameter pipes, a 1½-inch dauber up through 3-inch pipe, and a natural-bristle brush, swab, or roller half the pipe diameter on pipes 4 inches and up. Do not allow daubers to dry out.

6 DO NOT mix primer with cement. Do not use thickened or lumpy cement. Cement should have the consistency of syrup or honey.

7 Do not handle joints until they are fully cured.

8 All colored cements and primers will leave a permanent stain.

If your water is acidic, use CPVC instead of copper, since copper will corrode in an acidic environment.

1 **PURCHASE PRIMER AND CEMENT.** There are one-step cements available for CPVC that eliminate the need to use purple primer. But check local codes carefully to determine if primer is required. In some localities, you will fail inspection if you don't use it. Purple primer leaves a permanent and recognizable stain on the pipe, so everyone will know whether or not you've used it. **Choose a dauber that is the proper size for the job.** Daubers should be half the diameter of the fittings being joined so the job will take less time and cause less mess.

2 PREPARE THE PIPE. Use a pipe cutter or a saw and miter box to ensure a square cut. Square cuts ensure that the pipe will seat properly in the fitting to produce a strong joint.

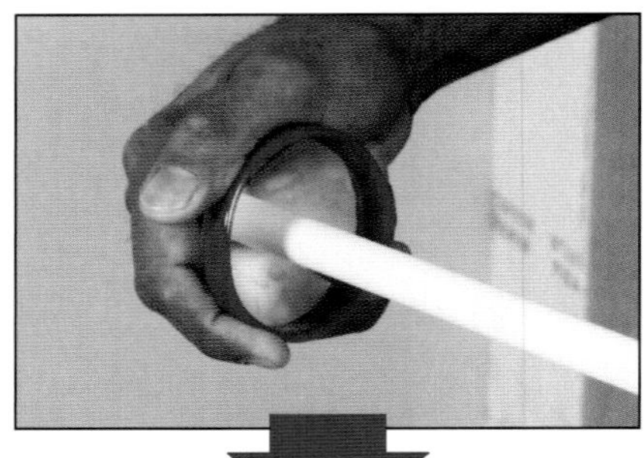

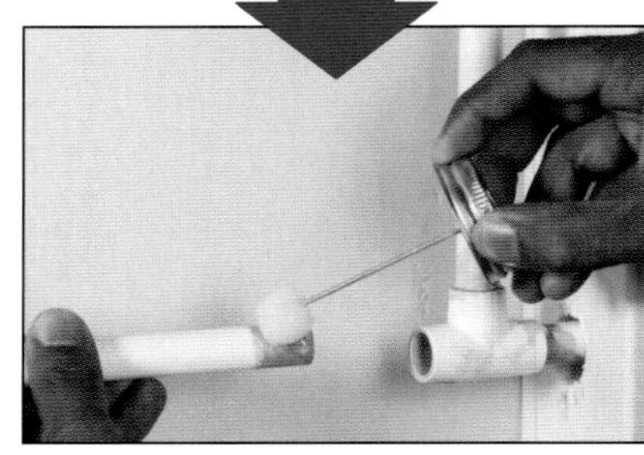

3 DEBURR THE PIPE. Removing burrs ensures even coverage with the primer and cement. Sanding can change the diameter and cause a poor fit. Test-fit the pipe to the fitting; it should seat snugly in the fitting.

COAT THE SURFACES WITH PRIMER. Apply an even coat of primer to the pipe and the fitting. Primer softens the pipe to help seat it and reacts with the cement to make a permanent bond.

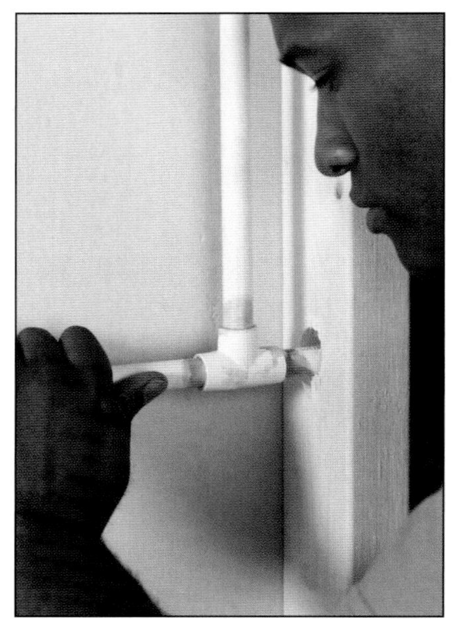

4 APPLY CEMENT AND ASSEMBLE THE PARTS QUICKLY BUT CAREFULLY. Apply an even coat of cement with a dauber to the pipe and fitting and insert the pipe all the way into the fitting until it stops. Twist a quarter of a turn to spread the cement evenly. Hold the pipe together for 30 seconds to prevent the heat from the cement from pushing the connection apart. Wipe off excess cement between the fitting and pipe with a clean rag.

CLOSER LOOK

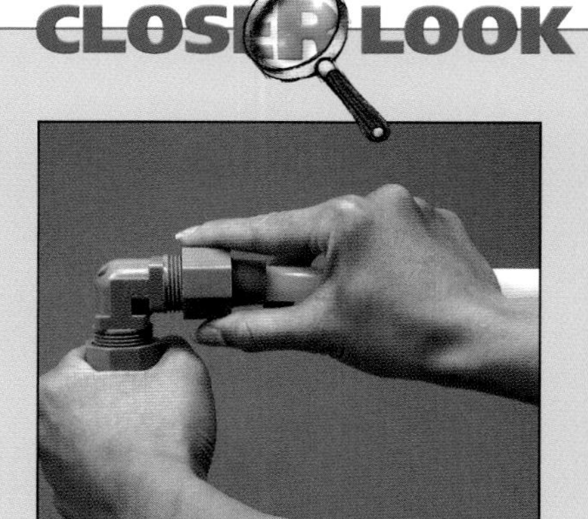

PEX FLEXIBLE PLASTIC PIPE.

PEX (cross-linked polyethylene) is a flexible plastic pipe used for hot and cold supply lines. While it is gaining wider national acceptance, PEX is primarily used in the southern United States and in parts of southern California. In these areas it can also be used to run the main supply line from an outside water meter into a slab home.

Good resistance to deterioration, heat, and the high pressure required for supply, plus ease of assembly, make PEX an ideal choice for do-it-yourselfers in areas where it is approved for use. (Check local codes.)

Easy to assemble, the system uses two types of fittings:

- Plastic compression-type fittings, which are tightened by hand and then given one full turn with pliers until snug. No tape or pipe compound is required.
- Brass ribbed fittings, which are permanently sealed to the pipe with a crimping tool and crimp ring. (This method requires special tools and is usually a professional installation.)

The pipe, which is available in rolls of 50 to 100 feet, is cut to length with a utility knife or a tubing cutter. It is flexible enough to turn corners that would require new connections and fittings with other types of plumbing. (Follow the manufacturer's instructions for maximum bends.)

Both the compression and crimped fittings are required by code to be accessible for inspection and repair and cannot be sealed in walls or ceilings. In order to make the fittings accessible they are often grouped in manifolds with access panels installed in convenient locations.

CONNECTING GALVANIZED AND BLACK IRON PIPE

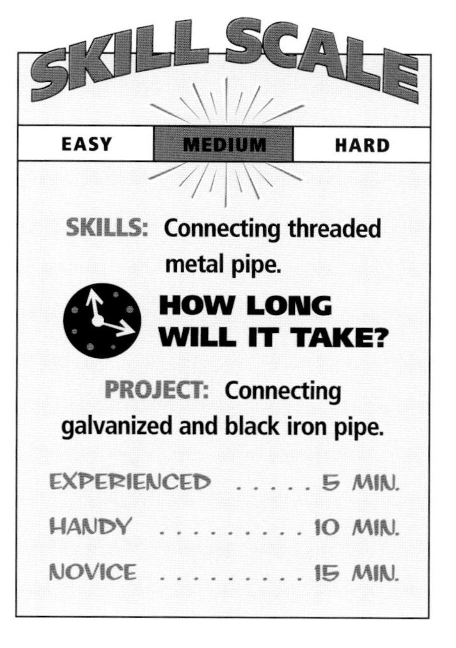

SKILL SCALE
EASY
MEDIUM
HARD
SKILLS: Connecting threaded metal pipe.
HOW LONG WILL IT TAKE?
PROJECT: Connecting galvanized and black iron pipe.
EXPERIENCED ..... 5 MIN.
HANDY .... 10 MIN.
NOVICE .... 15 MIN.

Many years ago galvanized pipe replaced lead in supply and drain applications, along with cast iron. ABS, PVC, CPVC, and PEX are now the most common choices. Black iron pipe is still used for gas lines.

Working with threaded galvanized or black iron pipe might seem simple, but it's a demanding job. You will need to keep track of what length and fitting goes where. It is also difficult to determine what lengths you need for the run. You may find that most of your time is spent running to your local home center. In the end, it is likely to be easier to measure carefully and have a home center or hardware store cut and thread the pieces for you.

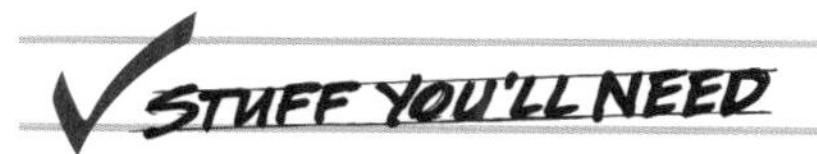

TOOLS: Two pipe wrenches

MATERIALS: Pipe, fittings, pipe compound, rag, liquid dishwashing soap, sponge

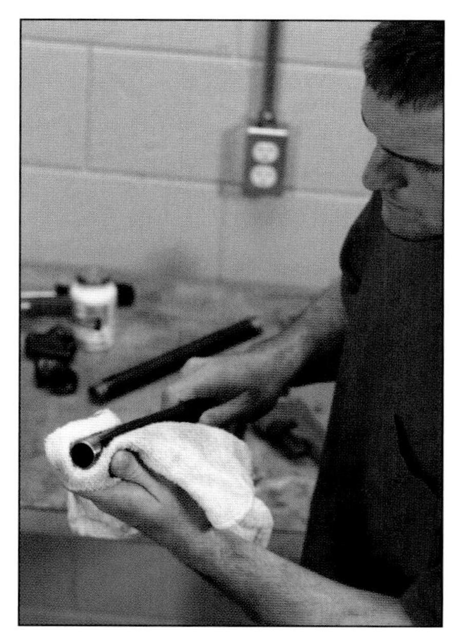

1 CLEAN THE THREADS OF THE PIPE AND FITTING. Use a rag to clean dirt and grease from the threads. This will ensure the best connection for the pipe and fitting.

2 APPLY PIPING COMPOUND TO THE PIPE THREADS. Connect the pipe to the fitting by holding the fitting in place with a pipe wrench and turning the pipe with the other wrench until it is snugly seated. Clean off any excess piping compound with a rag.

3 TURN ON THE GAS AND TEST THE LINE FOR LEAKS. If you are installing a gas line, it must be tested for leaks. Fill a sponge with liquid dishwashing soap and water. Apply it to the new fitting and look for bubbles. It's the same process used for finding leaks in a car tire. If there's a leak, bubbles will form on the surface and you'll have to refit the joint.

CHAPTER

5 THE DRAIN-WASTE-VENT (DWV) SYSTEM

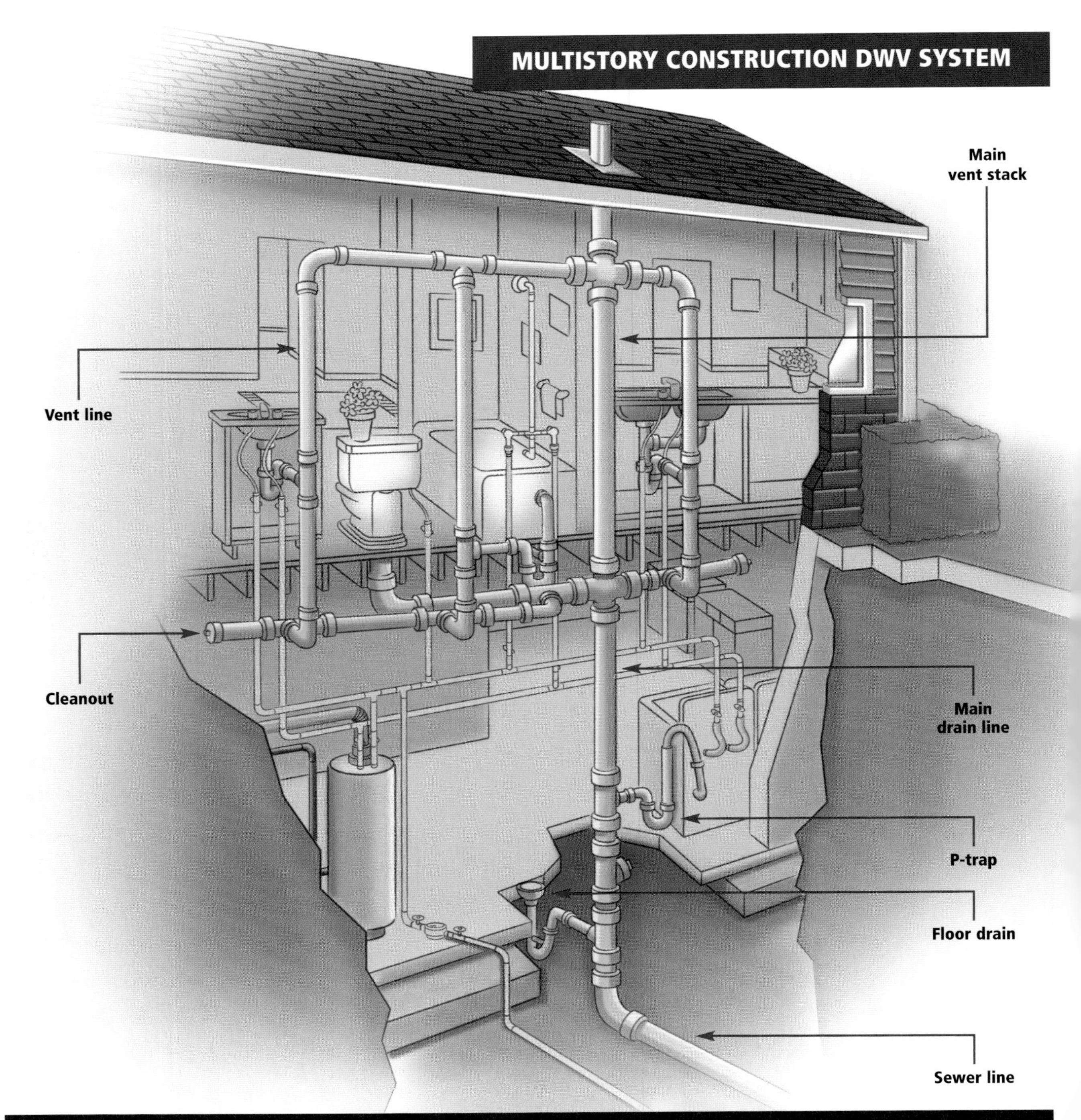

CHAPTER 5 PROJECTS

The DWV (drain-waste-vent) system removes liquids, gases, and solid waste from your home through a series of drains and vents.

WHERE THE SEWAGE GOES

Solid and liquid waste are drained away either to a septic tank on the property or into a municipal sewage system.

Gases produced from the waste must be prevented from entering and remaining in the house—that's the purpose of the vent system. The vent system also allows air into the drainage system so the wastewater and the gas can easily flow away from the house.

DRIVEN BY GRAVITY

Unlike the supply system, which uses water pressure to move water through the pipes, the drain system is installed at a slope and uses gravity to drain waste into the main sewage system. Water draining down a vertical pipe flows more quickly than from a horizontal one. Because of this rate of flow, deposits of grease and soap residues build up in horizontal pipes faster than in vertical pipes. To prevent deposit buildup, all horizontal pipes must be set and run at an angle called the "fall." The standard formula for calculating the fall is ¼ inch per foot over the course of the run. An 8-foot run, for instance, would have a fall of 2 inches from beginning to end.

WHO'S RESPONSIBLE?

The public sewage system is connected to your home through the main drain line. The main line serves as the trunk for the branch lines. The branch lines are tied to the main system by a large vertical pipe that leads down to the main drain and up through the roof to the main stack vent. You are usually responsible for all repairs on your side of the property line, which includes the service line. The municipality is responsible for repairs on the lines in the public rights-of-way.

DRAINING AWAY

When wastewater leaves a sink or plumbing fixture, it enters a curved section of pipe called a trap. The trap is a U-shaped section of pipe installed between the fixture drain and the drain line that forces water to drain against gravity. They are often called P-traps because of their shape when installed. As the sink, tub, or toilet drains, water is forced through the trap. When the fixture is empty, the trap holds the water, creating a seal to keep sewer gases from entering the house.

Bends are necessary in the branch lines because of the constraints of home construction. Elbows with gradual sweeping curves move the wastewater more smoothly through the system and prevent deposits from building at the elbows.

Cleanouts along the drain lines allow you to auger out blockages. There is probably a main cleanout where the drain exits the house. In slab construction you may find the main cleanout outside next to the house.

VENTS MAKE IT WORK

Without a constant air supply a vacuum would be created in drain lines and waste would not flow from the house. Air enters drain lines from the vent stack system that extends up through the roof. Each plumbing fixture should have its own vent line tied into the central vent system.

Air pressure also keeps sewer gases from breaking the seal made by wastewater in the drain traps. If the water barrier in the trap is broken, sewer gases rich in unpleasant smells and potential health and fire hazards can enter the house. Proper venting is rigorously enforced in plumbing codes by inspection and fines for violation.

Like drain lines, vent lines can become clogged. If the drain line gurgles or still seems sluggish after augering the main or branch line for a blockage, head up to the roof and auger down the vent to clear out the obstruction.

ANATOMY OF A P-TRAP

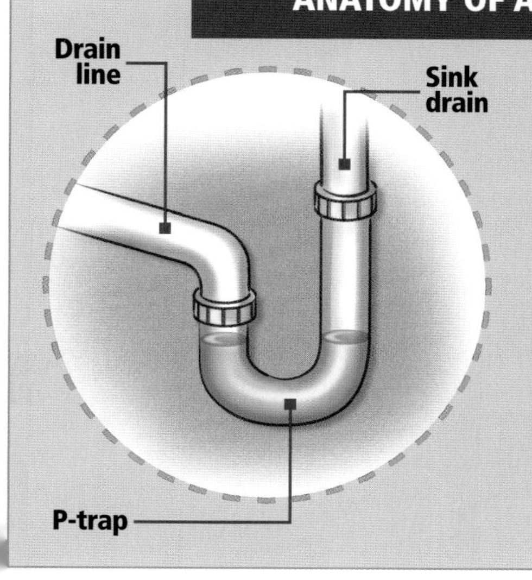

P-traps are required by code wherever there is an open drain line that flows into the DWV system such as from a sink, bathtub, or shower. It is constructed so that a water seal will form in the curve of the trap preventing backflow of air or gas from the sewer line while permitting free flow of liquids through the system.

USING ESCUTCHEON PLATES.

Escutcheon plates fit around the drain pipe covering the hole where the drain line enters the wall or floor. Where plumbing is visable, such as under a wall-hung lavatory, they add a finishing touch. Using escutcheon plates inside cabinets isn't always required but do help seal the hole. Two basic types are available in both metal and plastic.

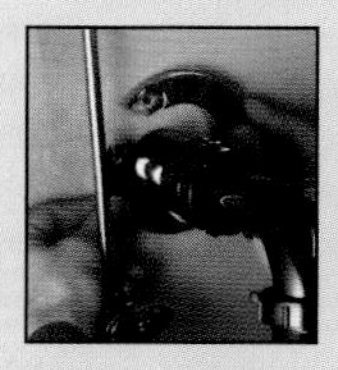

◄ Split-ring escutcheon plates fit around a pipe that is already in place.

◄ Solid-ring escutcheon plates slip onto the pipe before the drain is assembled.

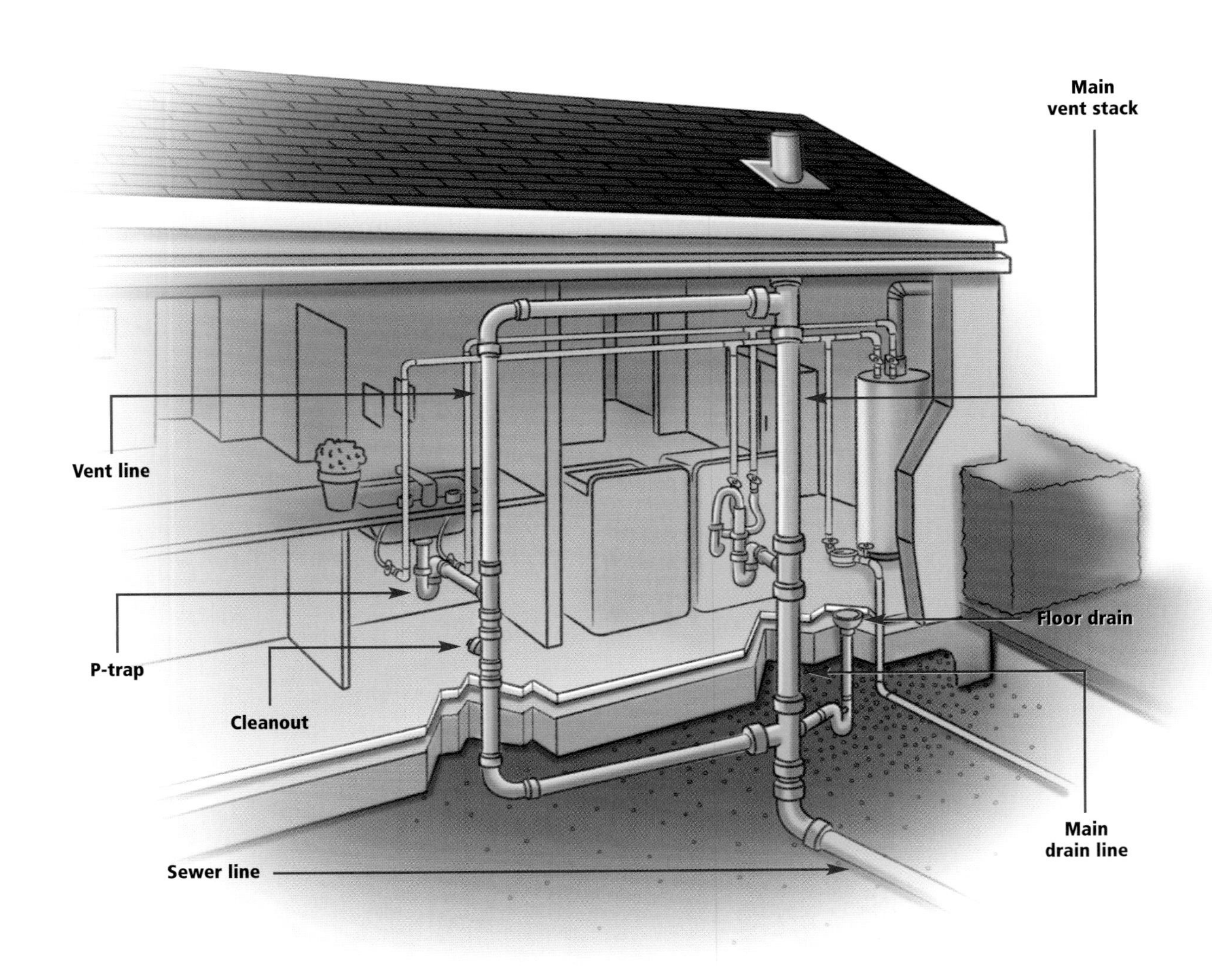

DWV PIPES AND FITTINGS

Different plumbing fixtures need different sizes of drain pipe because of flow requirements. Bathroom sinks have drainage flows requiring the pipe to be 1¼ inches in diameter; toilets require 3- to 4-inch lines; and sinks need 1½-inch pipes. The pipe size of the main drain is identified by its inside-diameter (ID). Standard is usually 4 inches.

TYPES OF DRAIN PIPES

The type of pipe used in the drainage system varies, depending on the age of the house. Vitrified clay, bitumized fiber, cast iron, ABS, and PVC all have been used for the main sewer line. Vitrified clay can be damaged by tree roots, a major cause of replacement of main lines and sewer lines in older homes. Older homes have cast iron for the stacks and galvanized steel pipe for the runs.

The durability of cast iron makes it an attractive choice for the drainage system stack, but it's difficult to work with. Galvanized steel tends to corrode over time and requires more frequent replacement than other materials.

Newer homes use either ABS or copper pipe for DWV runs because they are virtually maintenance-free.

3" ABS cap

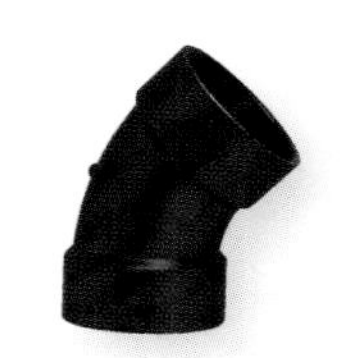
3" ABS 45° elbow

3" ABS Y fitting

3" ABS 90° long sweep elbow

ABS cleanout and plug

3" ABS coupling

3" ABS 90° elbow

ABS reducer

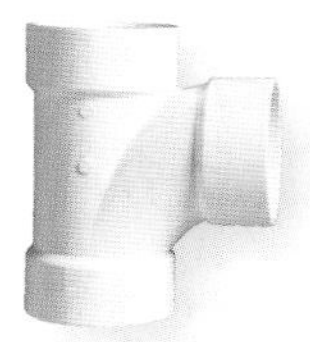
3" PVC T

PVC cleanout and plug

3" PVC coupling

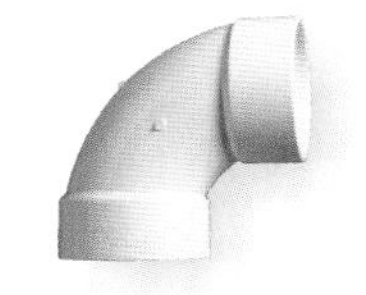
3" PVC 90° elbow

PVC reducer

3" PVC cap

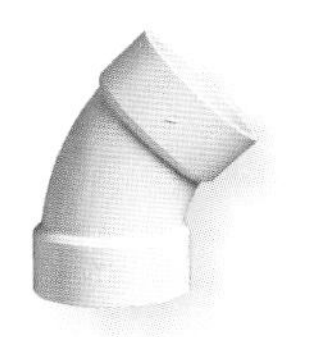
3" PVC 45° elbow

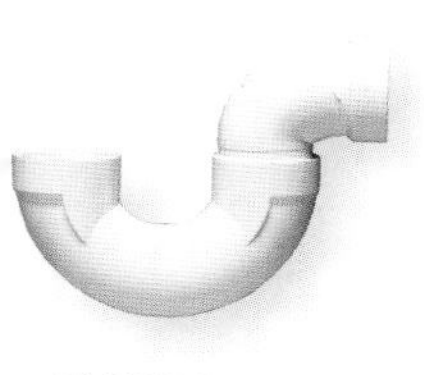
2" PVC P-trap

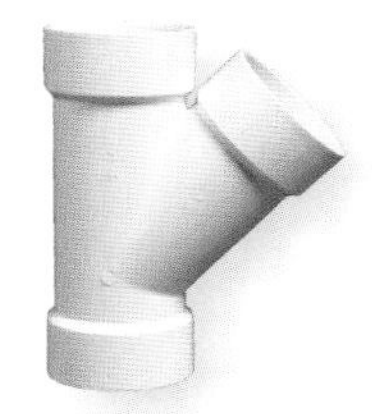
3" PVC Y fitting

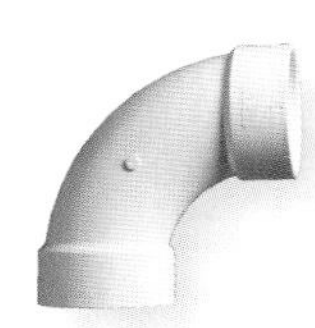
3" PVC 90° long sweep elbow

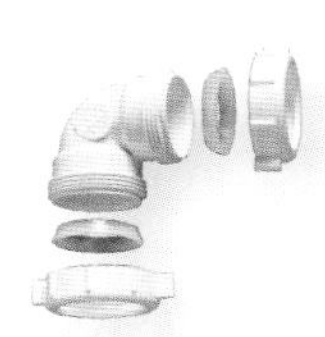
PVC slip elbow

PVC slip joint coupling

S-to-cast iron adapter

3" PVC pipe

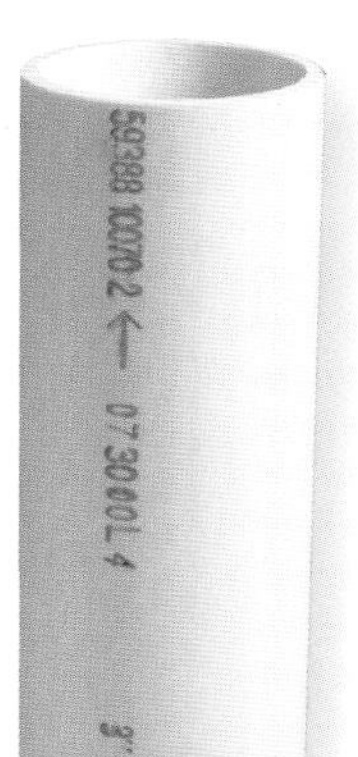

3" ABS pipe

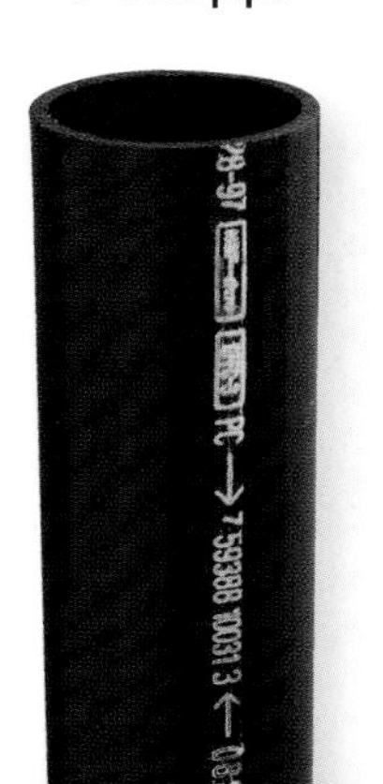

3" Cast iron pipe

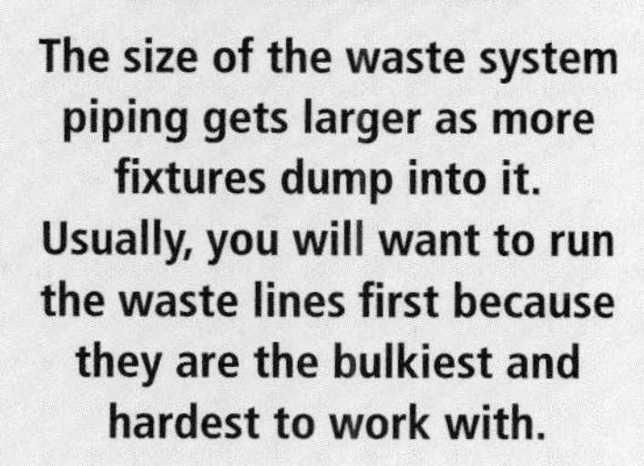

LAST THINGS FIRST.

The size of the waste system piping gets larger as more fixtures dump into it. Usually, you will want to run the waste lines first because they are the bulkiest and hardest to work with.

REPLACING AND REPAIRING CAST-IRON PIPE

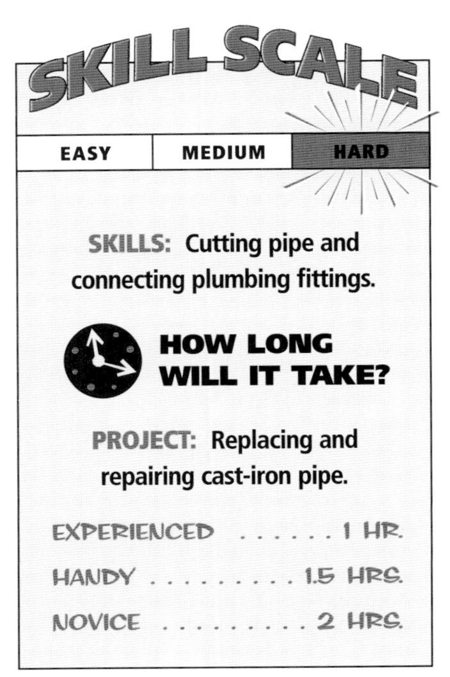

SKILL SCALE
EASY
MEDIUM
HARD
SKILLS: Cutting pipe and connecting plumbing fittings.
HOW LONG WILL IT TAKE?
PROJECT: Replacing and repairing cast-iron pipe.
EXPERIENCED ...... 1 HR.
HANDY .... 1.5 HRS.
NOVICE .... 2 HRS.

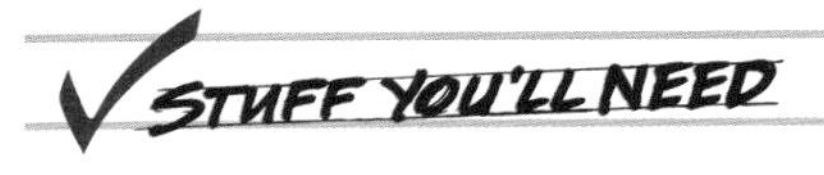

✓ STUFF YOU'LL NEED

TOOLS: Screwdriver, reciprocating saw or cast-iron snap cutter, riser clamps, ratchet wrench and sockets, chalk

MATERIALS: Neoprene sleeves, coupling bands, ABS pipe, 2x4s

Cast iron was the mainstay of drain and vent systems until the introduction of ABS. The mechanics of repair are not difficult to master, but cast-iron pipe is awkward, heavy, and difficult to cut. You will need a reciprocating saw or you can rent a special cast-iron snap cutter, which is more efficient. Sections of damaged pipe are commonly replaced with ABS that is lighter and easier to handle than cast iron.

SUPPORT THE PIPE

It is extremely important to support the pipe while you are making the repair. Riser clamps must be attached above and below the damaged section. These clamps must rest securely on wooden frames on the floor and be framed or chained to the ceiling joists above.

SAFETY ALERT!

BEFORE YOU BEGIN, CLEAN THE DRAIN LINE.

Flush the toilets and run some water through the fixtures to clean out the drain. Alert your family not to use house fixtures like toilets or sinks while you are making this repair. Waste in the lines is a potential biohazard. Wash your hands and change clothes when you're finished.

1 SHUT OFF THE WATER MAIN AND MARK THE PIPE. Use chalk to mark at least 6 inches above and below the damaged section. Build a support frame of 2x4s as shown in Step 2 and install riser clamps above and below the damaged section to support the pipe.

2 CUT OUT THE DAMAGED PIPE. Wrap the chain of the cast-iron cutter around the pipe. Align the cutting wheels with the chalk line. Tighten the chain and cut. After both ends have been cut, remove the section.

3 CUT THE REPLACEMENT PIPE. Measure and mark out the length of the replacement ABS pipe and cut a section 1 inch shorter than the span to be replaced.

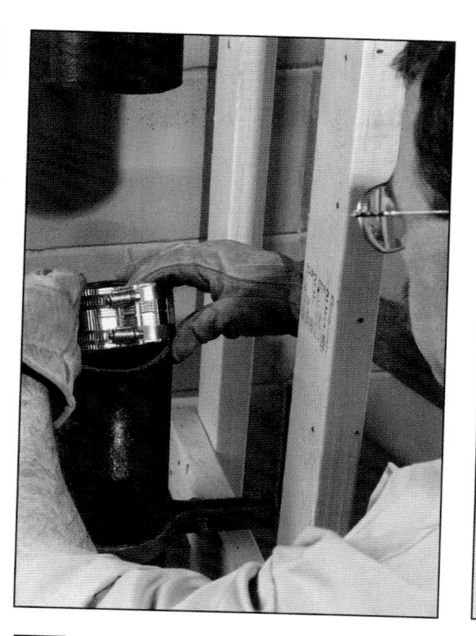

4 PLACE THE COUPLING BAND. Slip a banded coupling with a neoprene sleeve over each end of the cast-iron pipe. Make sure the cast iron is seated tightly against the rubber separator ring molded into the sleeve.

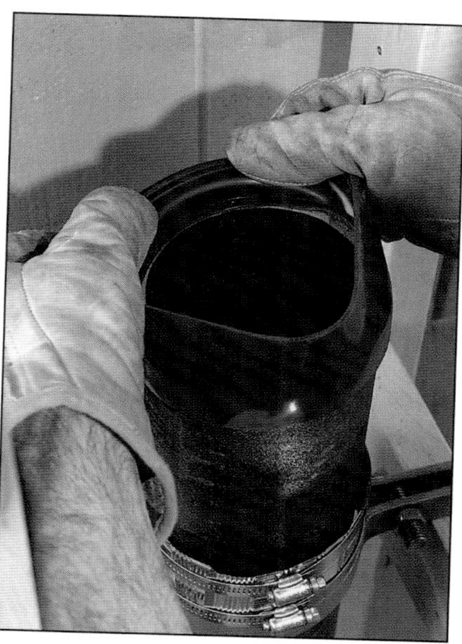

5 FOLD BACK THE SLEEVE. Expose the ends of the cast iron by folding back the neoprene sleeve on both sections.

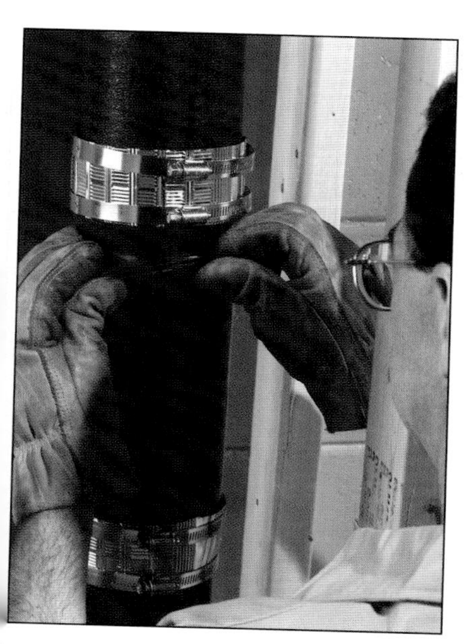

6 POSITION THE NEW SECTION OF PIPE. Align the ABS pipe with the cast-iron pipe. Roll the neoprene sleeves over the plastic pipe.

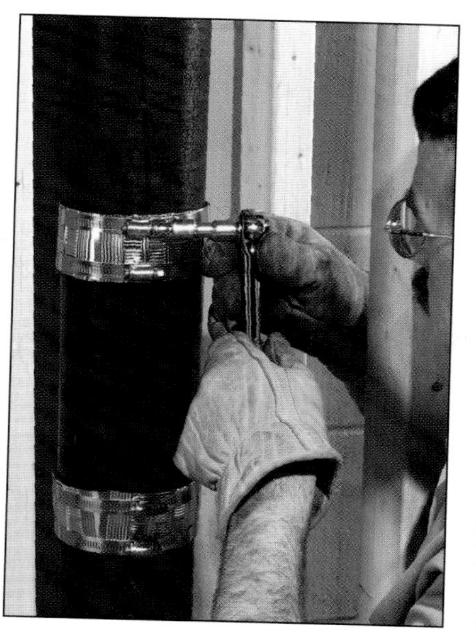

7 SLIDE THE BANDS OVER THE SLEEVES. Once in place, tighten the bands with either a screwdriver or a socket and ratchet wrench until snug. Remove the riser supports when the pipe is secured in place.

8 TEST THE SECTION. Run water through the drain system and check for leaks. Tighten the bands as necessary.

CONNECTING ABS PIPE

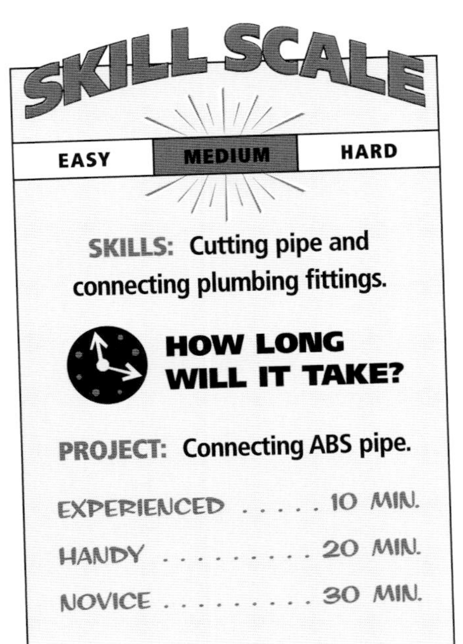

SKILL SCALE

EASY | **MEDIUM** | HARD

SKILLS: Cutting pipe and connecting plumbing fittings.

HOW LONG WILL IT TAKE?

PROJECT: Connecting ABS pipe.

EXPERIENCED . . . . . 10 MIN.
HANDY . . . . 20 MIN.
NOVICE . . . . 30 MIN.

ABS was the first rigid plastic pipe approved for use in drain, waste, and vent systems. It is easy to cut, lightweight, and very rigid.

✓ STUFF YOU'LL NEED

TOOLS: Tubing cutter, hacksaw or miter saw and box, deburring tool

MATERIALS: ABS pipe and fittings, cleaner, cement, dauber, rags

USE THE RIGHT DAUBER.
Using the right size dauber is the key to successful connecting. Because the cement dries very quickly, speed is of the essence. You should use a dauber that's no less than half the diameter of the pipes and fittings so you can minimize the time it takes to apply the cement and make the joint.

LOOKS CAN BE DECEIVING.
It's easy to assume all plastic pipe is the same, but that kind of thinking can get you in a lot of trouble. Each type is a different composition of materials and requires its own blend of cleaners and cements to make a proper bond. If you're in a situation where you have to join ABS and PVC, you'll need a special transition fitting to make the hookup.

1 **CHOOSE THE RIGHT SOLVENTS AND DAUBER.** Read the label to make sure the solvents are for use with ABS pipe. The dauber should be half the diameter of the pipe being joined to reduce the time for application.

2 **USE THE RIGHT CUTTING TOOL FOR THE JOB.** Tubing cutters are perfect for cutting small-diameter pipe. Cut larger pipe with a hacksaw, abrasive disk, or miter saw. Be sure to make straight cuts—they allow for proper seating into the fitting.

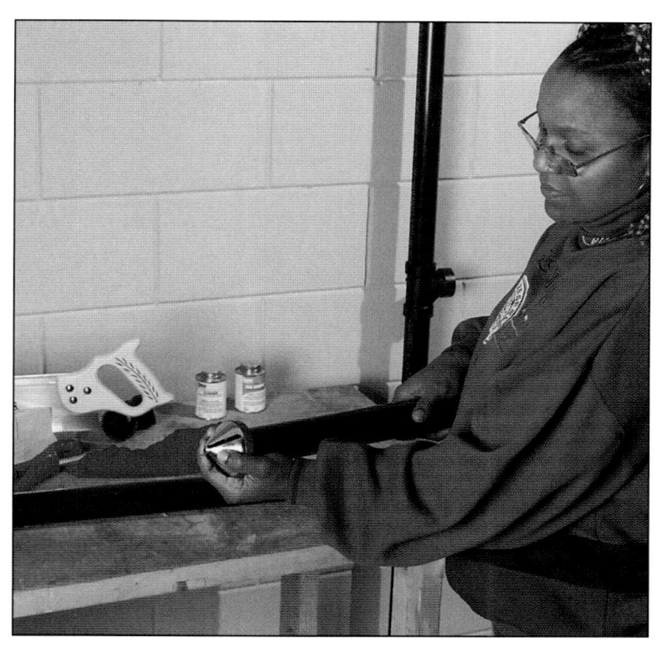

3 **DEBURR THE CUT PIPE.** Use a deburring tool to remove burrs and shavings. If you don't have a deburring tool, a knife or emery cloth will work. Dry-fit the pipe and fittings. Make sure the proper fall of ¼ inch per lineal foot run is made to allow proper drain flow. A fitting that is glued crooked can throw off a whole run. Discover potential problems by dry-fitting.

4 **APPLY THE SOLVENTS TO THE FITTING AND PIPE.** Apply the cleaner to the pipe surface to remove ink, oils, grease, and dirt. Allow the cleaner to dry. Coat the surfaces with ABS cement. Don't puddle the cement on—too much cement can weaken the pipe wall. Push the connections together with a twisting motion until properly seated.

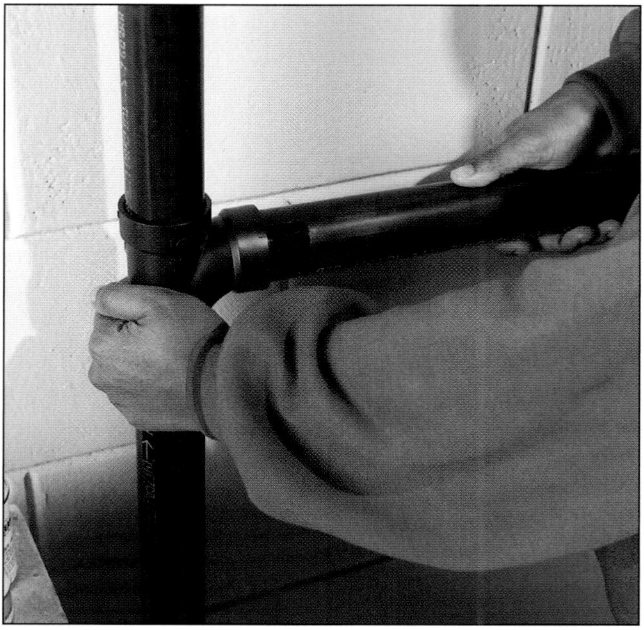

5 **HOLD THE CONNECTION FOR 30 SECONDS.** The solvent produces heat that may cause the connection to push apart. The cement melts the ABS pipe to fuse the connection together. Wipe away excess cement with a rag to prevent weakening of the ABS pipe walls.

CONNECTING PVC PIPE

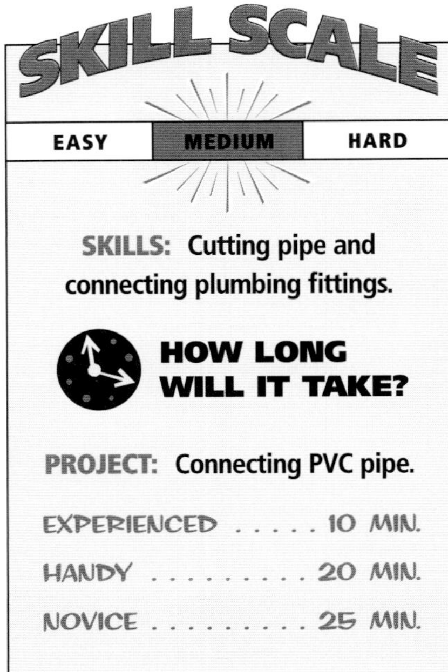

SKILL SCALE
EASY
MEDIUM
HARD
SKILLS: Cutting pipe and connecting plumbing fittings.
HOW LONG WILL IT TAKE?
PROJECT: Connecting PVC pipe.
EXPERIENCED . . . . . 10 MIN.
HANDY . . . . 20 MIN.
NOVICE . . . . 25 MIN.

✓ STUFF YOU'LL NEED

TOOLS: Tubing cutter, hacksaw or miter saw and box, deburring tool or emery cloth, torpedo level, utility knife

MATERIALS: PVC pipe and fittings, primer, cement, rags

In some areas you can use rigid PVC pipe for drain, waste, and vent systems. The pipe is available with inside diameters (ID) of 1½, 2, 3, and 4 inches.

- Use 1¼-, 1½-, and 2-inch ID ABS or PVC for sink drains and lavatories.
- Use 1½- and 2-inch ID pipe for tubs and showers.
- Use 4-inch ID for toilets.
- Drain lines and vent stacks can use 2-, 3-, or 4-inch ID.

DRY-FIT ALL THE CONNECTIONS BEFORE THE FINAL ASSEMBLY. Dry-fit the connections before applying primer and cement—once you've cemented the pipe, it can't be changed. Check the fall with a torpedo level. There should be a ¼-inch fall for each lineal foot of run.

1 USE THE RIGHT PRIMER AND CEMENT FOR THE JOB. The right primer and cement guarantee good connections. The right size dauber is important as well. Daubers should be half the diameter of the pipe to be joined to reduce application time.

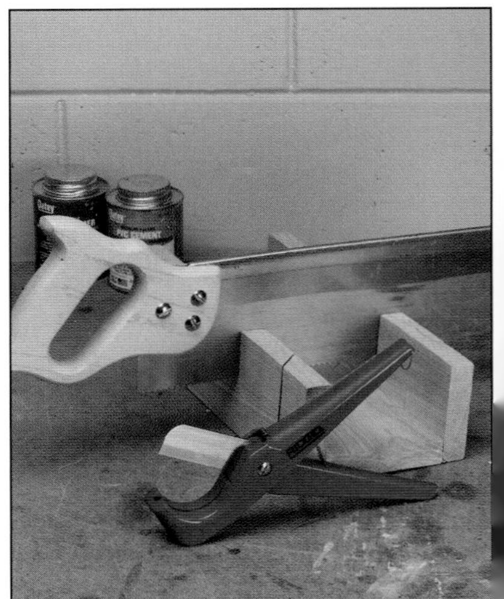

2 CUT THE PVC SQUARE. A straight cut is necessary for a good connection. There are PVC tubing cutters specifically for small-diameter pipes. Cut large-diameter pipes with a miter saw, hacksaw, or ratchet-type saw.

3 DEBURR THE CUT. After each cut, deburr the pipe with a knife, emery cloth, or rag to ensure that it seats snugly into the fitting.

4 APPLY PRIMER TO THE CONNECTIONS. The primer softens the ends, preparing it for application of the cement.

5 COAT THE PIPE AND FITTING WITH CEMENT. While the primer is still damp, quickly apply an even coat of cement to the surfaces. Apply a thin coat; too much cement can weaken the pipe and destroy the fitting.

6 WORK QUICKLY TO CONNECT THE PIECES. Insert the pipe into the fitting with a quick push and a quarter-turn to seat. Hold the connection for 30 seconds to prevent the heat produced by the cement from pushing the connection apart. The cement melts the surfaces and forms a secure bond. Wipe away excess cement, which can weaken the joint, with a rag.

TIPS FOR CONNECTING RIGID PLASTIC PIPE:

1 Use the correct primers and cements for the pipe you're installing. CPVC, PVC, and ABS are not interchangeable without transition fittings.

2 Cure time depends on the cement used, the size and tolerance of the pipe and fitting, and the air temperature. You will weaken the bond by trying to speed or retard the cure.

3 Keep the lid on cements and primers when not in use.

4 Stir or shake cement before using.

5 Use a ¼-inch dauber on small-diameter pipes, 1½-inch dauber on up through a 3-inch pipe, and a natural-bristle brush, swab, or roller one-half the pipe diameter on pipes 4 inches and up. Do not allow daubers to dry out.

6 DO NOT mix primer with cement. Do not use thickened or lumpy cement. Syrup or honey is the right consistency.

7 DO NOT handle joints until they are fully cured.

8 All colored cements and primers will leave a permanent stain.

CHAPTER

6 INSTALLING TUBS

Removing or installing a bathtub is a big job. It requires plumbing, carpentry, drywalling, electrical, and sometimes, tiling skills. Most bathrooms have limited space, so you'll probably have to remove other fixtures, such as the toilet and lavatory sink, to give yourself enough room to work. You may want to consider hiring professionals to do some or all of the work.

PLAN FOR REMOVAL

You've heard about the guy that built a boat in his basement? Well, before you take out the old tub, know how you're going to get it out the door, down the hall, down the stairs and out into the yard. Measure to make sure it will fit through doors and hallways. Call your local waste removal service to arrange for its disposal. Have friends help carry it out—a bathtub is too heavy and awkward for one person to handle.

TUB FACTS

Tub material determines the difficulty of installation. Acrylic and fiberglass tubs range from 60 to 120 pounds and some need to be installed on a

CHAPTER 6 PROJECTS

bed of mortar. Steel and cast-iron tubs don't need mortar but they can weigh more than 300 pounds, so make sure the floor will support the weight.

The National Kitchen and Bath Association (NKBA) categorizes bathtubs according to installation type: recessed, corner, freestanding, and platform or drop-in.

Recessed tubs have a finished front with unfinished sides that are concealed by the surrounding alcove walls.

Corner installations have three finished sides facing outward and the two unfinished sides concealed by the adjoining walls.

Claw-foot tubs are freestanding and can be placed anywhere that space allows.

Drop-in tubs are installed in a prebuilt platform and offer many design options by coordinating the color of the tub with the material of the platform structure.

MATERIAL CHOICES

Bathtubs constructed of **acrylic**, **fiberglass**, and **plastic composites** are durable, attractive, and a good alternative to traditional cast-iron construction. They weigh less, come in a wide range of colors, and have a nonporous surface that's easy to clean. Bathtubs constructed of acrylic, fiberglass, and plastic composite materials are easily molded into a variety of shapes and sport features such as foot and back supports and headrests.

Cast iron is strong and durable and offers contemporary as well as traditional choices (such as the claw-foot style) in design.

Bathtubs are also made of **enameled steel** and **cultured marble**. Today's models feature a wide range of metal trim and colors. It's not unusual for the same model tub to be available in more than 15 colors. White is usually the less expensive option.

SPECIAL FEATURES

Some manufacturers offer special surfaces designed to limit bacterial growth. **Antibacterial surfaces** can be especially in demand in homes with small children.

Whirlpool tubs provide basic bathing, but they also feature a motorized pump that draws water from the bathtub, mixes it with air, and injects it back into the bathtub through strategically placed jets to massage and relax tense muscles after a hard day of plumbing.

COMMON MISTAKES DO-IT-YOURSELFERS MAKE.

1. Ignoring or violating local code restrictions.
2. Using pipes that are too small for the job.
3. Connecting copper to galvanized pipe without a brass or dielectric fitting between them.
4. Not using pipe tape or compound at threaded joints.
5. Not leveling the fixture controls.
6. Not installing access and shutoff valves.
7. Not aligning tubing into compression fittings or stop valves.

REMOVING AN OLD BATHTUB

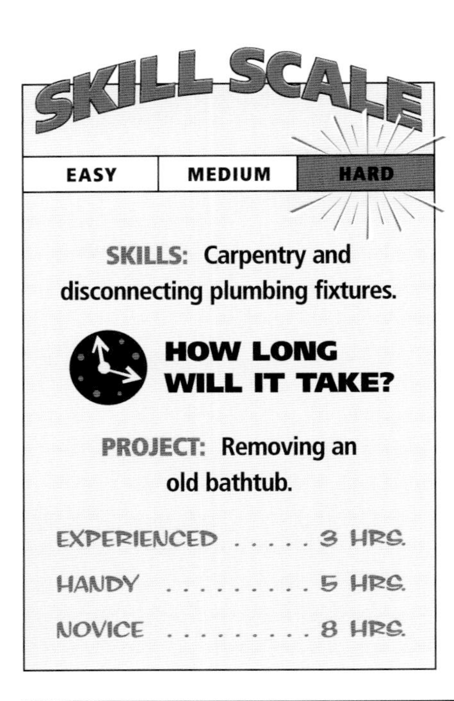

SKILL SCALE
EASY
MEDIUM
HARD
SKILLS: Carpentry and disconnecting plumbing fixtures.
HOW LONG WILL IT TAKE?
PROJECT: Removing an old bathtub.
EXPERIENCED ..... 3 HRS.
HANDY .... 5 HRS.
NOVICE .... 8 HRS.

STUFF YOU'LL NEED

TOOLS: Crowbar, pry bar, keyhole saw, screwdriver, pliers, shower stem socket set, gloves, safety glasses

MATERIALS: 1x4 slides

Make sure you have plenty of room to work before you tackle tub removal. You will need a clear space at least 3 feet deep when pulling the tub straight out from the wall. You need to be able to work around the tub when it's away from the wall. It's a good idea to remove other plumbing fixtures such as the sink or vanity and toilet.

MAP YOUR ROUTE

Measure the width of all door openings and the hallway along the path for removing the tub. Not only is this important for removing the old tub, but it's essential for bringing in the new one. There's nothing worse than having torn up your bathroom to remove the old tub, and discovering that the new one will not fit through the hall or doorways.

THE ORDER OF BATTLE

- Turn off the water at the main water shutoff. Drain the water supply lines by opening a faucet below the tub level.
- Remove the faucet handles, spout, and drain.
- Cut away at least 6 inches of drywall above the tub on all sides. Remove the screws or nails holding the tub flange to the studs. If you find a galvanized strip along the tub flange, use a flat bar or pry bar to remove it.
- Lift up on the front edge of the tub with a pry bar and slide a pair of 1×4s beneath the tub.
- Pull the tub away from the wall using the 1×4s as a skid. You'll need help carrying the tub, especially if it's cast iron. If you're not saving it or reusing it, cover the tub with a tarp and break it into pieces with a sledgehammer. Fiberglass and polymer tubs can be cut into pieces with a reciprocating saw.

Give yourself plenty of room to work when you're removing a tub. Cut away at least six inches of drywall above the tub.

INSTALLING A SUPPLY CONTROL RISER

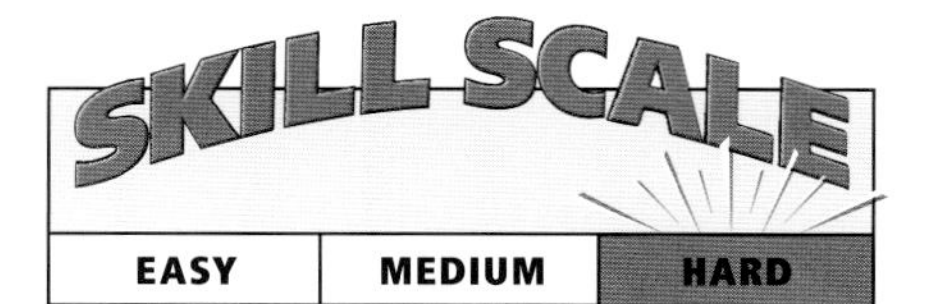

SKILLS: Carpentry skills and connecting plumbing fixtures.

HOW LONG WILL IT TAKE?

PROJECT: Installing supply risers and a bathtub.

EXPERIENCED . . . . . 6 HRS.
HANDY . . . . 8 HRS.
NOVICE . . . . 12+ HRS.

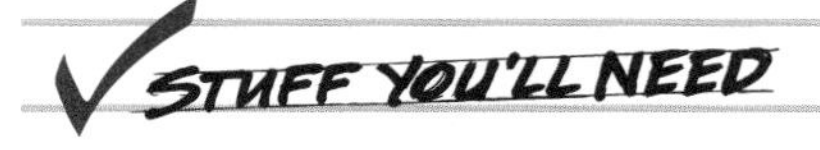

TOOLS: Propane torch and striker, tubing cutter, soldering kit, fiberglass flame barrier, hammer, electric drill, spade bits, screwdriver, saber saw, ABS cutter, tape measure, carpenter's level, pencil, gloves, safety glasses

MATERIALS: Copper supply pipe, copper fittings, integral stops, copper tube straps, faucet assembly, shower tee, solder, flux, ABS drainpipe, ABS fittings, ABS P-trap, tub overflow and drain assembly, ABS cleaner and cement, bathtub, galvanized nails, 2x4 blocking, 1x3 ledger strips, silicone caulk

Sketch an installation plan of the layout for your supply control riser so you can break the process into manageable phases. (Use the illustration below as a guide.) Use the drawing to make up a tool and materials list. You may find it easier to install the system in the bathroom and then connect the supply lines and the drain downstairs.

REPLACING AN OLD SYSTEM

If you're replacing older galvanized pipe with copper, and the interior wall structure isn't water damaged or otherwise deteriorated, the riser holes and the blocking will most likely be in place.

NEW CONSTRUCTION

For a new installation, determine where you want the tub, then where you'll tap into the water supply and drainage systems. Next, measure and sketch out both runs. If you're involved in new construction, you'll have to drill the riser holes and install blocking.

Adhere strictly to local codes and, if an inspection is required, make sure the inspector signs off before you close up the walls.

SUPPLY CONTROL RISER

Single-handle

Double-handle

Triple-handle

Standard spout height is 28" but it must be at least 6" above the finished top of the tub.

Standard shower height is 51" from the floor, but you can place it wherever you choose.

1 DRILL HOLES FOR THE RISERS. Drill riser holes through the stud wall into the basement. The diameter of the holes should be at least ¼ inch larger than the diameter of the riser so you'll have some flexibility when you're hooking up the supply lines. The type of faucet you're installing will determine the spread and placement of the riser holes.

2 INSTALL THE BLOCKING. Blocking for a tub spout should be centered 4 inches above the top of the tub. Faucets installed more than 6 inches above the spout need separate blocking. The blocking should be level, at the correct depth inside the wall, and toenailed firmly to the studs. Blocking for the showerhead is usually about 6 feet 6 inches above the floor, but you can adjust the height to suit your needs.

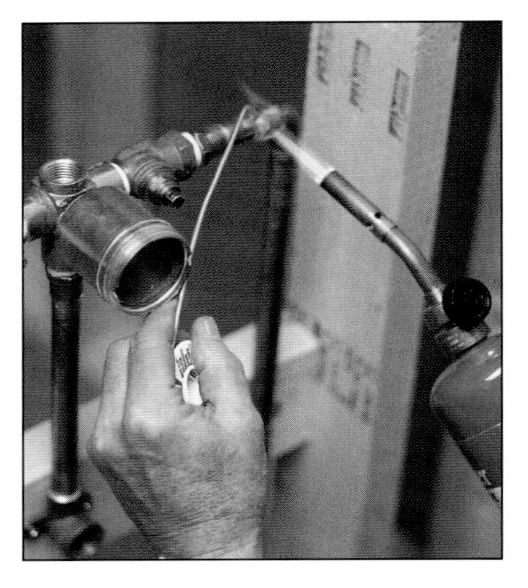

3 CUT THE RISER ASSEMBLY. Cut the pipe for the supply risers, integral stops, and tub spout. Fit the adapters onto the faucet assembly using Teflon® tape. Attach the pipe and fitting for the tub spout. Measure and cut the pipe for the supply and shower risers.

If local code requires an inspection, make sure you do it before you close up the walls around the risers and faucet.

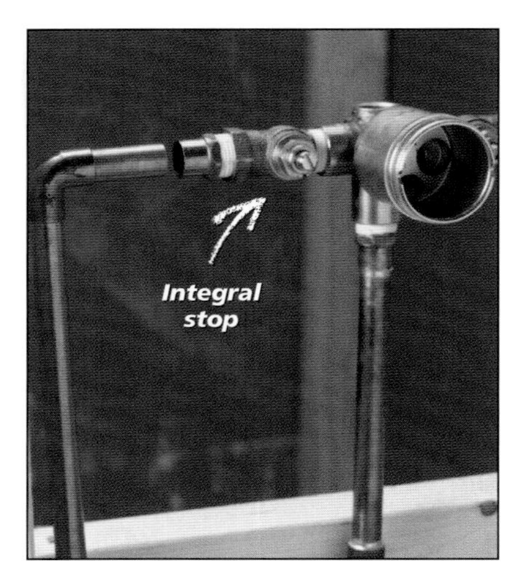

4 ASSEMBLE THE PIECES. Make sure the faucet controls are centered and level and the stops fit. Do a dry assembly as you go to make sure everything fits. Make any necessary adjustments and solder pieces into place.

5 SECURE THE SPOUT TO THE BLOCKING. After securing the spout to the blocking, connect all the pipes and fittings. (See pages 86–87 for soldering copper pipe and pages 89–90 for connecting CPVC.)

6 SECURE THE SHOWER RISER. Screw the fitting for the showerhead into its blocking.

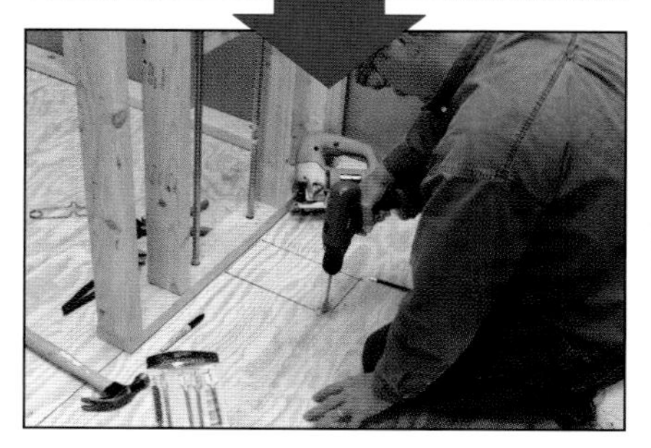

7 SECURE THE RISERS TO THE BLOCKING. Screw the risers to the riser blocking and install a dummy nipple in the tub spout to protect the threads and serve as a guide.

CUT AN ACCESS HOLE INTO THE SUBFLOOR FOR THE DRAIN. Measure and mark an opening in the floor 4 to 9 inches wide and 12 inches from the center of the end wall. Drill out the corners and cut out the opening with a saber saw.

OPTIONS FOR PROCEEDING

At some point you've got to test the system for leaks, and an inspection may be necessary before you install and finish the walls. After you've cut the access hole for the drain, you have a couple of options:

A You can install the tub and then hook it up to the drain and run the supply to the risers. (See "Installing a Tub on a Subfloor," pages 108–111.)

B If you're not ready to install the tub at this point, you can go below to run and connect the supply lines to the risers and run the drain line to the access hole. (See below.) Once the supply lines are hooked to the mains, test the system. Temporarily attach the faucet and spout and run water through the pipes to test for leaks at all the joints. Once you're satisfied the system is sound, finish the installation or schedule an inspection.

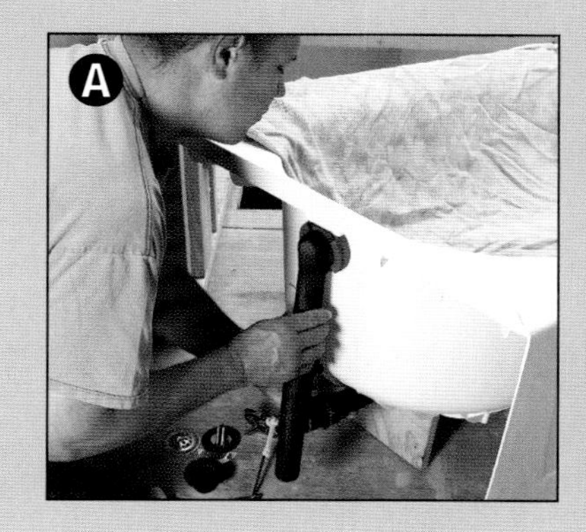

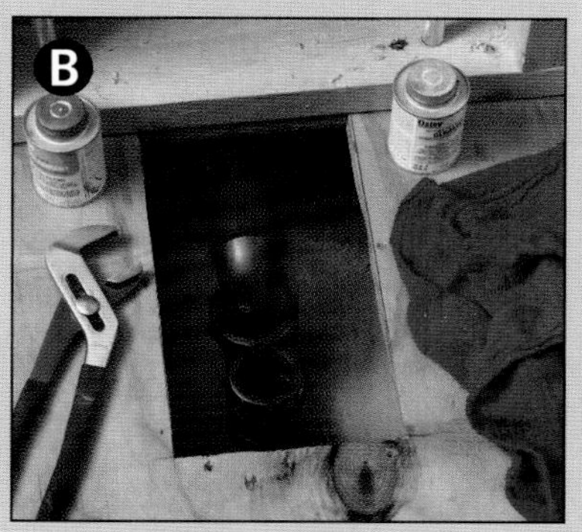

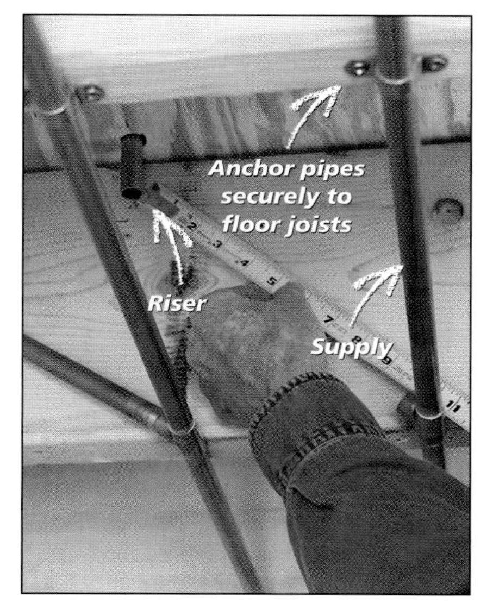

8 DOUBLE-CHECK THE MEASUREMENTS FROM THE PLAN. Cut pipe lengths. Test-fit each supply line. Mark adjustments on the piping. Take the runs apart and make the changes. Anchor the pipes securely to the joists.

9 SOLDER THE PIPE AND FITTINGS. Secure each run with hangers to prevent stress on the fittings and reduce line vibration. Run the hot water supply to the right side of the riser and cold water to the left.

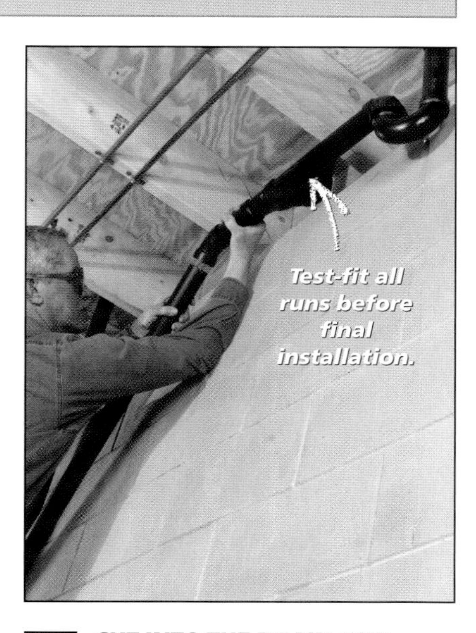

10 CUT INTO THE DRAIN AND VENT STACK SYSTEM AND RUN PIPES TO THE DRAIN. (See pages 98–101.) Cut the runs and test-fit all the pieces before you glue them. Mark each connection with a black mark for easy reassembly. For horizontal runs, slope the drain and vent pipes ¼ inch per foot. When satisfied with the setup, disassemble the pipes and cement them in place.

INSTALLING A TUB ON A SUBFLOOR

SKILLS: Carpentry and connecting plumbing fixtures and fittings.

HOW LONG WILL IT TAKE?

PROJECT: Installing a tub in an alcove on a subfloor.

EXPERIENCED . . . . . 3 HRS.
HANDY . . . . 4 HRS.
NOVICE . . . . 8 HRS.

STUFF YOU'LL NEED

TOOLS: Propane torch and striker, tubing cutter, soldering kit, fiberglass flame barrier, hammer, electric drill, spade bits, screwdriver, saber saw, ABS cutter, tape measure, carpenter's level, pencil, gloves, safety glasses

MATERIALS: Copper supply pipe, copper fittings, integral stops, copper tube straps, faucet assembly, shower tee, solder, flux, ABS drainpipe, ABS fittings, ABS P-trap, tub overflow and drain assembly, ABS cleaner and cement, bathtub, galvanized nails, 2x4 blocking, 1x3 ledger strips, silicone caulk

Bathtubs are designated either left-handed or right-handed. If you're facing the tub and the drain is on the left side, it's a left-handed model.

Installing a tub isn't rocket science, but it does require solid plumbing, carpentry, and sometimes, tiling skills. Before installing a tub, review everything that affects the project. Make sure you've qualified yourself for the job and are comfortable attempting it. Rather than have a contractor take over a halfway-completed project, it's better to consider hiring one before you begin. This is also one job where it's great to have a helper to hold and heft.

Place a blanket or piece of old carpet in the bottom of the tub to protect it while you are working. Dropped tools can easily crack or chip the finish.

WORK SMARTER

PLAN AHEAD.

Install an access panel behind the faucets and supply pipes so you won't have to remove the wall to work on them. Include shutoff valves or integral stops on the supply plumbing run. They isolate the plumbing from the rest of the house. Check local codes regulating installation of scald guards. They prevent someone from being scalded in the shower when cold water pressure changes as a toilet is flushed. Bathtubs are heavy. Have someone help with the installation. Measure the width of doors and passages. Make sure you not only can get the old tub out, but that the new one also fits into the bathroom.

1 FRAME THE WALLS AND RUN THE SUPPLY CONTROL RISER. Frame the walls so the alcove opening is just large enough to slide the tub into place. Leave a gap of ⅛ inch or less at the head and foot of the tub. Install plumbing for the risers, faucet, and showerhead (see pages 105–107).

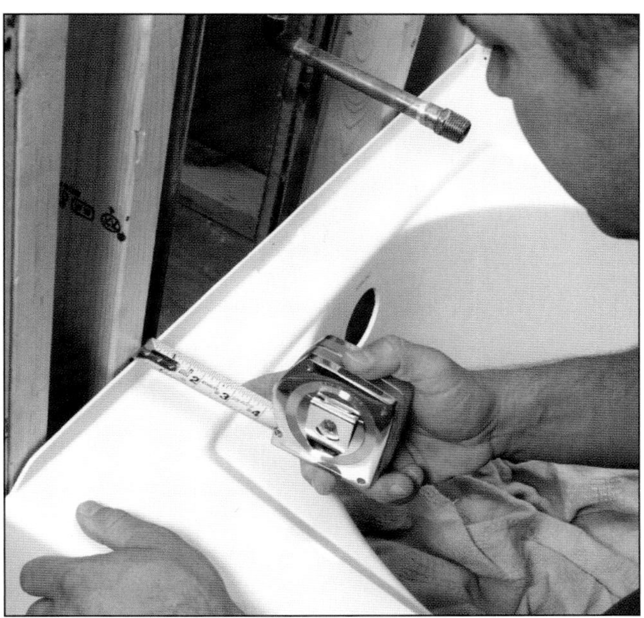

2 TEST-FIT THE TUB. After you've run the supply and cut the access hole, slide the tub into the alcove for a test fit.

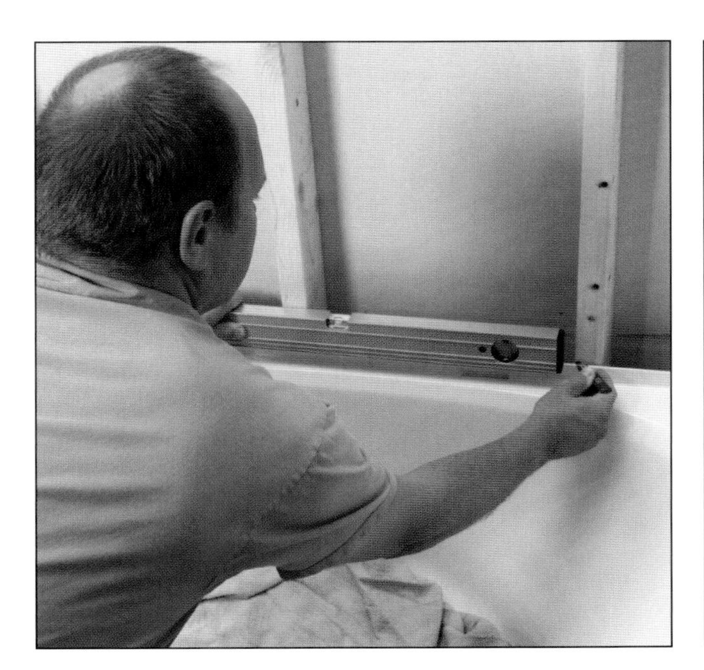

3 CHECK FOR LEVEL. Lay a carpenter's level on the tub to check for level. Shim the tub to level it. Don't use wood shims for a tub because they will rot when exposed to moisture. Flattened copper tubing or plastic shims work well. Once level, mark the top of the nailing flange at each stud.

4 DETERMINE THE LEDGER BOARD MOUNTING HEIGHT. Ledger boards are required for fiberglass and polymer/acrylic tubs. Measure the distance from the top of the nailing flange to the underside of the tub rim inset. Subtract that figure—about 1 inch—from the marks on the wall studs and mark the ledger board mounting height.

5 **CONNECT THE P-TRAP.** Install the drainpipe and P-trap if they have not already been roughed in (see page 107). Cut an access hole into the subfloor. Make it 4 to 9 inches wide and extending 12 inches from the center of the end wall. Connect the 1½-inch P-trap below the floor level so the slip nut fitting is centered directly under the overflow and drainpipe on the tub.

6 **DRY-FIT THE TUB DRAIN AND OVERFLOW.** Following the manufacturer's instructions, assemble the drain system so you can measure and trim the drain tailpiece to connect with the P-trap. (See the "Tub Drain Assembly" sidebar, page 111.)

TILING THE WALLS.

Planning to tile the walls? Tiles need a solid surface that can support their weight. Also, moisture may get behind the tiles, so the surface needs to withstand moisture damage. Use cement backerboard for the substrate; in most places it's required by code.

7 **TRIM THE DRAINPIPE.** The tub drainpipe will slide into the pressure fitting on the P-trap. It may be necessary to trim the pipe so it fits smoothly. Check the manufacturer's instructions for installation.

INSTALLING A TUB ON A SUBFLOOR (CONTINUED)

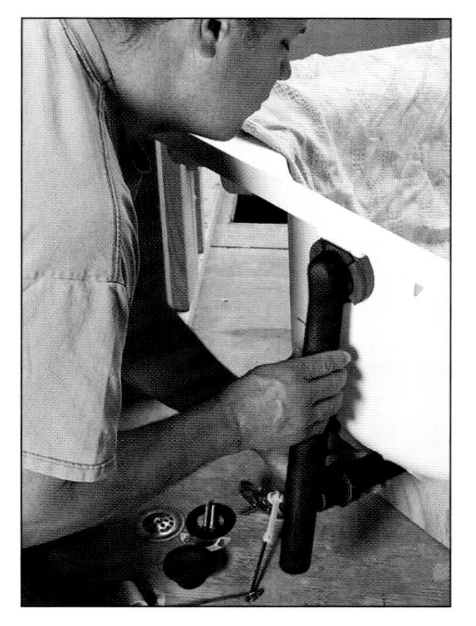

8 CONNECT THE OVERFLOW AND DRAIN TO THE TUB. Connect the overflow and drain to the tub and set the tub in place.

9 INSTALL LEDGER BOARDS. Cut the ledger boards to fit the alcove. Then use wood screws or galvanized nails to attach them to the studs following the marks you made in step 4. Install the boards in sections if necessary to make room for any structural braces at the ends of the tub. Double-check for level.

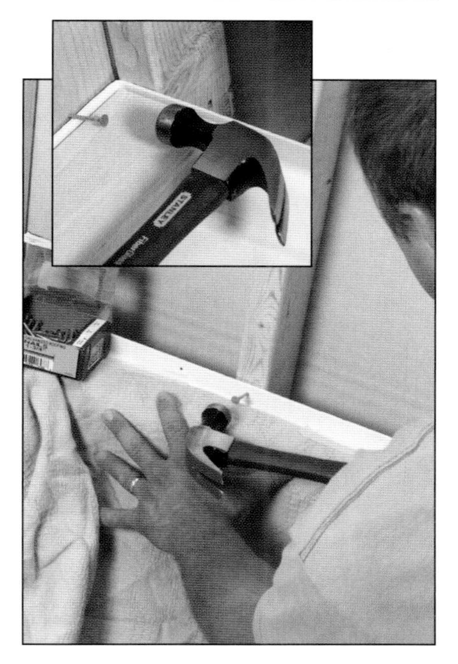

10 SET THE TUB INTO PLACE. Seat the tub on the ledgers. The tub must sit firmly on the ledger strips and the drain must fit smoothly into the P-trap. Nail through the predrilled holes using galvanized nails. (If there are no holes, drive the nails so the nailhead anchors the tub to the studs; see above inset.) Protect the tub with cardboard placed underneath a towel or rag.

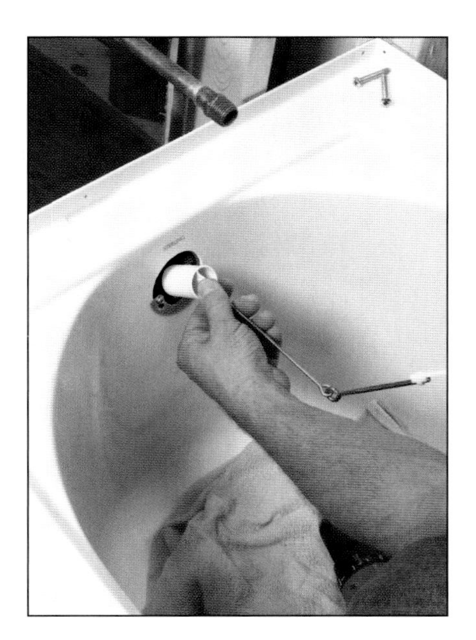

11 INSERT THE DRAIN PLUG LINKAGE. Install the drain linkage through the overflow opening. Attach the overflow cover plate to the mounting flange with screws. Test the system for leaks and schedule an inspection if necessary before you finish the walls and tile.

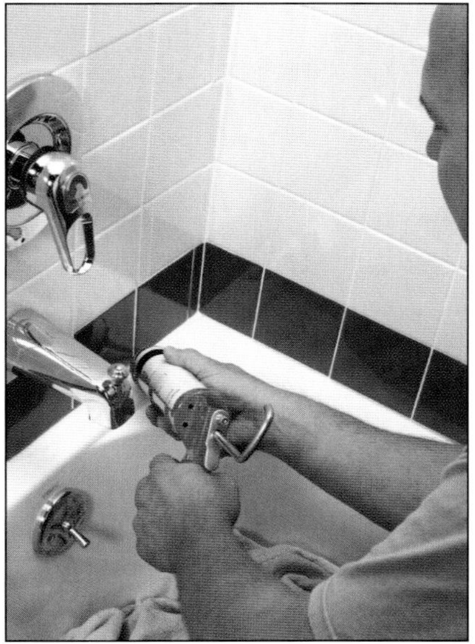

12 CONNECT THE FAUCET HANDLES AND TUB SPOUT. Use tub-and-tile caulk to seal around the faucet handles and tub spout. Apply a bead of caulk around the edge of the tub. Turn the water on and check for leaks.

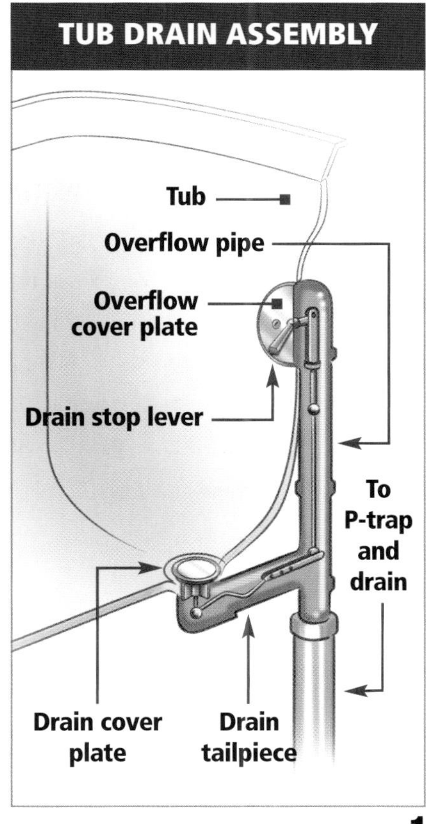

INSTALLING A TUB ON A SLAB

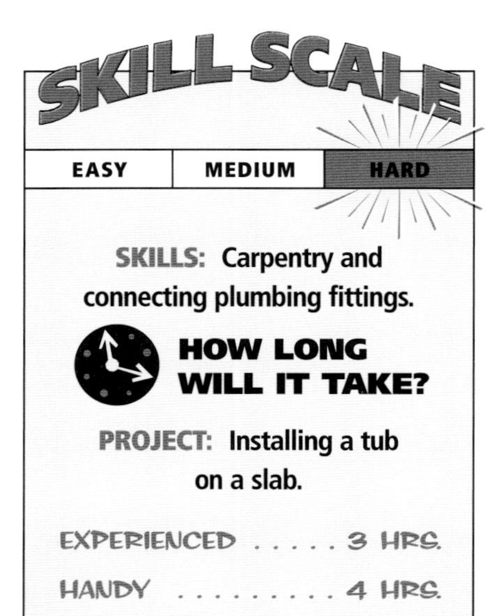

SKILL SCALE
EASY
MEDIUM
HARD
SKILLS: Carpentry and connecting plumbing fittings.
HOW LONG WILL IT TAKE?
PROJECT: Installing a tub on a slab.
EXPERIENCED . . . . . 3 HRS.
HANDY . . . . 4 HRS.
NOVICE . . . . 8 HRS.

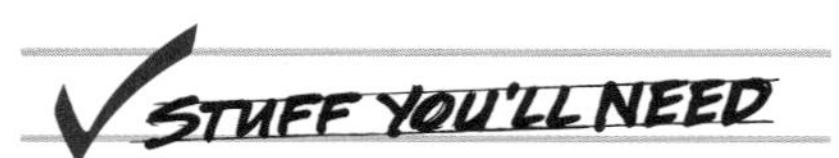

✓ STUFF YOU'LL NEED

TOOLS: Propane torch and striker, tubing cutter, soldering kit, fiberglass flame barrier, hammer, screwdriver, saber saw, ABS cutter, tape measure, carpenter's level, pencil, gloves, safety glasses

MATERIALS: Copper supply pipe, copper fittings, integral stops, copper tube straps, faucet assembly, shower tee, solder, flux, ABS drainpipe, ABS fittings, ABS P-trap, tub overflow and drain assembly, ABS cleaner and cement, bathtub, galvanized nails, 2x4 blocking, 1x3 ledger strips, silicone caulk, antiscald faucet, plumber's putty

Slab installations present unique problems. You don't have access from beneath, which limits where you can place the tub. The drain lines should be installed and roughed in before the slab is poured. If not, cut the slab to allow installation of the drain line—a messy and difficult job for the average do-it-yourselfer. Consider hiring a professional contractor if you need to install the drain line.

WILL IT FIT?

Measure the doorways in the house before purchasing a tub so you will know you can get it into the bathroom. NOTE: Bathtubs are heavy and awkward. Have someone help you with the installation.

Check local codes. You may have to install scald guards. They prevent a person from being scalded in the shower when a toilet is flushed anywhere in the house.

1 FRAME THE AREA FOR THE TUB. Using the measurements of the tub and the manufacturer's recommendations, frame the enclosure. Leave enough space to slide the tub into place and still maintain ⅛ inch of clearance at the head and foot of the tub. Test-fit the tub by sliding it into the alcove.

2 INSTALL THE SUPPLY PIPES. Mount the riser piping to a 2x4-inch crossbrace mounted between the studs. Attach the antiscald faucet body assembly and showerhead to the water supply pipes (pages 106–107). Install shutoff valves on the hot and cold water supply lines. In the installation above, the escutcheon plate will allow access to integral shutoffs. If that's not the case, install regular shutoffs on the risers and put in an access panel behind the faucets so you can reach the shutoffs for repairs.

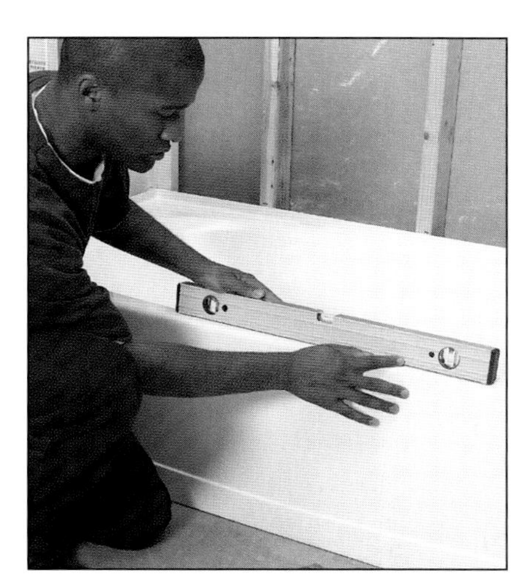

3 LEVEL THE TUB. Slide the tub into place and use a carpenter's level to make sure it's level. Shim with plastic or metal shims if necessary. Never use wood shims; they will rot in moist environments. Use a carpenter's level to mark the top of the nailing flange at each stud. Slide the tub back out.

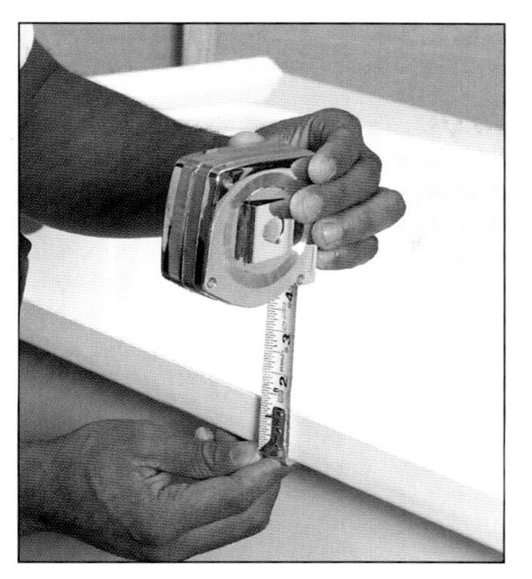

4 MEASURE THE FLANGE. Use a tape measure to find the distance from the top of the nailing flange to the underside of the tub. Subtract the measurement from the marks on the wall studs to determine the mounting height of the ledger board, which is required for fiberglass or polymer/acrylic tubs.

5 INSTALL THE LEDGER BOARD. Measure the length for each ledger board. Cut the strips and attach them to the wall studs just below the mark for the underside of the tub rim. You may need to install the ledger board in sections if you need to make room for any structural braces at the ends of the tub. Check the boards for level.

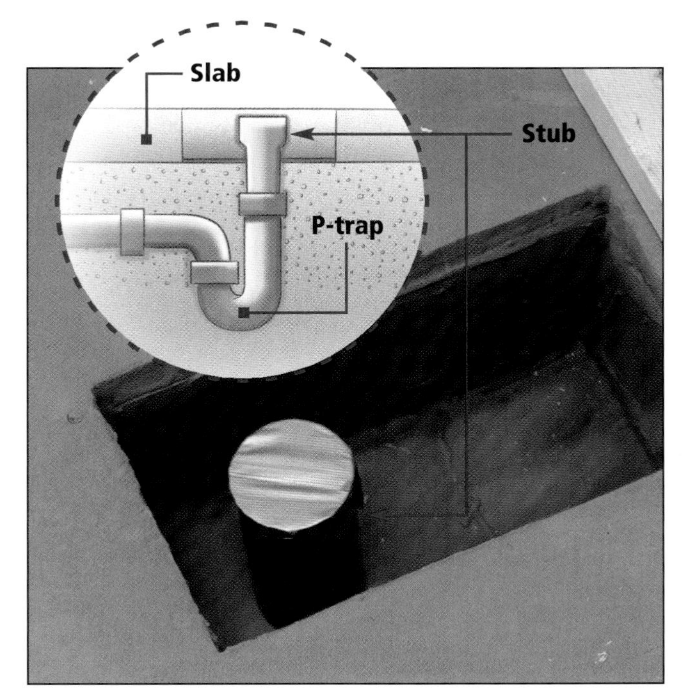

6 ANATOMY OF THE STUB. The P-trap for the drain is buried underneath the slab. All you will see is the stub rising out of tar or concrete, which helps keep insects and vermin from coming up from underneath the slab. Running the drain line beneath the slab and running the vent line is usually a job for a professional. The drain stub is wrapped in duct tape to keep debris out of the line—don't remove it until you're ready to install the tub. The illustration shows how the drain line gets to the pit.

7 CONNECT THE OVERFLOW PIPE. Install the slip nut fitting centered under the overflow pipe of the tub. Adjust the height of the overflow unit to fit the drain and the overflow opening of the tub. Follow the manufacturer's instructions for installing any washers and gaskets for the overflow.

INSTALLING A TUB ON A SLAB (CONTINUED)

8 CONNECT THE OVERFLOW. Apply a ring of plumber's putty to the bottom of the flange. Insert the pipe through the washer, then through the drain opening in the bathtub. Screw the fitting into the drain tailpiece and tighten until it is snug.

9 INSTALL THE DRAIN LINKAGE. Insert the drain plug linkage into the overflow opening. Attach the overflow cover plate, with the long screws provided, into the mounting flange of the overflow pipe. Adjust the plug linkage as directed by the manufacturer.

10 SLIDE THE TUB INTO PLACE. Check for level and shim, if necessary. Press down on the tub rims until they touch the ledger boards. Nail the tub rim flanges to the studs. Attach the flange by nailing through the predrilled holes with galvanized nails. If the flange does not have predrilled holes, drive the nails so the heads cover the rim of the flange. Before you install the enclosure or tile the wall, check the system for leaks and get any necessary inspections.

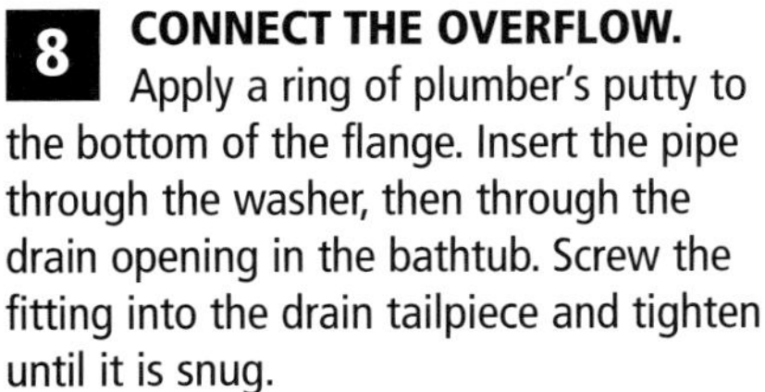

Protect the surface of the tub. Place an old piece of carpet or an old blanket over it to prevent chipping or cracking from dropped tools.

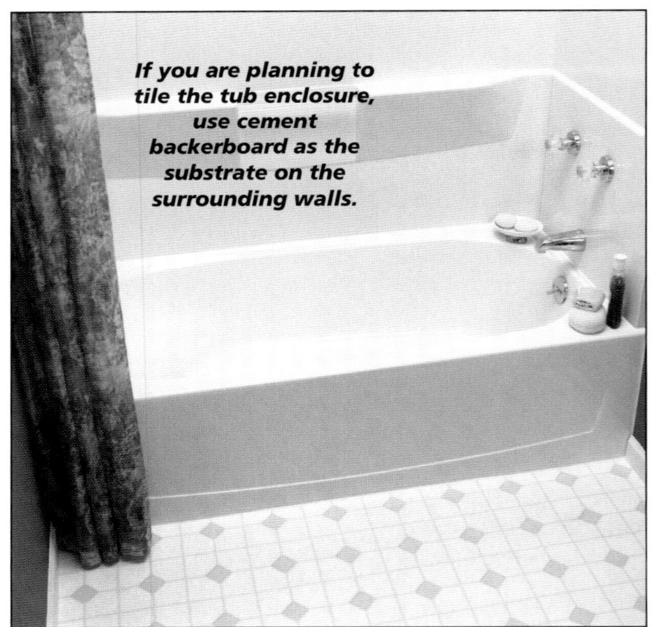

If you are planning to tile the tub enclosure, use cement backerboard as the substrate on the surrounding walls.

11 INSTALL AND FINISH THE ENCLOSURE. (See pages 118–120.) Install the faucet handles and tub spout. Caulk as necessary. Turn on the water supply to test for leaks. Tighten fittings if necessary.

INSTALLING A TUB SPOUT

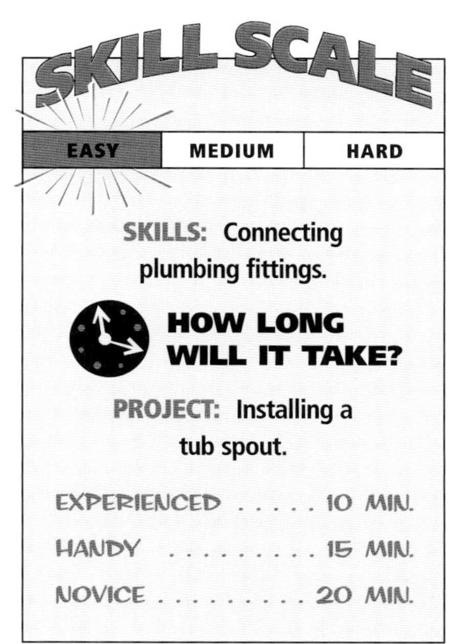

✓ STUFF YOU'LL NEED

TOOLS: Water-pump pliers or allen wrench

MATERIALS: New spout, rags or towel, pipe compound, silicone caulk

Tub spouts come in two basic types—screw-on and set-mounted. Make sure you get the right type.

LOOK FOR THE SETSCREW

Examine the underside of the spout. If there is a hole in the bottom, it's mounted with a setscrew. Slide the fitting over the pipe until it's snug with the wall. Then tighten the setscrew with a small screwdriver or allen wrench. If there is no hole under the spout, screw it until it's hand-tight.

1 DISCONNECT THE OLD SPOUT. Place an old towel or piece of carpet in the bottom of the tub to prevent chipping. Check the underside of the spout to see if there is a setscrew. If that is the case, loosen the screw and slide off the spout. If there isn't a setscrew, use a pipe wrench to unscrew the spout from the threaded pipe. Be gentle; twisting too hard can potentially bend the spout nipple, especially if it's copper.

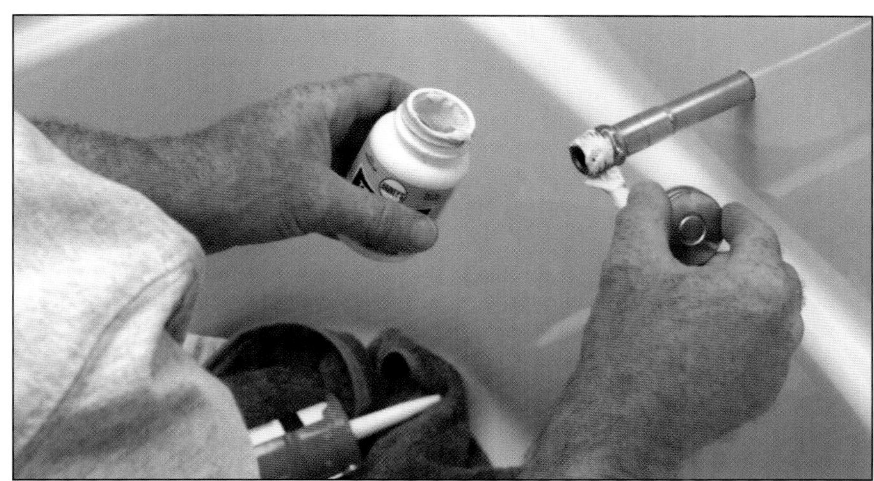

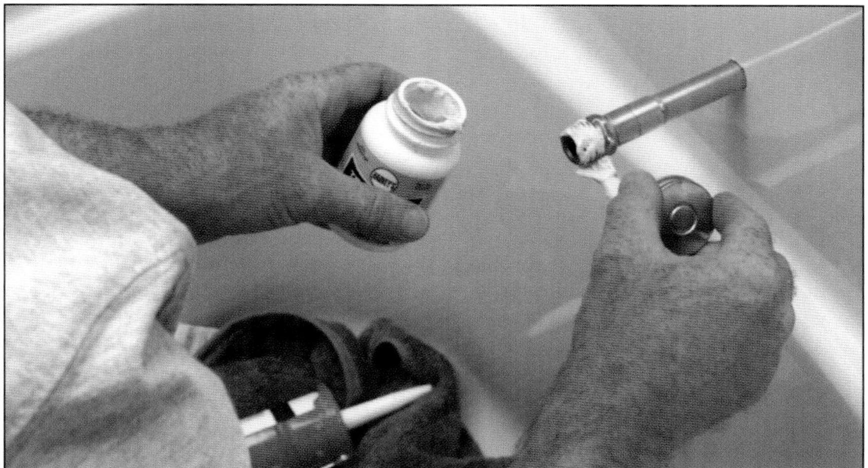

2 APPLY PIPE COMPOUND. For threaded spout pipes, coat the threads with pipe compound to ensure a good connection for the fixture.

3 INSTALL THE NEW SPOUT. Slide the new spout over the pipe and snug to the wall. If it has a setscrew, tighten it. If the pipe is threaded, screw on the spout hand-tight. If further tightening is required, use taped water-pump pliers.

4 APPLY A BEAD OF CAULK. Run silicone caulk around the spout or escutcheon plate if there is one.

CHAPTER

7 INSTALLING SHOWERS

Successful shower installations require careful planning, adequate time to do the job right, and a combination of plumbing and carpentry skills. If you want to install a light and vent fan, you will also need to do some electrical work.

THE BIG PICTURE: FUNCTION AND DESIGN

Think about the design of the entire bathroom before you begin working. If you have the space, you may want to install a separate tub and shower. Tubs, showers, and tub/shower combinations come in a variety of shapes and sizes. Check out the options and consult with the experts at your local home center before finalizing your bathroom design. You may find that there are options for shapes and sizes that will provide better use of space and allow for some additional amenities.

CHAPTER 7 PROJECTS

To get a "perfect bead" when using silicone caulk around a shower, smooth the caulk line with a small spoon dipped in soapy water.

CHECK LOCAL CODES

Check local codes before beginning any shower or tub installation. You may need to have an inspector check your work at certain points in the project.

PLANNING AND PREP

Proper preparation is key for the successful installation of any shower enclosure. Before beginning, prepare the walls.

- Wipe the wall surfaces with a damp cloth to remove soap scum and debris.
- All surfaces must be square, level, smooth, and dry before installing enclosure panels.
- Plaster can pose problems because of rough or uneven textures. Prepare a plaster surface for adhesives by sanding the plaster smooth. Use a rotary sander and medium-grit paper.
- You can install an enclosure over old ceramic tile, but remove any loose tiles first. Remove plastic tiles or old wallpaper.
- Seal any wall surface with two coats of primer or all-purpose enamel paint. This provides a better surface for the adhesive, allowing a stronger bond.

ALLOW FOR EXPANSION

Tub and shower enclosures expand and contract slightly. Follow the manufacturer's instructions. Allow an extra ⅛-inch space for expansion for every 5 feet of enclosure length. Allow 1/16 to ⅛ inch of space at the bottom, between the panels and the base.

The NKBA (National Kitchen and Bath Association) Guidelines for Shower Design

1. Make sure the interior dimensions of an enclosed shower include at least 34x34 inches of usable space. Measure from wall to wall. Grab bars, controls, and movable and folding seats don't reduce the measurement; but built-in seats do. You can settle for 32x32 inches, but it will make the shower uncomfortable for some users.
2. Design shower doors so they open into the bathroom—not into the stall—to avoid crowding the space in the shower.
3. To reduce the risk of falls, avoid installing steps for climbing into the shower.
4. Install a pressure-balancing/temperature regulator or a temperature-limiting device for the showerhead to prevent scalding.
5. Design the shower so you can reach the controls from inside and outside the stall. Put the controls 38 to 48 inches above the floor, and above the grab bar if there is one. Locate the controls between the showerhead and the stall door. You can install a hand-held showerhead instead of a fixed one. Place a hand-held model no higher than 48 inches above the floor in its lowest position.
6. Install only laminated safety glass with a plastic interlayer, tempered glass, or approved plastic for any clear face of a shower enclosure or partition that reaches to within 18 inches of the floor.

NEEDED: UMBRELLAS IN THE BASEMENT.

I learned the hard way how important it is to check a new installation for leaks. I thought that I checked out my new shower pretty extensively. I looked at the supply risers, the faucets, and the P-trap. I ran some water through the system. But I soon found out I wasn't thorough enough. The area around the drain flange wasn't sealed properly, and it was leaking around the base. It was a couple of days before I went downstairs and saw the stains spreading across my basement ceiling. I had to take the drain assembly apart and refit it. Always check every possible part of the system for leaks before you consider the job finished. And, of course, if I had had an inspection, somebody might have caught my mistake.

INSTALLING A TUB AND SHOWER SURROUND

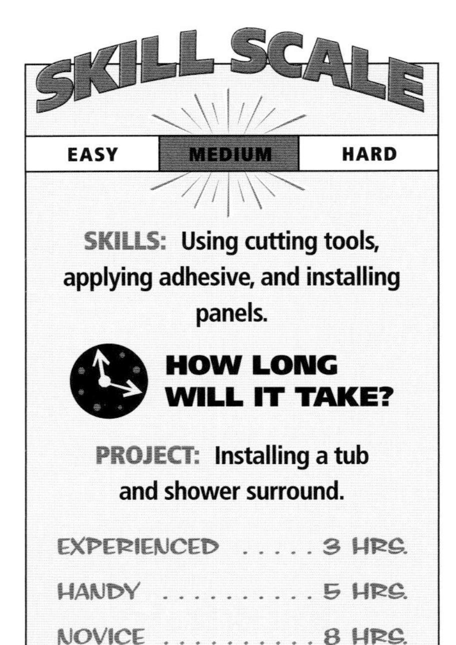

SKILL SCALE
EASY
MEDIUM
HARD
SKILLS: Using cutting tools, applying adhesive, and installing panels.
HOW LONG WILL IT TAKE?
PROJECT: Installing a tub and shower surround.
EXPERIENCED . . . . . 3 HRS.
HANDY . . . . 5 HRS.
NOVICE . . . . 8 HRS.

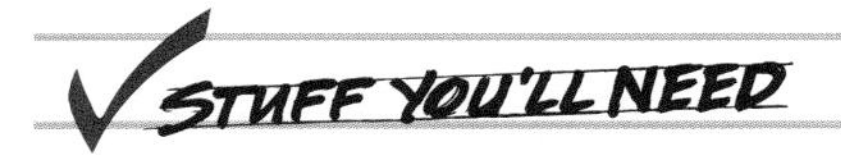

STUFF YOU'LL NEED

TOOLS: Tape measure, pipe wrench, screwdriver or hex wrench set, bucket, sponge, hole saw, utility knife, carpenter's pencil, screwdriver, caulking gun

MATERIALS: Tub enclosure, detergent cleaner, cardboard for template, double-stick tape, tub surround adhesive, rags, silicone caulk, expanding foam

Shower surrounds can be installed directly over securely fastened ceramic tile. Loose ceramic tiles—and any plastic tiles—must be removed and the walls sanded smooth before installation.

DEALING WITH WINDOWS

In many cases the surround will cover a window in the bathroom that is the primary source of ventilation. If that's the case, you'll need to install a vent fan or make a cutout in the surround so you have access to the window. (See "Providing Ventilation for the Bathroom," page 119, for more information.)

PICK YOUR FEATURES

When selecting a surround, consider features such as built-in soap and shampoo caddies and towel bars that provide space for bathroom items.

1 PREPARE THE ENCLOSURE. Turn off the hot and cold water valves or shut off the main water supply. Remove the faucet, escutcheons, filler spout, and any fittings mounted to the wall, such as towel bars and soap dishes. Remove soap film and dirt from the wall surface with a detergent cleaner. Wipe dry. Depending on the condition of the surface, it's often a good idea to prep it with a stain-blocking primer to create a good bonding surface with the tub surround adhesive. (See page 121.) Allow the surface to dry completely before installing the tub surround.

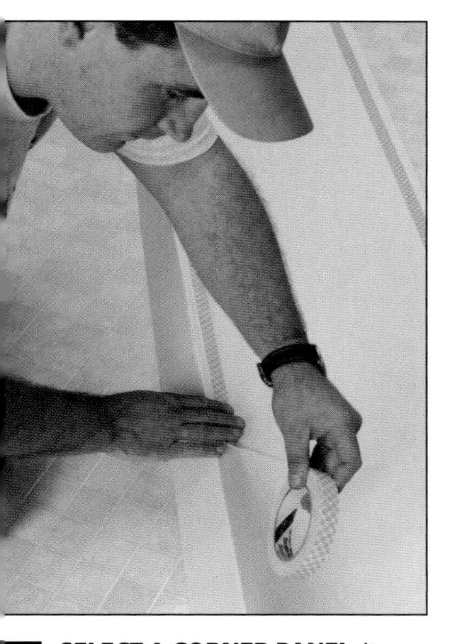

2 **SELECT A CORNER PANEL.** It doesn't matter which corner you art in. Select a corner panel and test-fit it. emove the panel and lay it on the floor ith the surface that goes next to the wall cing up. Apply a pressure-sensitive, ouble-sided, 1-inch tape along the vertical dge of the panel.

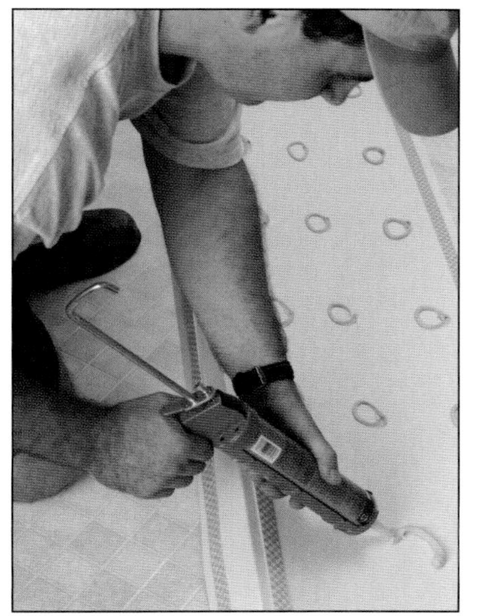

3 **APPLY ADHESIVE TO THE PANEL.** Use a caulking gun to apply the adhesive. Apply dots of the adhesive at intervals of no more than 12 inches along the vertical length of the panel. Rows of adhesive near the edge should be set back from the edge by 3 inches.

USE THE RIGHT ADHESIVE!

Check the manufacturer's instructions to make sure you use the recommended kind of cement for the panels. Some adhesives are too "hot" and will melt into the panel, showing through on the finished side of the enclosure.

4 **INSTALL THE FIRST CORNER PANEL.** Position the panel. Press it rmly onto the wall, then pull it back bout 6 inches for a few minutes (or per e manufacturer's instructions) to let the dhesive set up. This will provide a ermanent bond. Push the panel back in ace. Apply pressure with your hands, up nd down, and side to side, making firm ontact with the wall. Repeat this ocedure in the opposite corner.

5 **MEASURE THE OUTLETS.** Create a template to cut the openings into one end panel for the faucet and spout piping. Use a tape measure to determine the height from the bottom of the panel and the distance of each fitting from the inside edge of the panel. Make the template from a piece of cardboard. Cut the openings into the cardboard.

BUYER'S GUIDE

PROVIDING VENTILATION FOR THE BATHROOM.

A new shower surround will often cover a window which is the primary source of ventilation for the bathroom. If that's the case, you'll need to install a vent fan or make a cutout in the surround to allow access to the window.

Kits are available with all the materials you'll need to do the framing job, or you can trim it out with a primed, painted, and sealed wooden frame.

6 **TEST-FIT THE TEMPLATE.** Lay the cardboard template against the wall and make sure the template openings line up with the pipe outlets. If they don't, remeasure them and make a new template.

7 **CUT THE OPENINGS IN THE PANEL.** Mark the positions of the openings on the panel by laying the template over the back of the panel—drill from the back of the panel to minimize chipping. Place a piece of scrap wood beneath the panel under the markings for the openings to produce a cleaner edge to the hole. Test-fit a final time before cutting to make sure the markings line up with the openings. Use a hole saw to cut the openings.

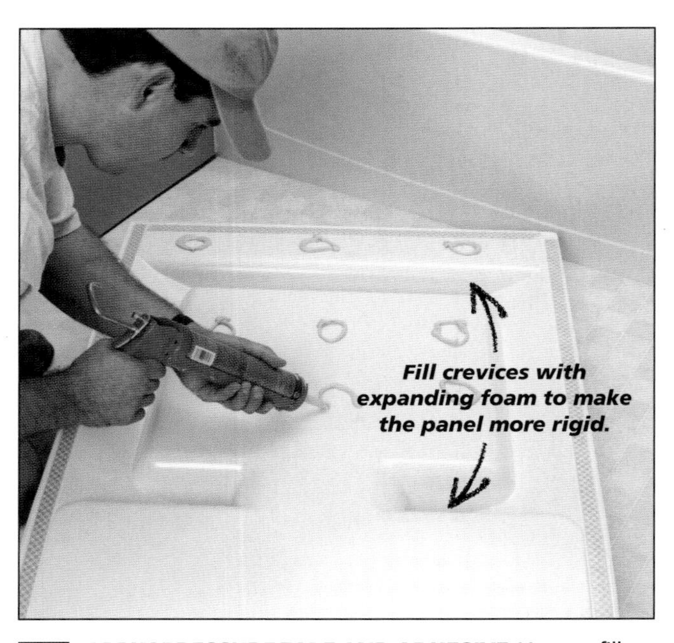

Fill crevices with
expanding foam to make
the panel more rigid.

8 **APPLY PRESSURE TAPE AND ADHESIVE.** You can fill the crevices in the panel with expanding foam to make them more rigid. Apply carefully; too much foam can expand and crack the panel. Position the panel onto the wall. Apply pressure by hand—side to side—to mount. Install the remaining panels.

9 **SEAL WITH A BEAD OF CAULK.** Use a caulking gun to apply a quality tub and tile sealant. Seal along the outside edges of the panel and around the fixtures. To get a "perfect bead," use a small spoon dipped in soapy water to smooth the caulk line. Do not use for at least 24 hours or follow the manufacturer's instructions.

INSTALLING A SHOWER ENCLOSURE

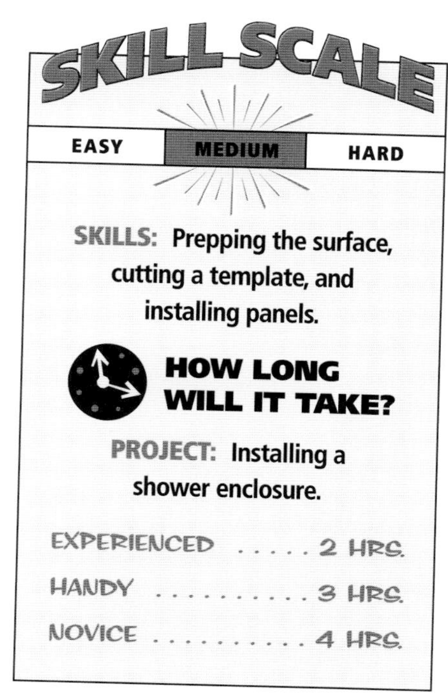

SKILL SCALE
EASY
MEDIUM
HARD
SKILLS: Prepping the surface, cutting a template, and installing panels.
HOW LONG WILL IT TAKE?
PROJECT: Installing a shower enclosure.
EXPERIENCED ..... 2 HRS.
HANDY .... 3 HRS.
NOVICE .... 4 HRS.

Preformed enclosures are easier and faster to install than tiling a shower surround. Between prepping, painting, and mounting the enclosure, you can expect to spend a weekend completing this project. You might have to make a cutout to gain access to a window for ventilation. (See "Providing Ventilation for the Bathroom," page 119, for more information.)

BIG ENOUGH TO FIT

Before selecting a new enclosure, measure the existing space and purchase the largest enclosure that will fit. If possible, avoid shower surrounds smaller than 34×34 inches—people need to turn around comfortably.

TOOLS: Roller and roller pan, utility knife, straightedge, tape measure, pencil, drill, carbide hole saw, caulking gun, carpenter's level

MATERIALS: Shower enclosure, waterproof drywall, primer, cardboard template, panel adhesive

1 PREPARE THE WALLS FOR THE PANELS. Waterproof drywall or marine plywood are the best subsurfaces for a shower enclosure. If you're installing in an existing shower area, shut off the water supply and remove all fixtures. Clean the surface with denatured alcohol to ensure a good bond for the enclosure panels. Prime or water-seal raw plaster, wallboard, or spackling before you install the panels.

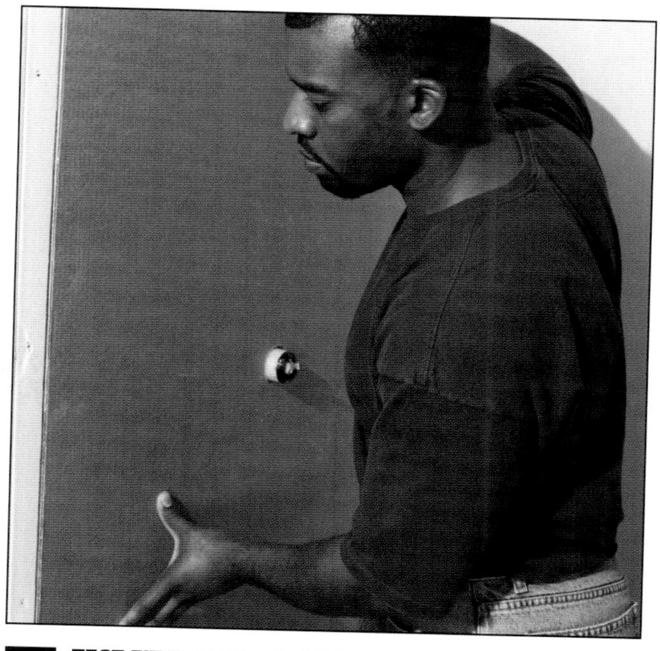

2 TEST-FIT THE ENCLOSURE. Use the cardboard shipping carton to make a template. Mark the front of the template so you know which side is which. Draw a vertical line from the base corner of the shower to an inch below the ceiling. Measure the center of the showerhead pipe and faucet valve stems from the top of the mounting surface along the vertical line. Transfer these measurements to the template. Cut holes in the cardboard large enough to fit over the fittings. Place the template on the wall and over the fittings. Align the front edge with the vertical line. Line up the holes to check the fit. Mark again as necessary.

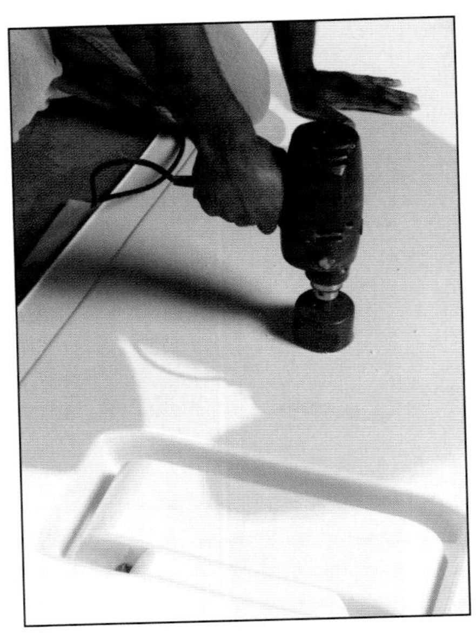

3 CUT THE PANEL OPENINGS. Lay the template over the end panel with the exposed surfaces facing down. Mark the holes using the template. Drill the holes in the panel using a carbide hole saw. Make the holes large enough to allow clearance for the pipe but small enough so the escutcheon completely covers the hole. To avoid chipping the surface while drilling, drill from the back side of the panel. Test-fit the panel against the wall and check for accuracy by lining up the edge with the marked vertical line.

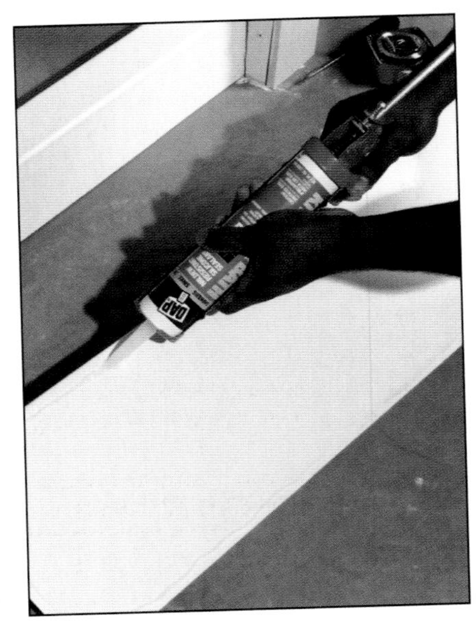

4 APPLY ADHESIVE TO THE PANEL. Use a caulking gun to apply adhesive to each corner of the panel. Apply adhesive following the manufacturer's instructions. Carefully position the panel and press it firmly onto the wall, then pull it back about 6 inches for a few minutes (or per the manufacturer's instructions) to let the adhesive set up. This will provide a permanent bond. Push the panel back in place and apply pressure with your hands, up and down, and side to side, making sure that all areas of the panel are in firm contact with the wall. Repeat this procedure in the opposite corner.

5 CONTINUE TO APPLY ADHESIVE TO THE WALL FOR EACH PANEL SECTION. Place a line of adhesive an inch below the top of the panel and in 1½-inch-diameter dots at 12-inch intervals, 3 inches from the side edges of the panel.

USE THE RIGHT ADHESIVE!

Check the manufacturer's instructions to make sure you use the recommended kind of cement for the panels. Some adhesives are too "hot" and will melt into the panel, showing through on the finished side of the enclosure.

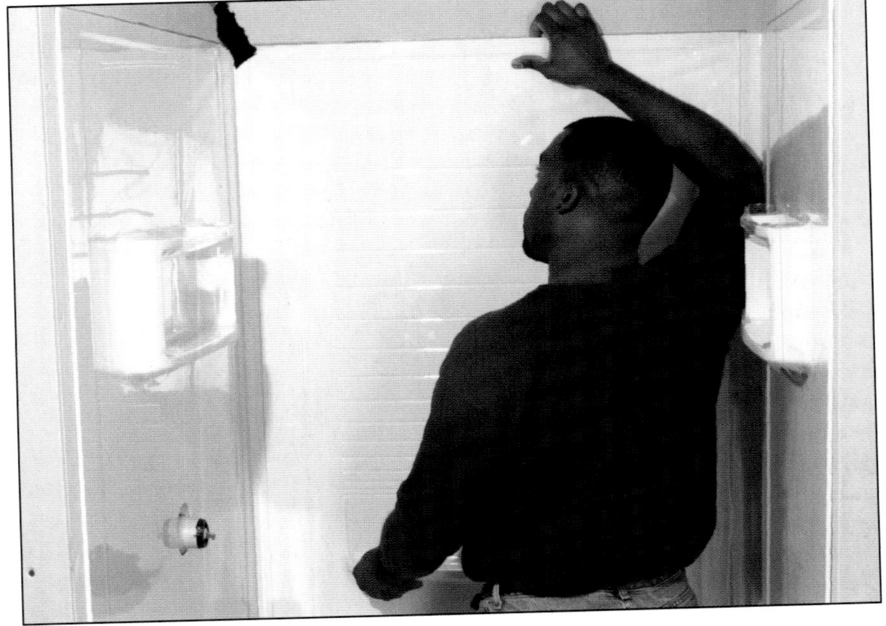

6 INSTALL THE REMAINING PANELS. Use a carpenter's level to line up the top of each panel and make sure it's level. Seal each panel (see step 4 above) to the wall by applying hand pressure over the surface, up and down and side to side, to ensure maximum contact. Remove excess adhesive. Don't use the shower for 24 hours. Install the fixtures after the adhesive has dried.

INSTALLING A SHOWER BASE

SKILLS: Assembling and hooking up a drain. Installing control riser, pan, and enclosure.

HOW LONG WILL IT TAKE?

PROJECT: Installing a fiberglass shower base and enclosure.

EXPERIENCED ..... 2 DAYS
HANDY .... 3 DAYS
NOVICE .... 5 DAYS

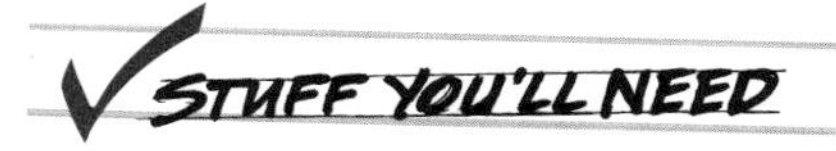

STUFF YOU'LL NEED

TOOLS: Tape measure, ABS tubing cutters, carpenter's level, screwdriver, power drill and bits, hammer, caulking gun, propane torch, spark lighter, 4-in-1 tool, hole saw

MATERIALS: Shower base, drain assembly, shower enclosure, copper pipe, copper fittings, shower fixtures, rags, ABS cleaner, ABS cement, galvanized nails, silicone caulk, panel adhesive, faucet and spout fittings, lead-free flux, lead-free solder, emery cloth, pipe joint compound, 2x4 for blocking

The first step in taking a shower in the basement is installing a shower base. The base fits over a drain flange which is roughed into the slab. Choose the base before you have the drainpipes installed to make sure it will fit. Allow space for the studs and framing as well as for the enclosure or tile that will finish the job. Shower bases and enclosures come in many colors and shapes.

POSITION IS EVERYTHING

The most critical part of the job is getting the drain positioned properly in the slab. Unless you've had some experience, this is usually a job better left to a professional.

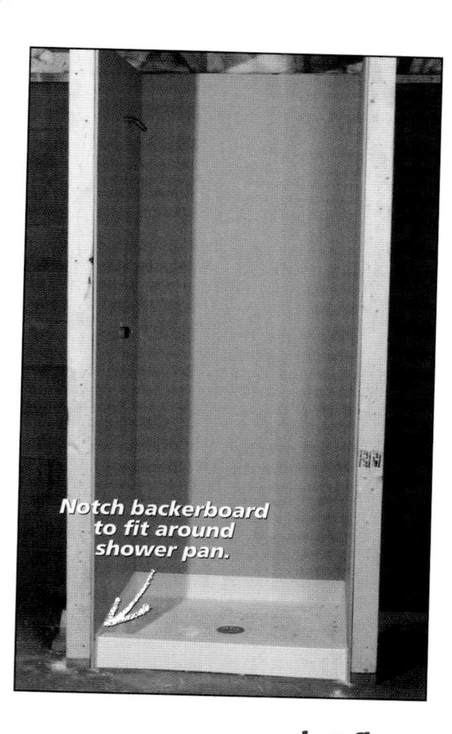

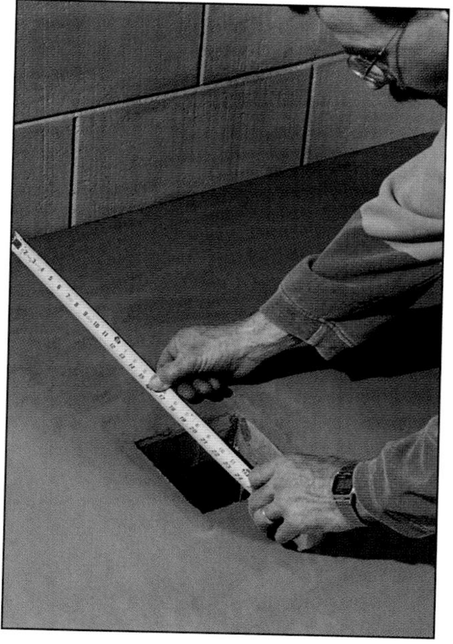

1 DECIDE ON A SHOWER BASE AND ROUGH-IN THE DRAIN. Choose a shower base, then have the drain roughed in to meet that dimension. Make sure there is enough room between the pan and the wall to allow for studs and framing. Check to make sure the floor is level. If not, you will have to level the area where the base will rest.

Some bases require a stand, underlayment, or a coat of thinset; some are freestanding. Follow the manufacturer's instructions for installation.

CLOSER LOOK

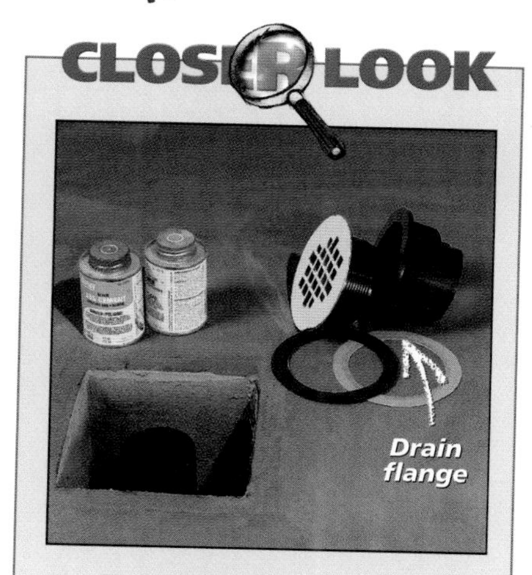

To install a shower base, you will need a drain flange, drain screen, ABS pipe cleaner and cement, and a shower pan.

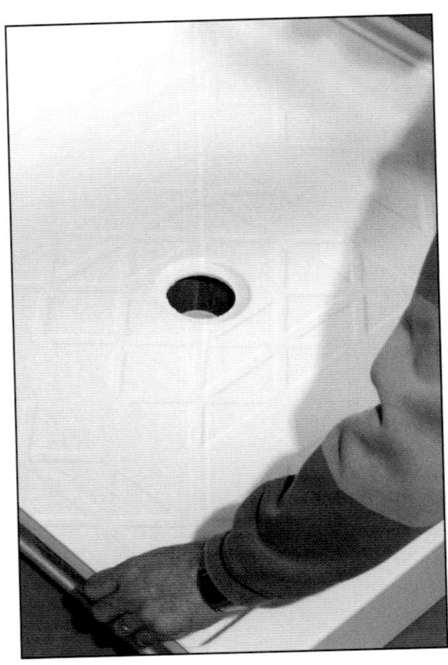

2 TEST-FIT THE SHOWER PAN. Verify that the shower base will fit and that the drain will be under the opening in the shower base. Be sure the floor is level and clean.

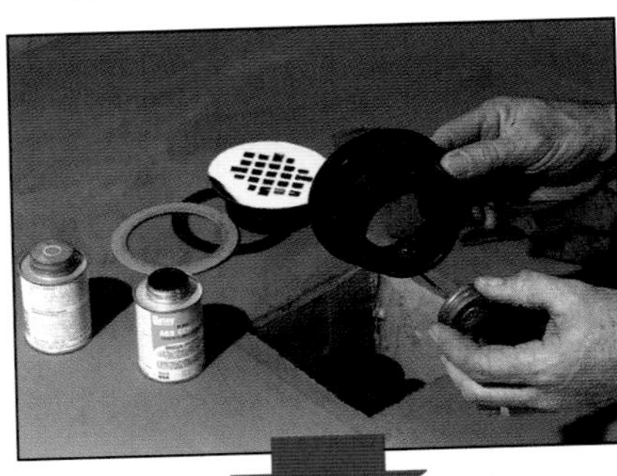

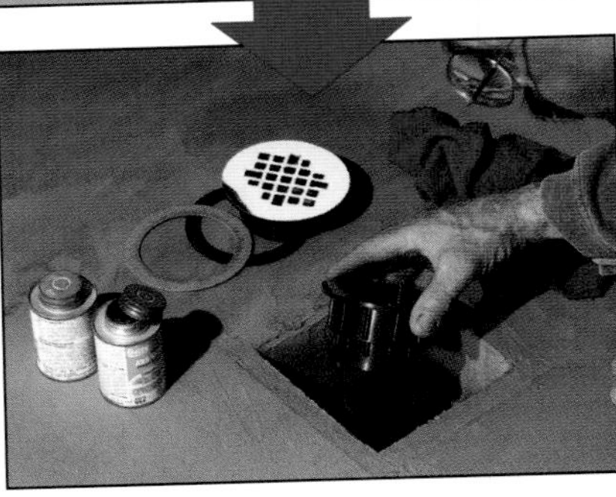

3 ATTACH THE FLANGE TO THE DRAINPIPE. Test-fit to make sure the flange will be flush with the surface of the shower base. Add an extension to the drainpipe if necessary. Apply primer and cement to the flange and the drainpipe, then attach.

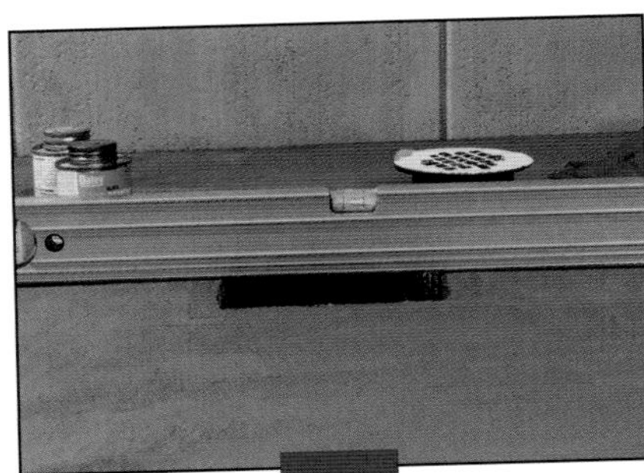

4 ADJUST THE TOP OF THE FLANGE SO IT'S FLUSH WITH THE FLOOR SURFACE. Allow the cement to dry according to the cement manufacturer's directions. Attach the gaskets and set the base in place.

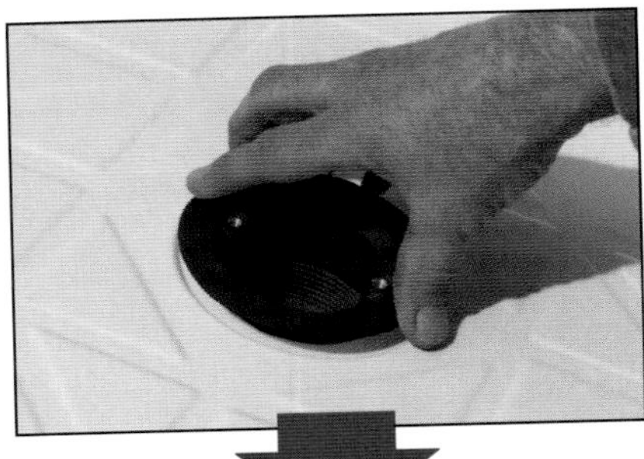

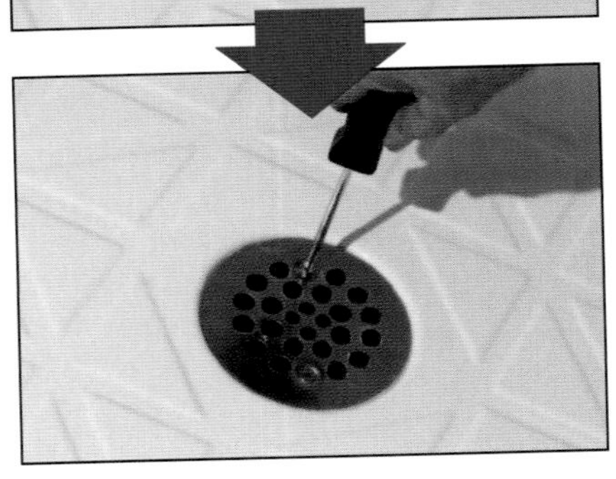

5 CAULK AND SEAT THE DRAIN BODY. Caulk the drain flange with silicone. Line up the holes in the drain body with the screw holes in the flange body. Press the drain body firmly into place and attach the drain screen with the screws that came with the assembly.

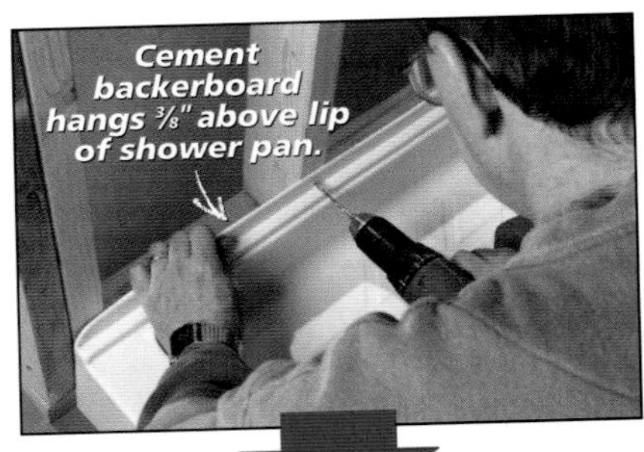
Cement backerboard hangs ⅜" above lip of shower pan.

6 PREDRILL HOLES INTO THE SHOWER BASE RIM. Check the base for level. Mark the rim at the center of each wall stud. Predrill a pilot hole for the galvanized nails that hold the base to the stud walls. Use rustproof galvanized nails, and shim as necessary to keep the shower base square. The cement backerboard will hang ⅜ inch above the shower pan so you can caulk.

Notch backerboard to fit around shower pan.

7 INSTALL THE SHOWER RISER AND THE ENCLOSURE. Install the shower riser. Test for leaks and get an inspection, if necessary, before you install the enclosure. Apply cement backerboard to the stud walls, tape, and fill as necessary. Prime and install a plastic enclosure or tile the walls.

INSTALLING A SHOWER PAN LINER

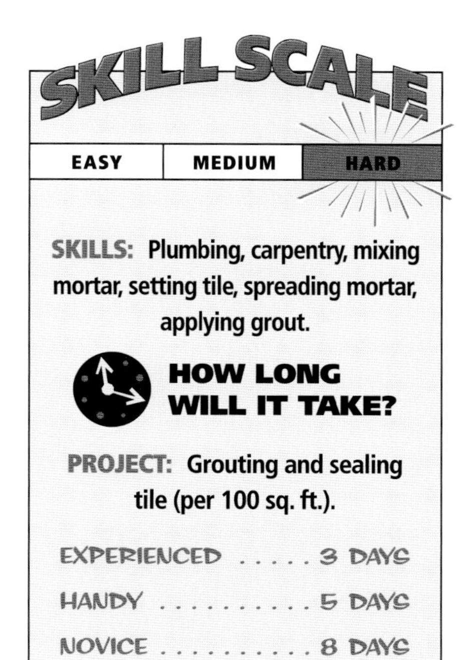

SKILL SCALE
EASY
MEDIUM
HARD
SKILLS: Plumbing, carpentry, mixing mortar, setting tile, spreading mortar, applying grout.
HOW LONG WILL IT TAKE?
PROJECT: Grouting and sealing tile (per 100 sq. ft.).
EXPERIENCED ..... 3 DAYS
HANDY .... 5 DAYS
NOVICE .... 8 DAYS

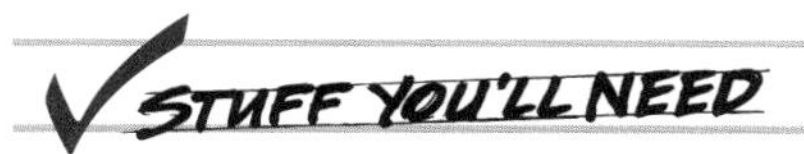

✓ STUFF YOU'LL NEED

TOOLS: Hammer, framing square, tape measure, 4-ft. level, carbide scorer, utility knife, adjustable wrench, scissors, stapler, ½-inch drill, mixing paddle, steel trowel, ¼-x¼-inch square-edge trowel, grout float

MATERIALS: ¾-inch exterior plywood, 2x4 studs, pressure-treated 2x4s and 2x6s, ½-inch cement backerboard, backerboard screws, dry mortar mix, CPE shower pan membrane, CPE primer, CPE solvent-weld, 4-mil poly, shower drain, mosaic tile, grout, Teflon tape, ABS drain assembly

The advantage of installing a shower pan liner is that you can create any shape shower stall you choose, and you can place the drain anywhere you want it on the shower floor. Shower pan liners are finished with tile. They can be installed over wood or concrete. Shower pan liners are made of either CPE or PVC. Installation is similar for both products, but follow the manufacturer's recommendations.

THE ORDER OF WORK

Professionals install the control and supply systems and install and tile the shower walls after they've put up the enclosure and before they mortar and tile the shower floor. That way, they won't damage the mortar bed or the tiles while they're installing plumbing and tile on the walls.

1 BUILD THE SHOWER ENCLOSURE. Frame in the shower walls with 2x4 studs. Frame the curb with 3 pressure-treated 2x4s on edge. End nail 2x6 pressure-treated blocking between the studs so you can staple or nail the shower pan liner into place.

2 MOUNT THE DRAIN ASSEMBLY IN THE FLOOR. Position the drain hole, mark it, and cut it out. Remove the top half of the drain assembly, but leave the bolts in place. You will use them later to mark the cutout for the drain hole in the liner.

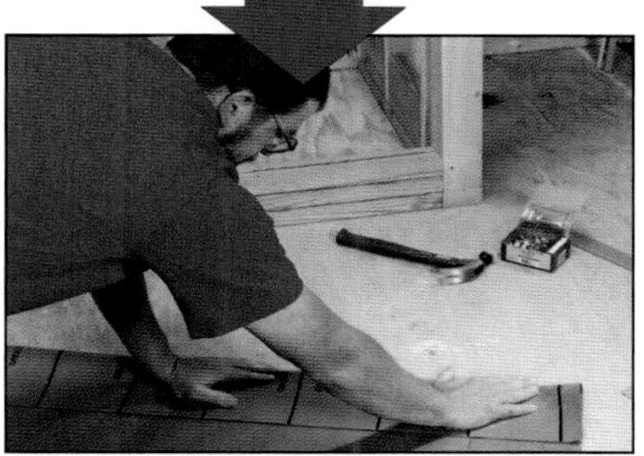

3 INSTALL THE SHOWER LINER. Lay out the shower dimensions with a marker and straightedge. Leave a minimum of 9 inches above the finished floor for the shower walls and 16 inches for the curb. Fold the liner to fit (see "Closer Look," page 126).

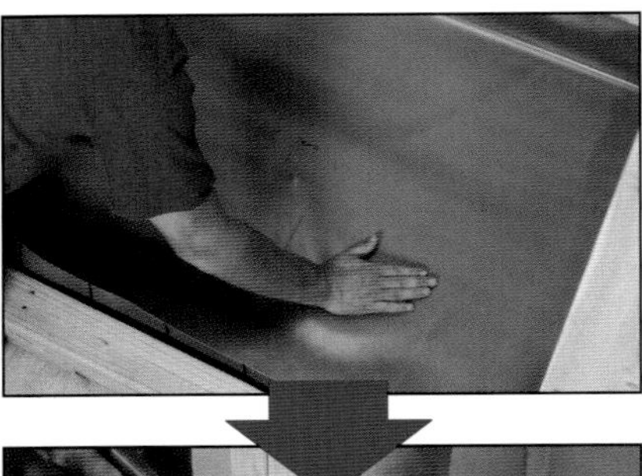

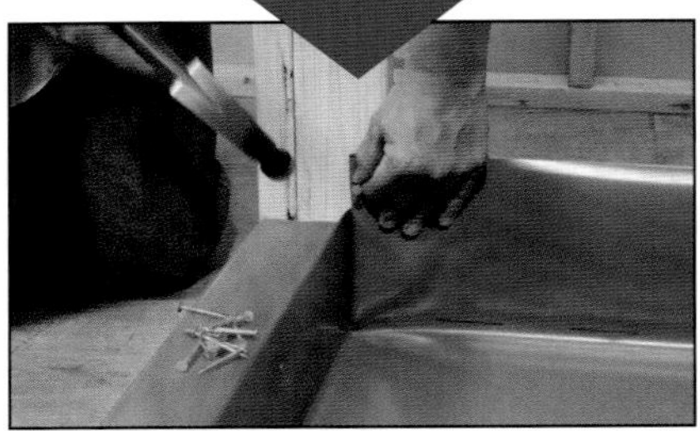

4 SET THE LINER IN PLACE AND FIT THE WALL CORNERS AND CURB. Working from the drain out, set the liner in the shower and nail it to the blocking ½ inch from the top. Make a "hospital corner" and fold into each wall corner. Fold the liner over the curb at least an inch; nail it in place. Eliminate wrinkles as you work to make a smooth surface.

CLOSER LOOK

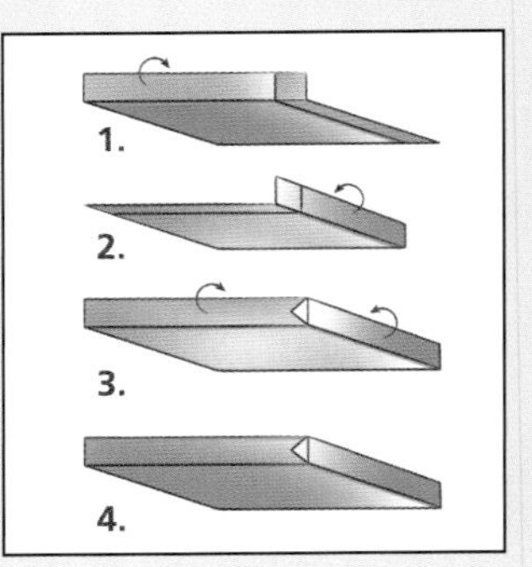

TO MAKE A HOSPITAL CORNER:

1. Fold and crease one flap.
2. Fold and crease the second flap.
3. Join both flaps at the corner and pinch them together to form a triangle.
4. Press the triangle to one side, then staple in place.

TO SEAM A CPE SHOWER PAN MEMBRANE:

1. Remove dust, oil, and grease.
2. Coat both the mating surfaces with the recommended cleaner/primer. Let dry.
3. Coat mating surfaces with solvent.
4. After 1 minute, line up surfaces and roll together.
5. Wait 5 minutes; then try to separate. If the seam opens, pull apart; repeat steps 3 through 5 until the seam holds.

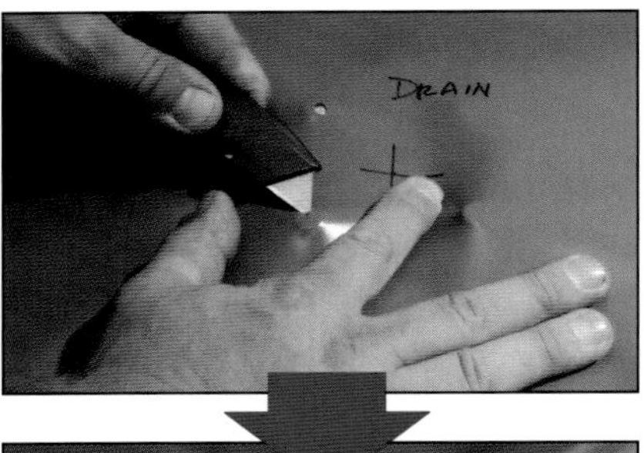

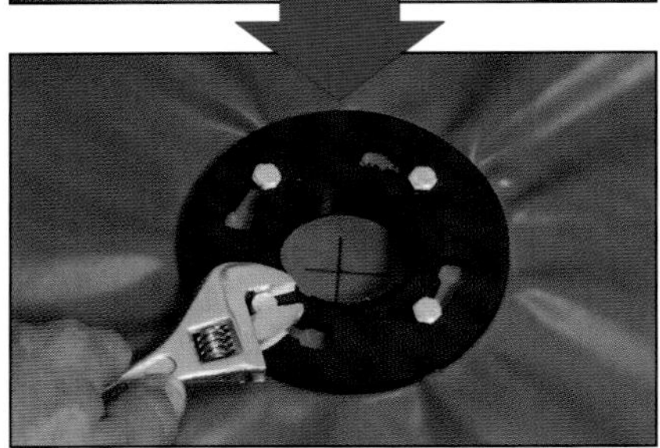

5 CUT HOLES FOR THE DRAIN BOLTS AND ATTACH THE CLAMP. Carefully locate the drain clamp bolts and make small cuts so you can stretch the membrane over the bolts. Attach and tighten the locking clamp to the drain flange with an adjustable wrench.

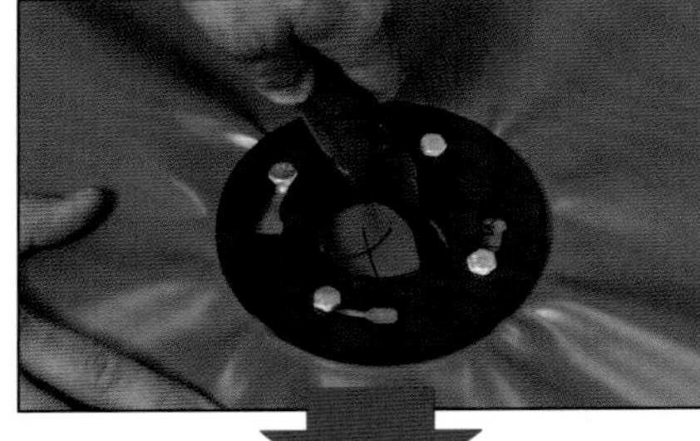

6 CUT OUT THE DRAIN HOLE AND SCREW IN THE DRAIN. Use a utility knife to cut away the membrane from the inside of the drain. Wrap the strainer's threads with Teflon tape and thread the fitting into the drain body.

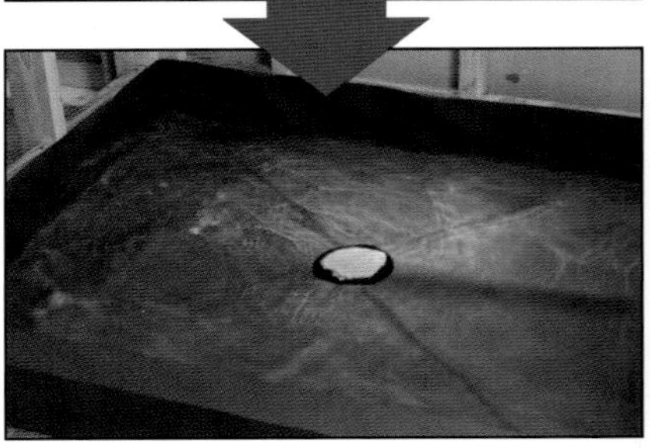

7 TAPE THE STRAINER AND TEST THE SHOWER PAN LINER AND SHOWER CONTROLS. Use layers of masking tape to cover the strainer and fill the pan to within 1 inch of the curb. Let stand for at least 20 hours, then check for leaks. Repair as necessary. At this point, test the control system for leaks; have the plumbing inspected if necessary.

8 STAPLE POLY OVER THE STUDS, ATTACH THE BACKERBOARD, AND TILE THE WALLS. Staple on 4-mil poly, overlapping the pan liner by at least 2 inches. Don't staple into the liner. Install ½-inch backerboard over the poly, leaving 1 inch at the bottom. Install backerboard up, over, and down the side of the curb. **Protect the shower pan lining and tile the walls before you continue with the floor.**

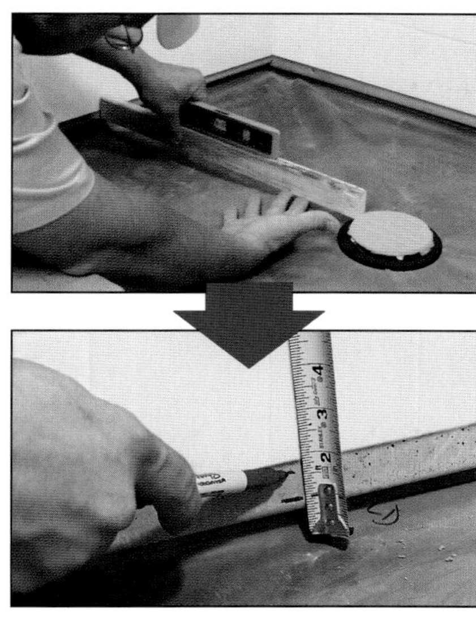

9 TRANSFER THE FLANGE HEIGHT TO THE WALL AND MARK FOR THE PITCH TO THE DRAIN. Use a level and a straight board to transfer the height of the drain flange to the wall. Add ¼ inch per foot for the slope to the drain and make a mark. Use your level to extend the mark around the entire enclosure as a guide for the mortar bed.

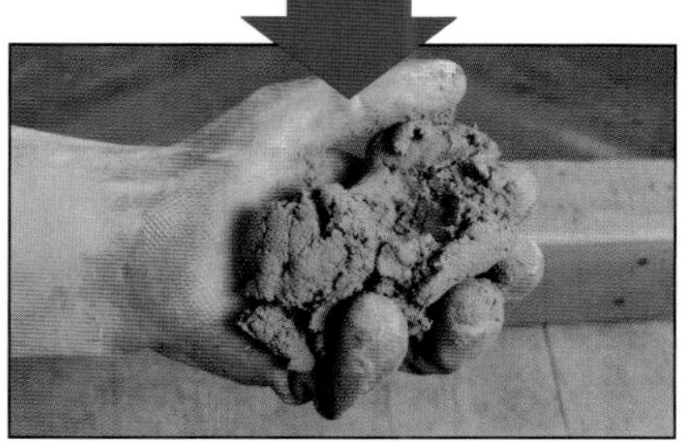

10 PROTECT THE WEEP HOLES AND MIX THE MORTAR. Use tile spacers or pebbles around the drain weep holes to prevent them from clogging. Mix dry mortar (one part cement, three parts fine sand) to a crumbly consistency that holds together.

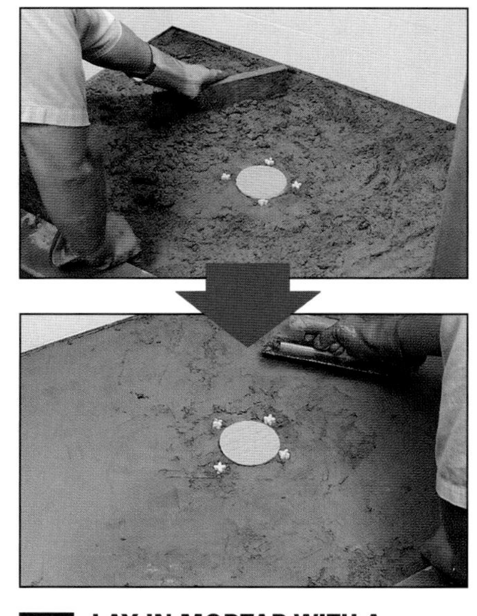

11 LAY IN MORTAR WITH A SLOPE OF ¼ INCH TOWARD THE DRAIN. Fill the pan with mortar to your wall marks and the bottom of the flange. Pack firmly and screed off excess. Finish with a steel trowel, removing clumps and smoothing the finish. Extend the mortar bed up, over, and down the shower curb.

12 APPLY THINSET AND INSTALL TILE. When the mortar base is thoroughly set (at least 24 hours), apply a layer of latex-modified thinset mortar and set the tile. Sheet tile is easier to work with and faster than installing singles, depending on the look. Let the tile set for 24 hours, then grout and finish.

THE ORDER OF WORK IS EVERYTHING!

Following the order of work in big projects is the key to success. When you're installing a tile shower with a shower pan lining, this is how it should go:

1. **Frame out the enclosure, install the supply, and set the drain. Test both.**
2. **Install the poly barrier and hang the backerboard.**
3. **Tile the shower walls.**
4. **Lay the mortar and install the tile floor.**
5. **Let it all dry. Take a shower.**

8 INSTALLING SINKS AND FAUCETS

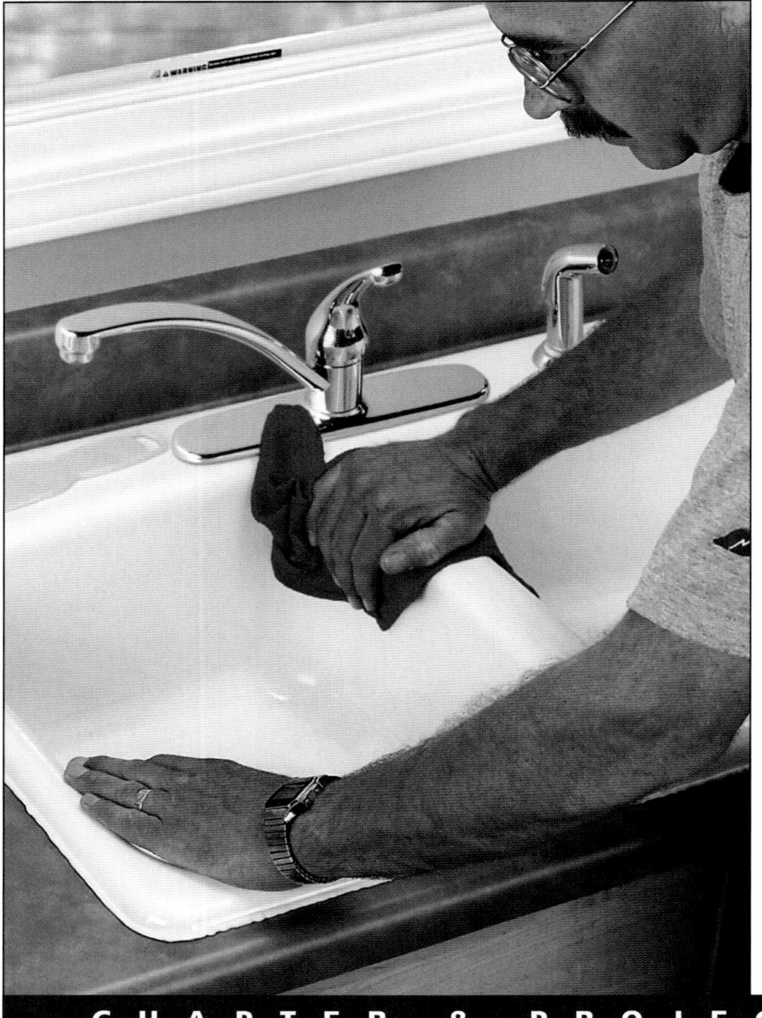

With the variety of stylish, high-profile sinks available today, it's easy to make any kind of design statement you wish.

New colors, shapes, and features have revolutionized design in both style and function.

MATERIAL CONSIDERATIONS

Materials now available include porcelain, fired clay, porcelain-coated cast iron and steel, stainless steel, and more. Even the old standard—stainless steel—is the focus of a resurgence in popularity with the perfecting of the brushed stainless-steel surface. Other new alternatives include quartz-acrylic and solid-surface drop-in sinks. Quartz-acrylic sinks are popular among many homeowners and designers because they are attractive, durable, and heat resistant. Solid-surface drop-in sinks are sought now because of their longer warranties and aesthetically pleasing, seamless look.

CHAPTER 8 PROJECTS

INSTALLATION TIPS FOR SINKS AND FAUCETS.

A new sink and faucet installation will do wonders for an older bathroom.

- **A** Do as much work above ground as possible. Install the faucets and drain on a single sink before you set it in place. It will make the job easier because space to work is scarce when connecting fittings from beneath a counter.
- **B** The drain holes on a double sink make perfect handholds for setting the sink into the countertop opening. Install the drain system once the sink is in place.
- **C** Square up the faucet body on the sink top before you go below and tighten it up.
- **D** Flexible, braided supply line is a great alternative to copper because of its ease of installation. Follow the instructions for hooking up compression fittings carefully to prevent leaks.
- **E** Don't use pipe compound on compression fittings; compound keeps the fitting from connecting tightly and can cause leaks.
- **F** Pipe compound and Teflon tape act as sealants, but they also lubricate the threads so you can fully tighten the connections. Don't use too much—no more than one pass with compound or two wraps of tape—or you may not get a good connection.
- **G** ABS is light, easy to use, and more flexible to work with than steel pipe for drain systems.
- **H** There are two kinds of joint compound—with and without Teflon. Compound with Teflon is better for general use because it can be applied on plastic, brass, copper, steel, ABS, PVC, and CPVC. Joint compound without Teflon can damage plastic pipe.
- **I** Plumber's putty is not recommended for use on cultured marble or plastic sinks or fittings; it can discolor surfaces and weaken fittings. Follow the manufacturer's instructions and/or use silicone products.
- **J** Metal strainer basket assemblies are superior to the plastic kind; they last longer and are less prone to leaking.
- **K** If the drain isn't exactly where you need it to be, you can usually combine 90-degree and 45-degree elbows and short pieces of straight pipe to make the connection. P-traps and extensions with flexible sections may also help.
- **L** If the stub coming out of the wall or floor for the drain is metal or an incompatible plastic and you want to install ABS or PVC, rubber transition fittings are available to make the connection.

USING ESCUTCHEON PLATES.

Escutcheon plates fit around the drain pipe covering the hole where the drain line enters the wall or floor. Where plumbing is visable, such as under a wall-hung lavatory, they add a finishing touch. Using escutcheon plates inside cabinets isn't always required but do help seal the hole. Two basic types are available in both metal and plastic.

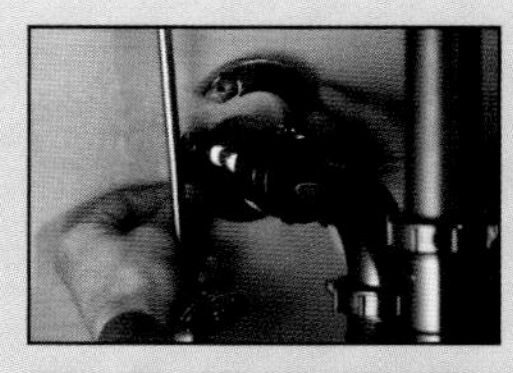
◀ **Split-ring escutcheon plates fit around a pipe that is already in place.**

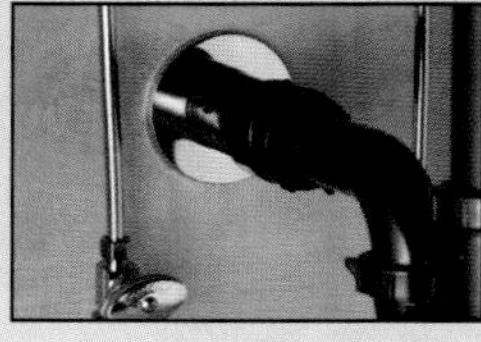
◀ **Solid-ring escutcheon plates slip onto the pipe before the drain is assembled.**

INSTALLING A COUNTERTOP SINK

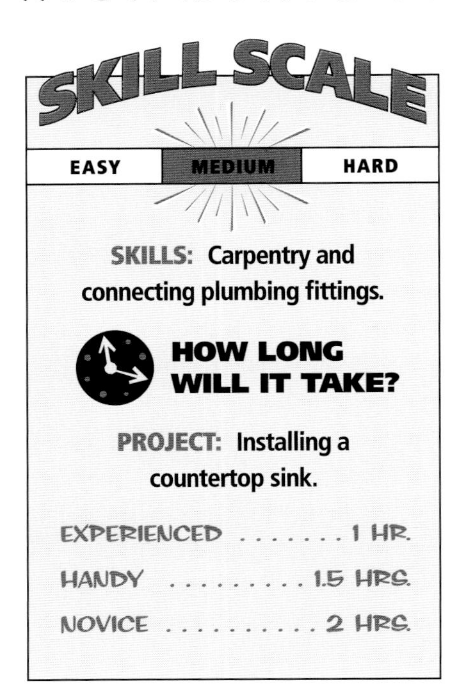

SKILL SCALE
EASY
MEDIUM
HARD
SKILLS: Carpentry and connecting plumbing fittings.
HOW LONG WILL IT TAKE?
PROJECT: Installing a countertop sink.
EXPERIENCED .... 1 HR.
HANDY .... 1.5 HRS.
NOVICE .... 2 HRS.

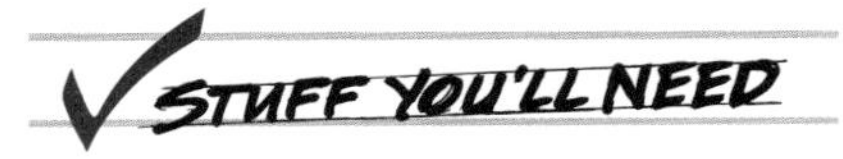

✓ STUFF YOU'LL NEED

TOOLS: Tape measure, scissors, carpenter's pencil, power drill and bits, saber saw, caulk gun, putty knife, screwdriver, utility knife

MATERIALS: Countertop sink, fixtures, cardboard template, masking tape, rag, silicone caulk

TOOL TIP

CUT IT SHORT.
High-speed cutting tools make short work of cutting holes in drywall and are a perfect alternative to using a saber saw for cutting out a countertop.

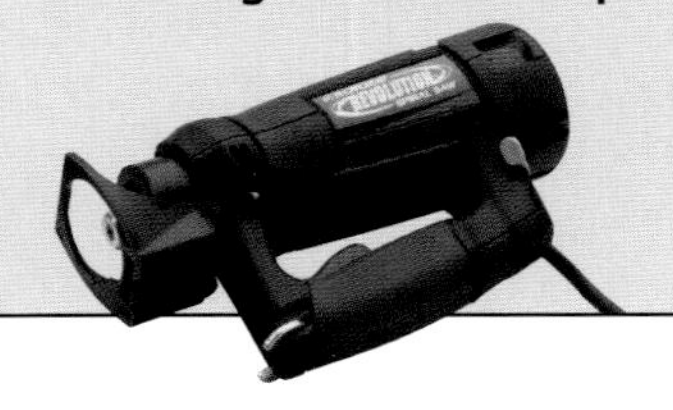

Sinks show their age over time. Cracks, scrapes, chipping, dents, and persistent stains take their toll. The time will come when it's time to install a replacement and perhaps go for a new look in your kitchen.

Choosing a sink style depends on your personal taste and the size of your kitchen. The colors, shapes, and features offered today give you a vast choice to meet your practical needs and design choices. Now is the time to get a fixture that really does the job.

CONNECT THE FIXTURES FIRST

Connect the faucets and other fixtures to the sink before you install it. It will be much easier to install them out of the cabinet. Wait to install the drain so you can use the drain holes as handholds when you set the sink in place.

1 CREATE A TEMPLATE. If the sink doesn't come with a template, create one by laying the sink facedown on a sheet of cardboard and drawing a line around the edge. Lift off the sink and draw a second line ¾ inch inside the first line. Cut along the second line with scissors.

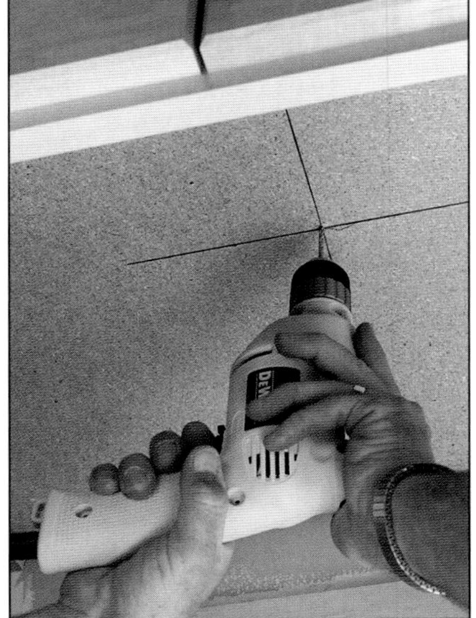

2 MARK THE CENTER OF THE CABINET FROM UNDERNEATH. Drill a hole large enough to fit a nail through it.

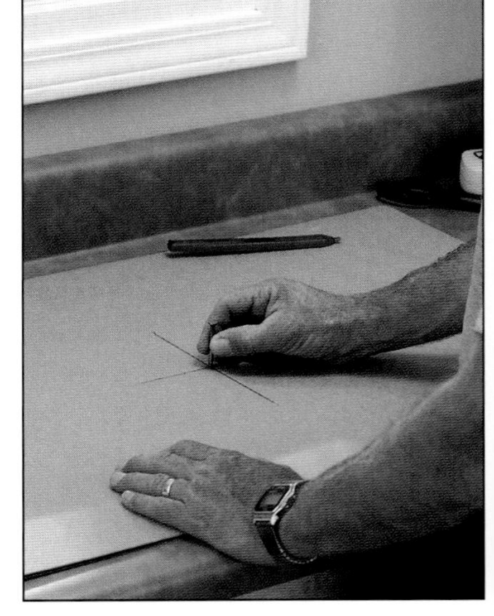

3 MARK THE CENTER OF THE TEMPLATE. Push a nail through the center of the template. Align the center of the template over the hole in the countertop. Push the nail through the countertop to anchor the template into place. Center the template and check to make sure the edges are square so the rim will lie entirely on the surface.

4 TRACE THE OUTLINE OF THE TEMPLATE ON THE COUNTERTOP. Remove the template and place tape over the line. Replace the template, center it, and draw the outline of the template on the tape. Remove the template.

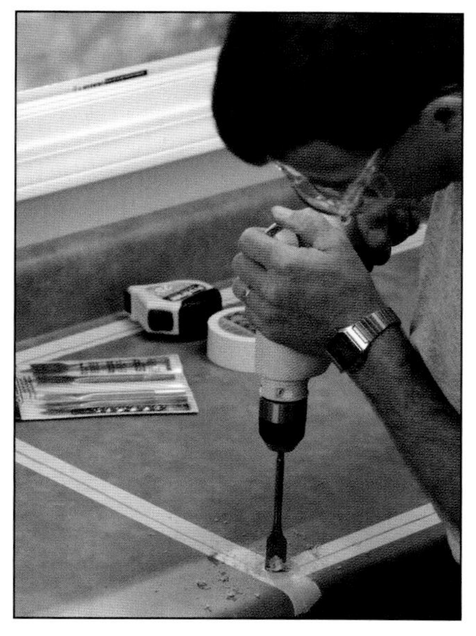

5 DRILL A ¾-INCH HOLE INSIDE THE CUTOUT LINE. Use a power drill and spade bit to make a starter hole for the saber saw.

6 INSTALL BRACES BENEATH THE COUNTERTOP. The braces will support the section to prevent it from binding while cutting.

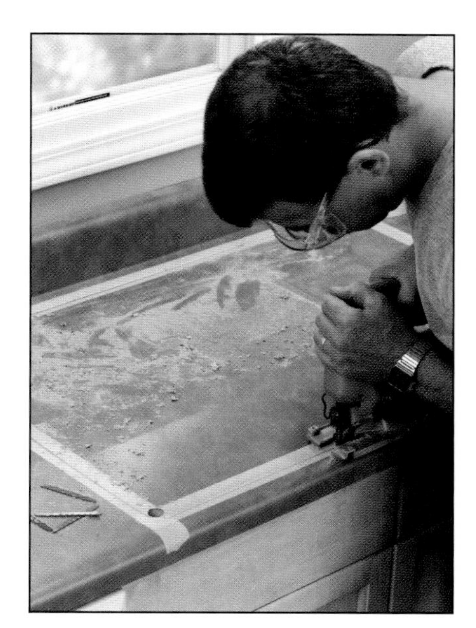

7 CUT ALONG THE LINE USING A SABER SAW. Use a blade designed to cut countertops without chipping the surface. Test-fit the sink in the opening. Trace the rim lightly on the countertop.

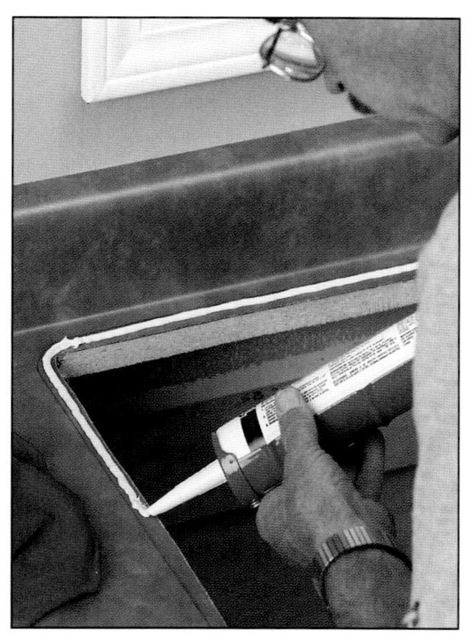

8 APPLY A BEAD OF SILICONE CAULK INSIDE THE LINE. Remove the sink and apply silicone caulk around the opening in a steady, continuous bead between the mark on the opening.

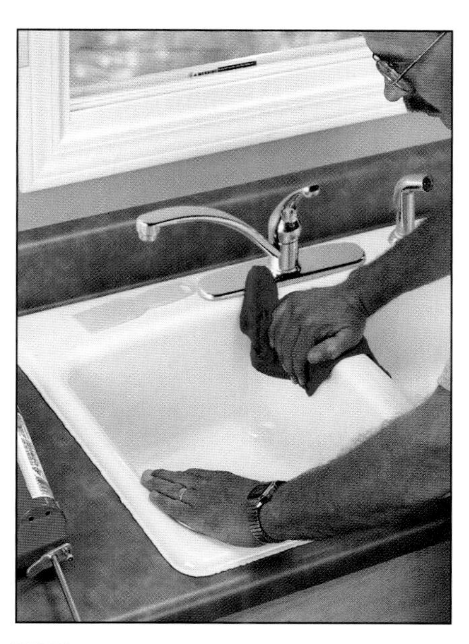

9 PLACE THE SINK IN THE OPENING. Press the sink firmly into the silicone caulking, then level and clean. Connect the faucets and the drains. Test for leaks.

COUNTERTOP SINKS (CONT.) INSTALLING A SELF-RIMMING SINK

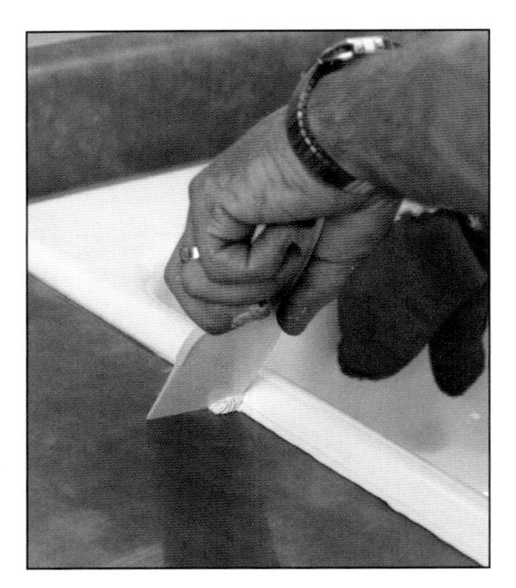

10 CLEAN AWAY EXCESS SILICONE. Trim silicone caulking with a plastic putty knife for a professional look.

1 INSTALL A SELF-RIMMING SINK. Follow the steps for installing a countertop sink through step 7, check for fit, then caulk the rim and lower the undermount sink into place.

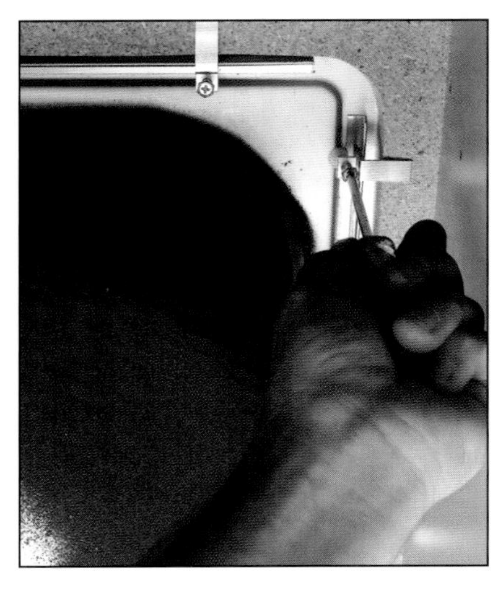

2 TIGHTEN THE CLAMPS BENEATH THE SINK. Align the clamps over the countertop; stagger-tighten with a screwdriver as you would the lugs on a tire, alternating so they tighten evenly.

REMOVING AN OLD SINK

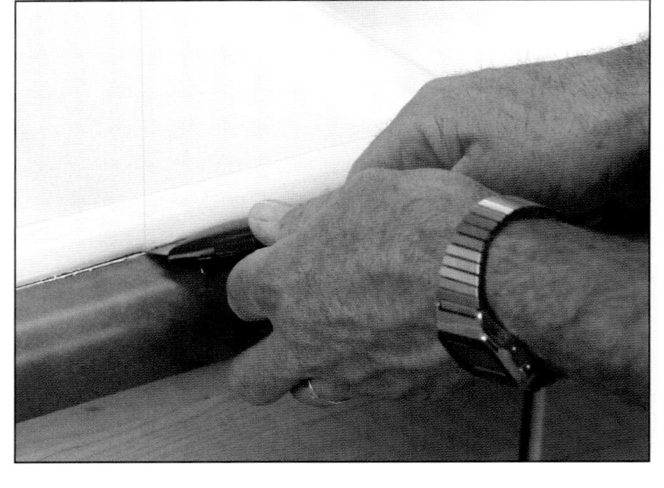

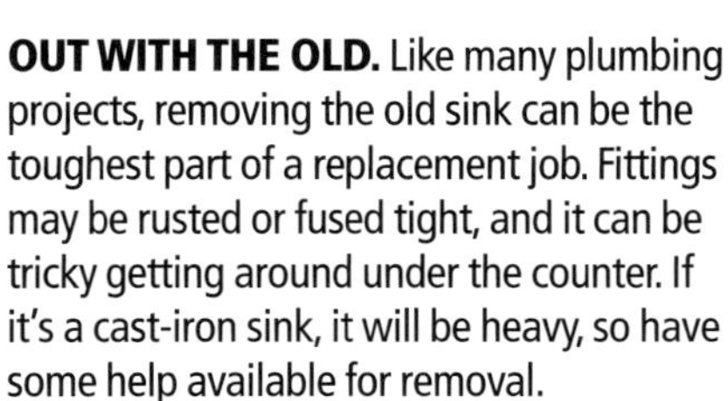

OUT WITH THE OLD. Like many plumbing projects, removing the old sink can be the toughest part of a replacement job. Fittings may be rusted or fused tight, and it can be tricky getting around under the counter. If it's a cast-iron sink, it will be heavy, so have some help available for removal.

1. Before you begin, make sure the new fixture will fit properly into the old hole.
2. Turn off the water supply valves to the hot and cold water faucets. Place a bucket beneath the drain trap. Loosen the slip nuts and remove the trap. (See "Installing a Single-Bowl, ABS P-trap," pages 146–147.)
3. Remove the bucket and place a shallow tray or rags beneath the sink supply lines to catch water that may remain in the water supply lines.
4. Remove the coupling nuts connecting the supply tube to the faucet tailpiece.
5. Disconnect additional plumbing for disposers (see pages 156–157), dishwashers (see pages 158–159), and sink sprayers.
6. Slice through the caulking around the rim using a utility knife.
7. Lift the sink from the countertop using the drain hole as a handhold.
8. Look for water damage to the countertop and clean any excess caulking from the rim before you install the new sink.

Call your local waste removal service for instructions on disposing the old sink.

INSTALLING A WALL-HUNG LAVATORY SINK

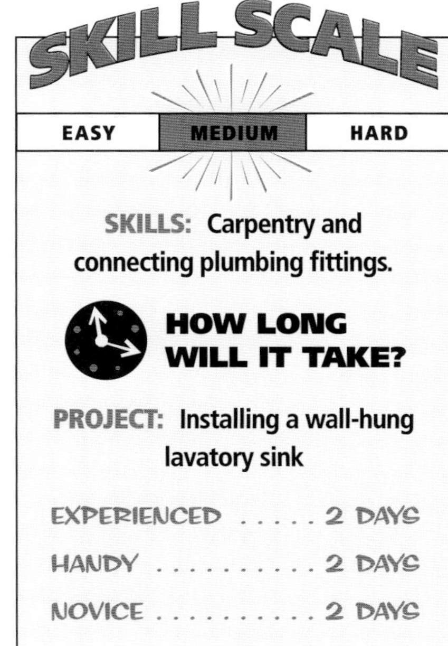

SKILL SCALE
EASY
MEDIUM
HARD
SKILLS: Carpentry and connecting plumbing fittings.
HOW LONG WILL IT TAKE?
PROJECT: Installing a wall-hung lavatory sink
EXPERIENCED ..... 2 DAYS
HANDY .... 2 DAYS
NOVICE .... 2 DAYS

If you install a wall-hung lavatory sink, you have to install wooden blocking between the wall studs to mount the support bracket for the fixture.

PICK YOUR HEIGHT

Lavatory sink heights vary between 30 and 38 inches, depending on individual needs. Position the blocking for the mounting bracket so the sink will hang at a height that is comfortable for you. Locate the studs and cut away the existing drywall so you can install the blocking. Run the hot and cold water supply and drain lines to the fixture.

Manufacturers will often include a template to help you align the height and position of the sink on the wall.

✓ STUFF YOU'LL NEED

TOOLS: Keyhole saw, hammer, tape measure, carpenter's pencil, power drill and bits, ratchet wrench and sockets, caulk gun, screwdriver, utility knife

MATERIALS: Wall-hung sink, mounting bracket, anchor bolts, 2x10 wood blocking, water-resistant drywall scrap, drywall tape and compound, paint, faucets and fittings, drainpipe and fitting, rag, silicone caulk

1 CUT AWAY A 16x16-INCH SECTION OF DRYWALL AND NAIL OR SCREW THE BLOCKING INTO PLACE. Use a keyhole saw to remove the drywall. Nail the 2x10 between the studs so it's flush with the leading edge of the studs. Cut a piece of water-resistant drywall to cover the hole. You may be able to use the piece you cut away. Finish the drywall and paint it. Once dry, attach the mounting bracket.

2 PREPARE TO SET THE LAVATORY SINK. Attach the faucet and drain before mounting the sink. Run the supply line with stop valves. Install the drain line and P-trap. Set the sink on the bracket. Secure it to the bracket with anchor bolts. Connect the supply line to the faucet and the drain to the P-trap. Open the valves. Check for leaks. Tighten leaking fittings if necessary. **Legs for most wall-mounted sinks can be purchased as an option.**

INSTALLING A PEDESTAL SINK

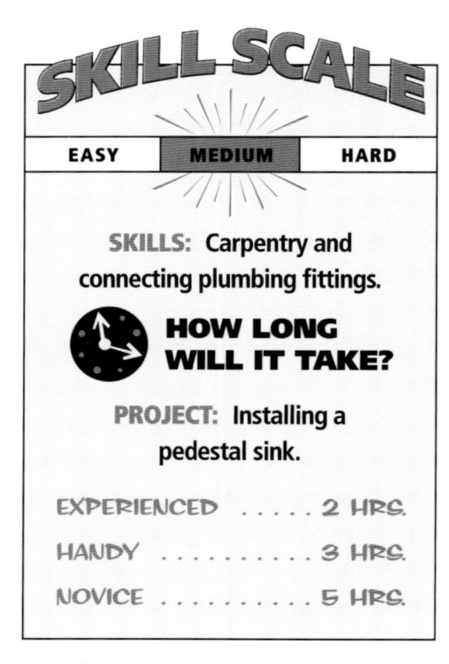

✓ STUFF YOU'LL NEED

TOOLS: Keyhole saw, hammer, tape measure, carpenter's pencil, power drill and bits, ratchet wrench and sockets, caulk gun, screwdriver, utility knife

MATERIALS: Wall-hung sink, pedestal, mounting bracket, anchor bolts, 2x10 wood brace, water-resistant drywall, drywall tape and compound, paint, faucets and fittings, drainpipe and fitting, rag, silicone caulk

The pedestal on a pedestal sink appears to be freestanding, but it's anchored to the floor with lag screws.

1 INSTALL THE MOUNTING BRACE. Remove a 16x16-inch square of drywall to expose the studs. Install a 2x10 mounting brace flush with the studs and nail it into place. Patch, finish, and paint the repair.

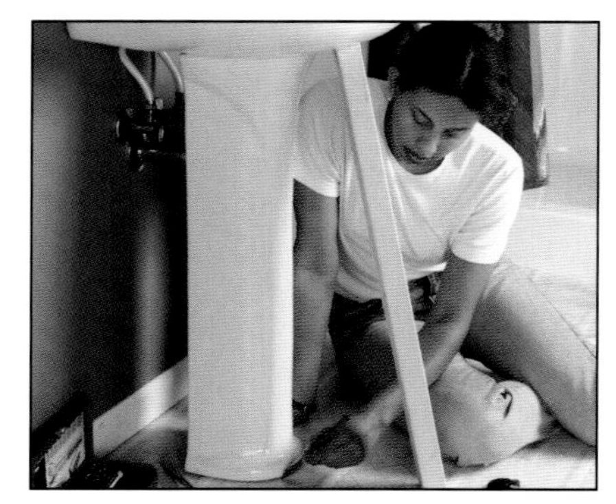

2 SET THE BASIN AND PEDESTAL IN PLACE. Install the fixtures and plumbing. Brace the basin with a 2×4. Outline the base of the pedestal on the floor with a pencil. Mark the wall with the location of the holes in the sink for the lag screws. Set the basin and pedestal aside. Drill holes in the floor for the lag screws.

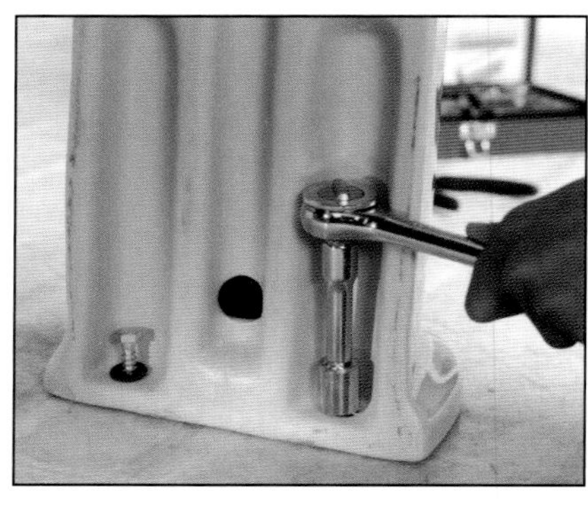

3 INSTALL THE PEDESTAL LAG BOLTS. Make sure they're snug. **Don't overtighten; you could crack the base.** Apply a bead of silicone caulk around the base. Install the faucet, then connect the supply lines and drain before you attach the basin to the wall so you'll have some play to make the hookup.

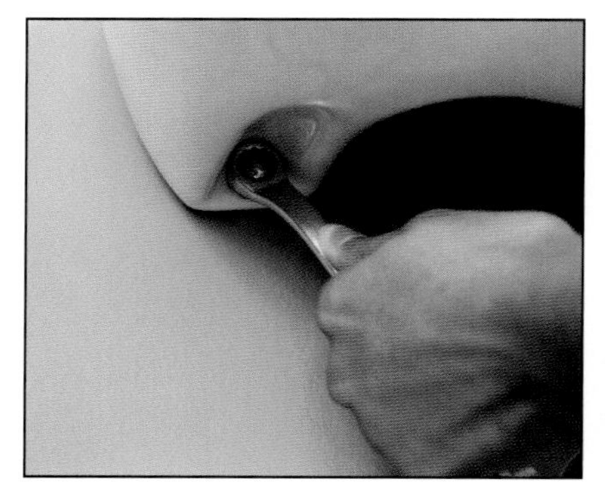

4 CONNECT THE BASIN TO THE WALL. Predrill a small hole through to the blocking, then insert and alternatively tighten the lag screws until just snug—don't overtighten. Caulk the back of the basin with silicone caulk. Test the system for leaks.

DIFFERENT MOUNTS.

Some sinks are mounted on a bracket rather than lag-screwing into a mounting brace. For this style, attach all the fixtures and plumbing before you hang it on the bracket. Mark for the pedestal holes, but complete the drain and supply hookups before you attach the pedestal to the floor.

INSTALLING FAUCETS

When you purchase a faucet, you get exactly what you pay for, so get the best you can afford. Purchase products and materials from businesses that back what they sell with good service and have a clear return policy in case problems arise. They have a presence in the community and a reputation to maintain. Older hardware stores and secondhand shops are potential sources of antique fixtures. Consider places that sell discontinued items, but keep in mind that it may be very difficult, if not impossible, to find replacement parts. Check sales and comparison shop; high- and medium-level fixtures can often be found at low prices.

SO MANY STYLES, BUT...

There are so many styles, finishes, and features to choose from that finding a match for your decorating tastes and functional needs is easy. Finishes vary. Chrome is popular because it is durable and easy to clean. New techniques in metallurgy have created brass finishes that are durable and easy to clean. Satin-nickel has gained in popularity because of its warm, soft color.

THE FAUCET STILL HAS TO FIT THE SINK

Regardless of how much you may love the style of a faucet, it still needs to fit your sink. Measure the distance between the tailpieces to determine the correct faucet size. Bathroom faucets are 4 to 12 inches apart on center. The spacing will vary for kitchen sinks by the number of holes in the sink. Three- and four-hole sinks are the most common. Some sinks offer only one hole for a single-control faucet, and as many as five for accessories, such as a hot-water dispenser, water filter, or soap dispenser.

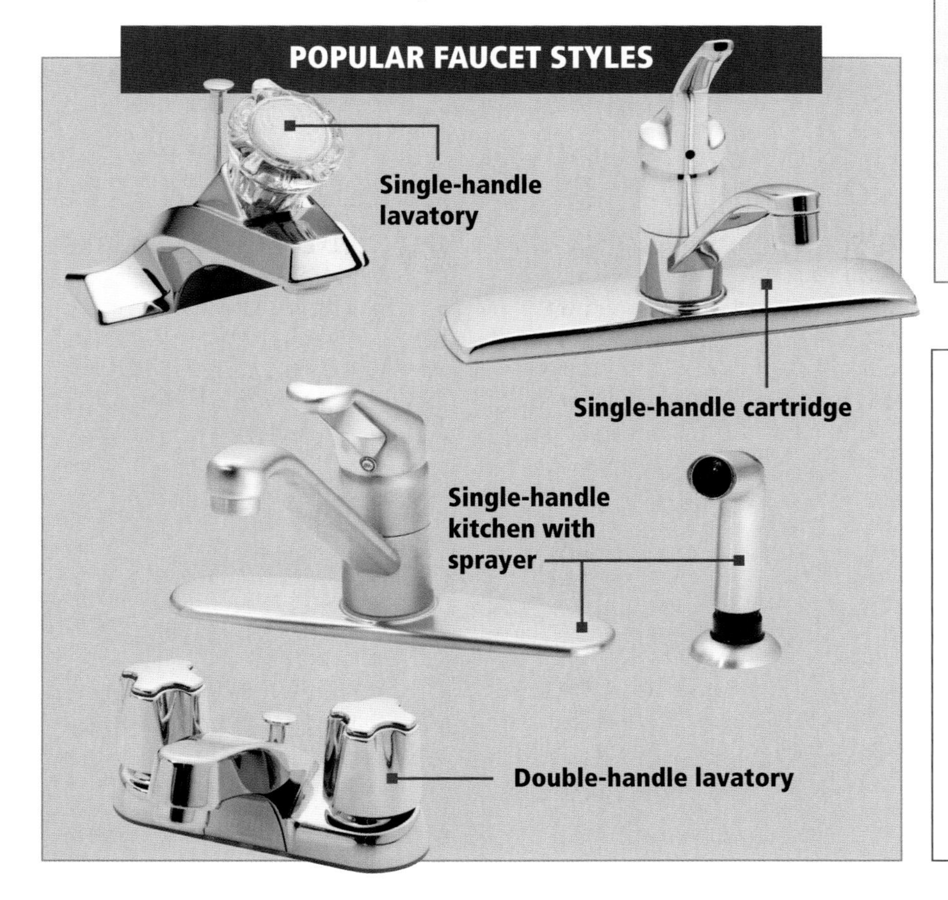

ONE HANDLE OR TWO?

It's a matter of taste. One- and two-handle faucets are popular and they function equally effectively, although there's now a trend toward one-handle models.

1. **Once you're used to them, single-handle faucets are usually easier to operate and balance the water better.**
2. **Another important advantage of single-handle faucets, especially ones with longer handles, is that you can turn on the water (when your hands are full) with a flick of the wrist or elbow.**
3. **Single-handle faucets are a little easier to install, take up less room on the ledge of the sink, and offer a wider swinging radius for the spout.**
4. **Double-handle faucets offer both contemporary and traditional styling. They fit well in almost any decor, and you can combine spouts and handles for custom looks.**

ANTISCALD DEVICES

Water over 49 degrees C can cause serious burns, especially to unsuspecting children. Antiscald devices limit the water temperature to the fixture. You can also add a programmable device to the faucet that shuts off the water before it reaches 49 degrees C.

INSTALLING A CENTER-SET FAUCET

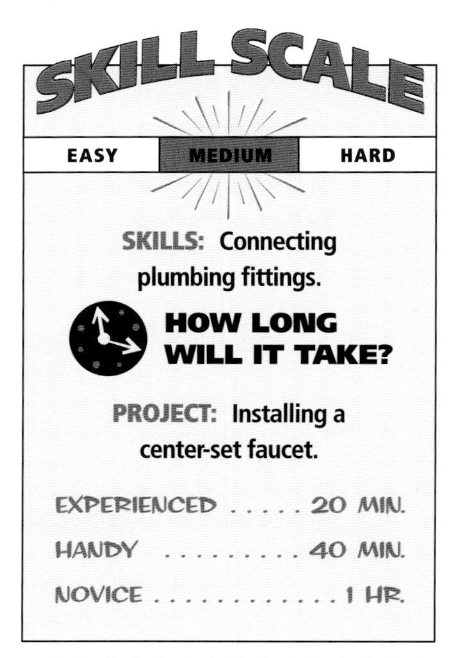

SKILL SCALE
EASY
MEDIUM
HARD
SKILLS: Connecting plumbing fittings.
HOW LONG WILL IT TAKE?
PROJECT: Installing a center-set faucet.
EXPERIENCED ..... 20 MIN.
HANDY .... 40 MIN.
NOVICE .... 1 HR.

TOOLS: Putty knife, adjustable wrenches, basin wrench

MATERIALS: Center-set faucet, gaskets, plumber's putty or silicone caulk

OUT WITH THE OLD IS THE HARDEST PART.

As opposed to other home improvement jobs, such as installing cabinets or hanging doors, the hardest part of almost any plumbing job is usually removing the old stuff. Corrosion and inaccessibility can make taking out an old faucet a real pain in the neck. Give yourself extra time to remove old fixtures—it will save on frustration and bruised knuckles.

Manufacturers are supplying center-set faucets in styles and colors to meet even the most discriminating decorating tastes and needs.

Many finish types are available. The old standby of chrome is popular because of its durability and ease of maintenance. Porcelain finishes have gained popularity. Polished brass and satin-nickel are also available.

Before you choose a center-set faucet, measure the distance between the tailpieces of the one you are replacing. Faucets are 4 inches on center up to 12 inches on center. Purchase one that will fit the opening in your sink.

Take your time when selecting a new faucet. And don't compromise quality. A new installation will provide years of service.

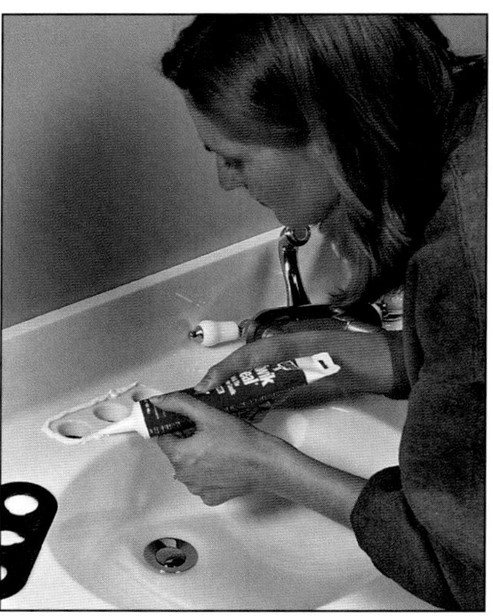

1 SEAL AROUND THE LAVATORY OPENINGS WITH PLUMBER'S PUTTY OR SILICONE CAULK. Apply a bead of silicone caulk or plumber's putty around the faucet openings.

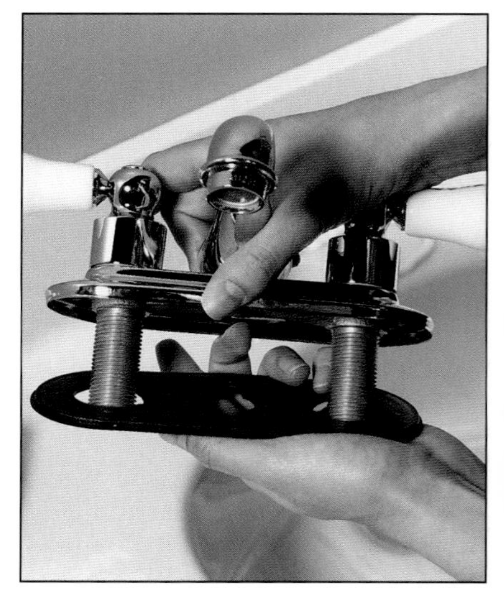

2 PLACE FAUCET GASKET. The gasket will set over the faucet tailpieces.

3 **SET THE FAUCET.** The tailpieces should fit into the hole spacing in the sink. If they don't, you purchased the wrong set and you'll need to return it.

4 **HAND-TIGHTEN THE BASIN NUTS.** Thread each, alternating from one nut to the other to draw the faucet body evenly over the gasket. Don't tighten all the way.

Don't use plumber's putty on cultured marble or other composite sinks. Putty will discolor the sink. Use silicone caulk instead.

5 **CENTER THE FAUCET BODY.** Measure or visually center the faucet body on the lavatory.

6 **HAND-TIGHTEN THE BASIN NUTS UNTIL THEY'RE SNUG.** Install the pop-up drain. (See "Installing a Pop-Up Drain," page 142.) Connect the water supply lines, turn on the water, and test for leaks.

Homer's Hindsight

INSTALL FAUCETS AND DRAIN FITTINGS BEFORE YOU SET THE SINK.

You would be amazed at the places I have installed a sink. Some of the bathrooms have been so small that it's a miracle there is any room to work at all. Save yourself a lot of grief. Install the faucets and as much of the water and drain piping as possible before you set the sink into the counter.

INSTALLING A WIDESPREAD FAUCET

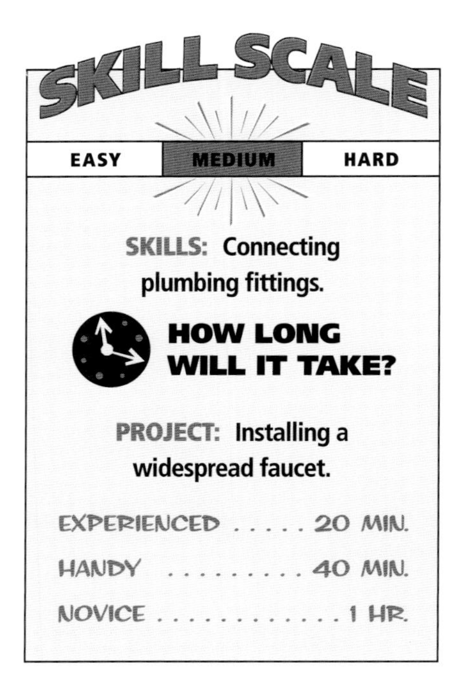

SKILL SCALE
EASY
MEDIUM
HARD
SKILLS: Connecting plumbing fittings.
HOW LONG WILL IT TAKE?
PROJECT: Installing a widespread faucet.
EXPERIENCED ..... 20 MIN.
HANDY .... 40 MIN.
NOVICE .... 1 HR.

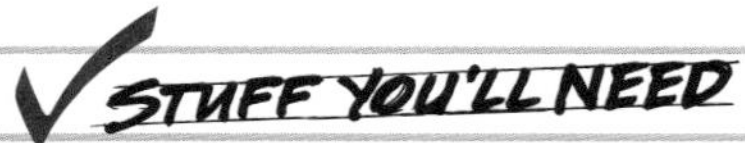

STUFF YOU'LL NEED

TOOLS: Plastic putty knife, adjustable wrenches, basin wrench

MATERIALS: Widespread faucet, gaskets, pipe compound or Teflon tape, plumber's putty

Style is a very important factor when remodeling. Faucets, like everything else, are either in or out of fashion. The most common faucet finishes include chrome, brass, nickel, and white. Choose the finish that will work with the fixtures in your bathroom or kitchen.

SEIZE THE MOMENT

Don't wait until your old faucet fails to replace it. If you're ready for a change, that's reason enough to tackle this project. The good news is that installing a new faucet isn't a difficult job.

After you've installed the faucet, install the pop-up drain. ***See "Installing a Pop-up Drain" on page 142.***

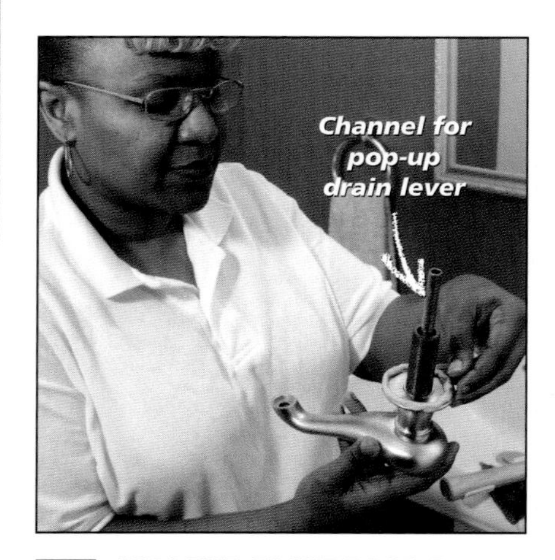

1 SEAL THE SPOUT BASE. Form plumber's putty into a rope and place it on the base of the spout. Press the putty against the base. Set the spout into place.

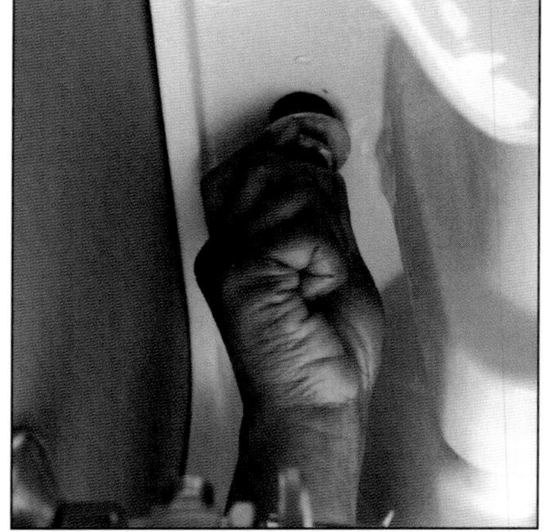

2 HAND-TIGHTEN THE BASIN NUT. Tighten the nut just enough to hold it in place. Don't overtighten, because you will need to center the spout on the sink in step 4.

3 THREAD THE TEE ON THE SPOUT. Center the tee so the outlets are approximately parallel to the back wall and line up with the faucets on either side. See step 7.

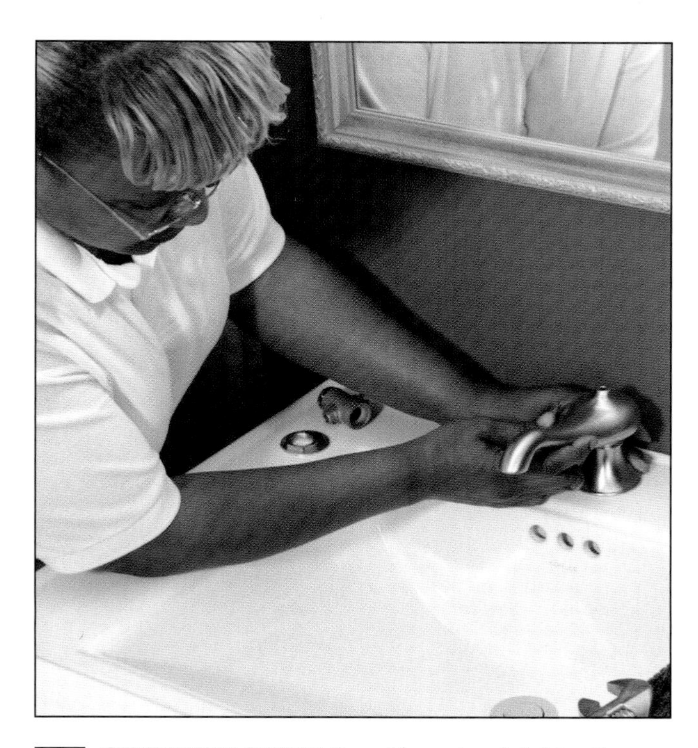

4 **CENTER THE SPOUT.** Once it's centered, tighten the spout from beneath the sink using a basin wrench.

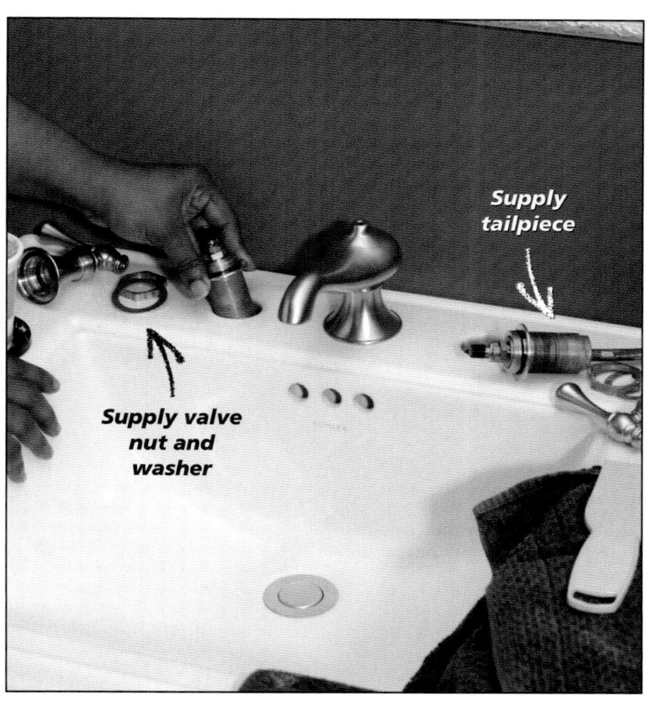

5 **PLACE THE FAUCET VALVES IN THE SINK.** Slide the washer over the threads from beneath the sink. Tighten the valve nut until snug.

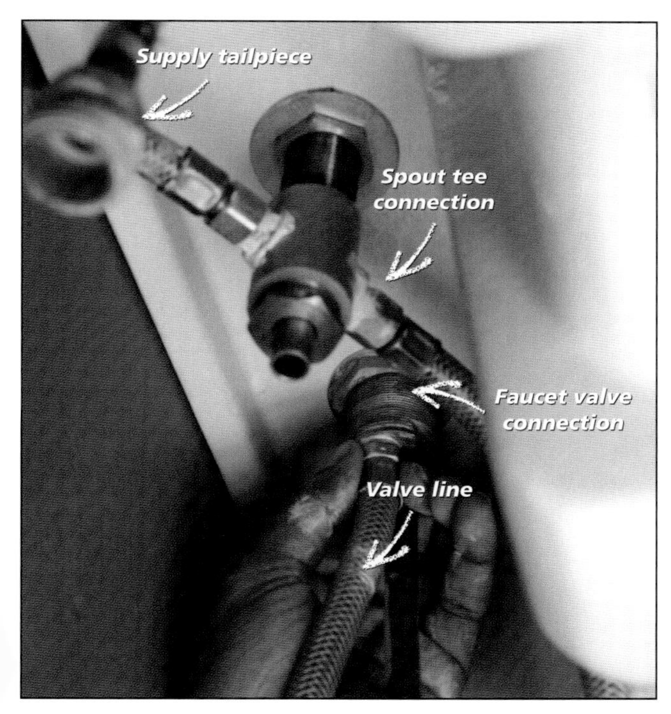

6 **CONNECT THE FAUCET VALVE LINES TO THE FAUCET VALVE.** Apply pipe compound or Teflon tape to the threads of the valve line, connect to the faucet valve, and hand-tighten. Apply pipe compound or Teflon tape to the other end of the valve line and hand-tighten onto the spout tee.

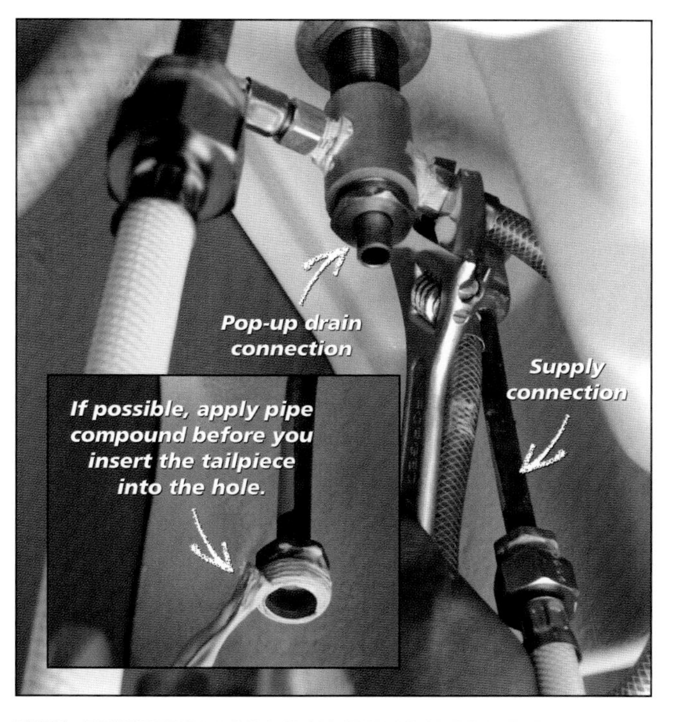

7 **FINISH THE ASSEMBLY AND TIGHTEN THE CONNECTIONS.** Apply pipe compound or Teflon tape to the threads of the water supply tailpieces (see inset) and connect the hot and cold water supplies to the hot and cold supply valves. When the entire assembly is complete, go back and tighten all the connections with an adjustable wrench. The last step is to connect the faucet handles to the valves. Apply a bead of plumber's putty to the faucet handle base, attach the handle, center it, and tighten the faucet nut below the sink with a basin wrench. Turn on the system and check for leaks.

INSTALLING VALVE STOPS AND SHUTOFFS

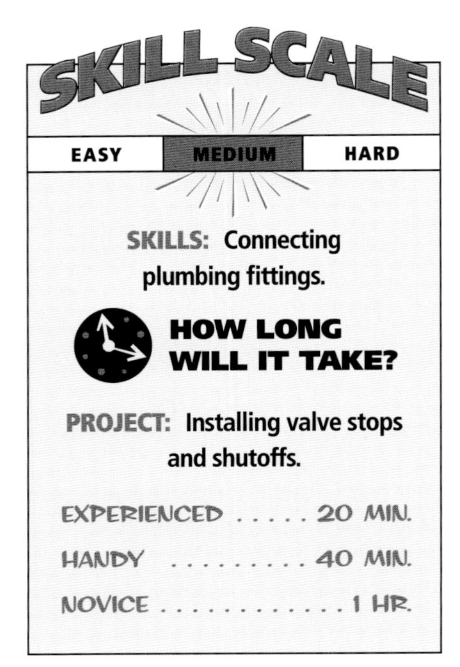

SKILL SCALE
EASY
MEDIUM
HARD
SKILLS: Connecting plumbing fittings.
HOW LONG WILL IT TAKE?
PROJECT: Installing valve stops and shutoffs.
EXPERIENCED ..... 20 MIN.
HANDY .... 40 MIN.
NOVICE .... 1 HR.

Shutoff valves are fitted in several different ways: by soldering, threading, or compression. Compression fittings come in a variety of combinations that will give you options for connecting supply lines. They will also give you control of the water supply near your fixtures so you don't have to go to the basement every time you need to make a repair. Compression fittings are easy to install and don't require pipe dope or compound—a compression ring makes the fitting watertight—as long as they are properly installed.

TOOLS: Mini hacksaw or tubing cutter, two adjustable wrenches, emery cloth

MATERIALS: Compression valve, compression fittings

VALVE STOPS AND SHUTOFFS

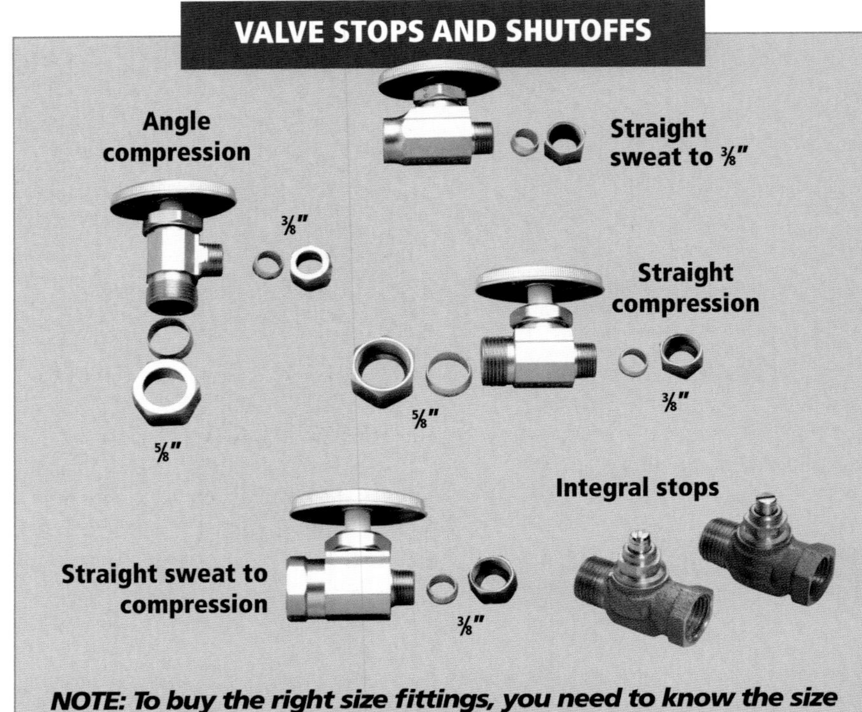

***NOTE:** To buy the right size fittings, you need to know the size of the pipe. See "Rigid Supply Pipes," page 84.*

Replacing old supply lines is generally a good idea, but if they're in good shape, you can reattach them and save a little time. If you have any doubts or difficulties, replace them with braided flexible supplies.

1 CUT THE SUPPLY PIPE. Turn off the main water supply valve. Open the faucets to drain the line. A tubing cutter makes a truer cut, but you can also use a mini hacksaw if the cutter won't fit. Leave enough room between the escutcheon plate and the cut to allow installation of the fitting. Cut slowly and gently. If it's out of round, the copper won't accept the compression fitting and will leak.

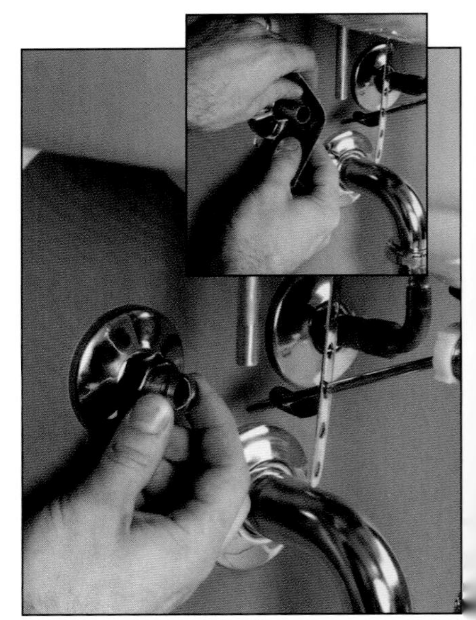

2 SLIDE THE COMPRESSION NUT OVER THE SUPPLY PIPE. Deburr the pipe with emery cloth. (See inset.) Slide the nut as far back on the pipe as possible to give yourself room to work.

3 PLACE THE COMPRESSION RING OVER THE END OF THE SUPPLY PIPE. The ring should completely cover the end of the supply pipe.

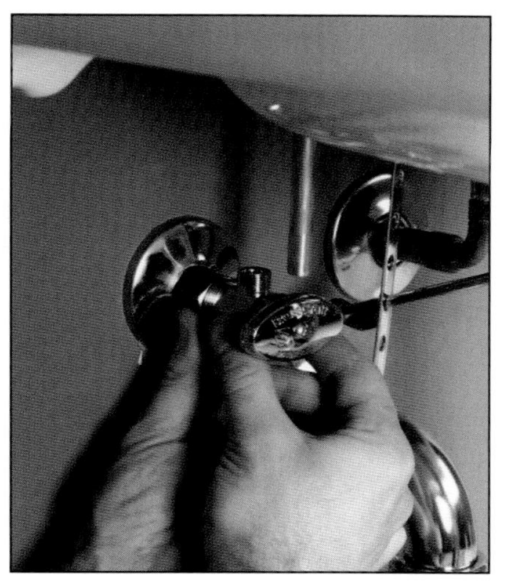

4 THREAD THE COMPRESSION VALVE INTO THE COMPRESSION NUT. The valve should slide squarely and snugly over the ring. Thread the compression valve onto the nut and hand-tighten. If the nut doesn't turn easily, add a tiny drop of oil to the threads. Don't use pipe compound; the fitting doesn't require it and it can actually make the fitting leak.

5 TIGHTEN (BUT DON'T OVERTIGHTEN) THE COMPRESSION VALVE TO THE NUT. Finish tightening the nut using one wrench to hold back the valve and keep it square and another one to turn the nut. Connect the other side in the same manner.

SEAL THE JOINTS.
If the supply pipe is galvanized steel it will be threaded. In that case you can install a regular threaded stop valve using Teflon tape or pipe compound to seal the joints.

Just because you don't recognize a fitting or pipe doesn't mean you can't find out what it is and work with it.

NEW

REAL-WORLD SITUATIONS

In a new installation, you call the shots—everything fits easily because it's made to order. That isn't always the case when you're replacing parts or adding to an existing system. And for reasons of aesthetics or design, you may choose to keep some of the original parts.

You may find that the supply pipe coming from the wall is copper, galvanized, CPVC, or PEX, all of which require different fittings.

In some cases, you may not have enough room for the new supply stops after you've cut away the old elbow. If the supply pipe is too short to take the fitting, you might need to knock away some wall to solder an extension on the supply pipe, or replace the supply line entirely.

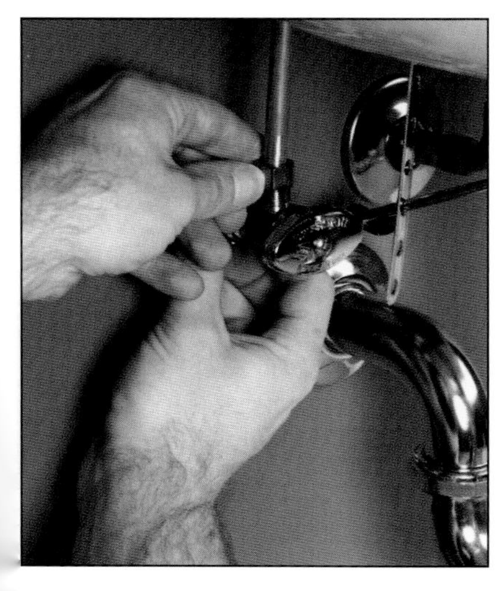

6 ATTACH THE SUPPLY LINES TO THE VALVE. If you are reattaching the old lines to the new valve, follow the same procedure you used to install the valve. If you are replacing the old supplies with new braided flexible lines, you will have to remove the old supplies from the faucets before you proceed.

7 TURN ON THE MAIN WATER SUPPLY VALVE. Turn on the main and open the supply valves. (See step 1.) Let the water flow through the faucet to clear air in the line, and test the fittings for leaks by feeling the fittings and examining them with a flashlight. Tighten as necessary until the joints are sealed.

INSTALLING A POP-UP DRAIN

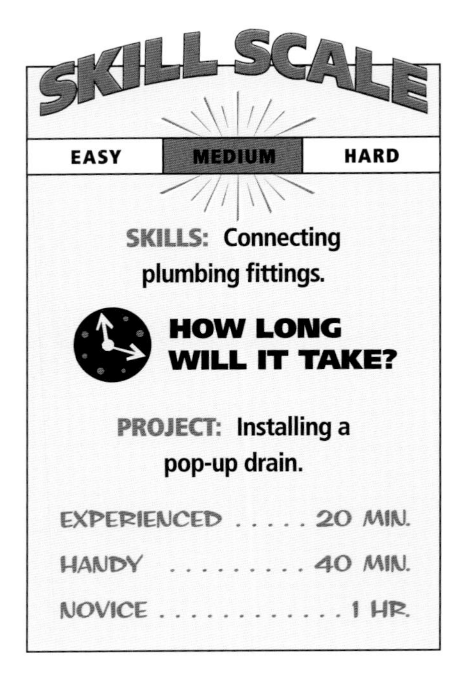

Unless your faucet can accept a pop-up drain, you will have to replace the fixture to get the convenience of a pop-up stopper.

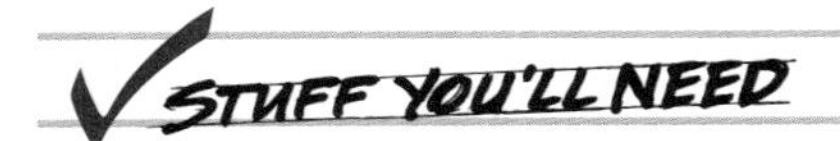

✓ STUFF YOU'LL NEED

TOOLS: Water-pump pliers, plastic putty knife

MATERIALS: Pop-up drain kit, plumber's putty, Teflon tape, silicone caulk, rag, 2x4 support

1 APPLY PLUMBER'S PUTTY. Use a 2x4 to support the sink while you're working. Cover the bottom of the flange with a rope of putty.

2 THREAD THE LOCKNUT ONTO THE DRAIN BODY. Then add the friction washer and beveled gasket.

3 INSTALL THE DRAIN BODY. Push the drain body up through the lavatory hole from underneath.

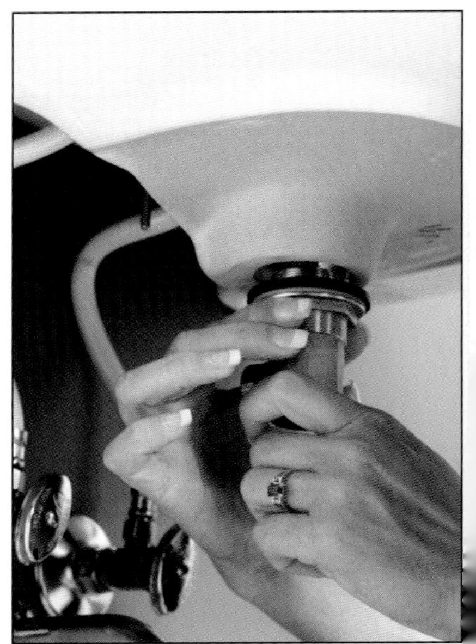

4 HAND-TIGHTEN THE DRAIN BODY TO THE FLANGE. Turn the drain body counterclockwise to tighten.

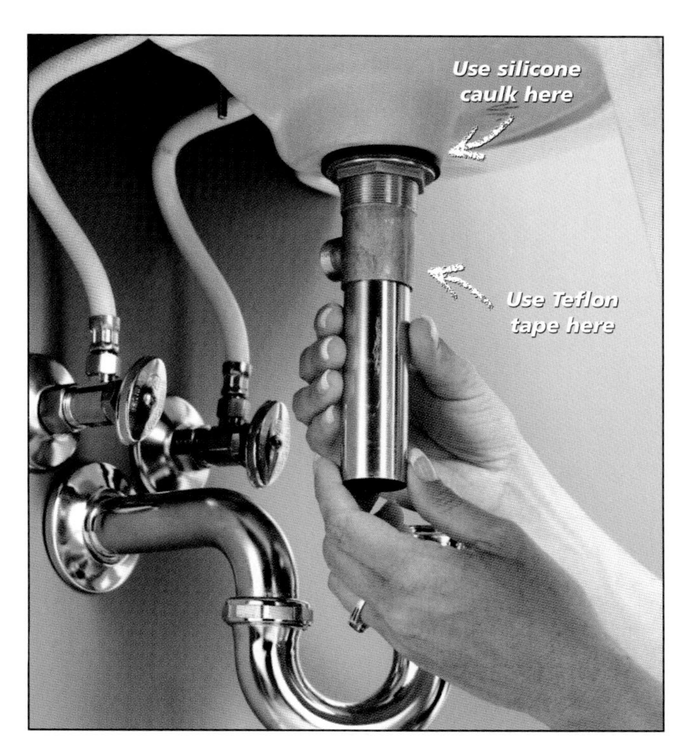

5 TURN THE DRAIN BODY TO THE REAR. This will line it up with the pop-up linkage. Tighten the locknut so the drain doesn't move. Apply silicone caulk at the drain and Teflon tape at the extension piece to seal the drain tube. Remove any excess putty with a plastic putty knife.

6 INSTALL THE DRAIN PLUNGER. Slide the plunger into the drain opening.

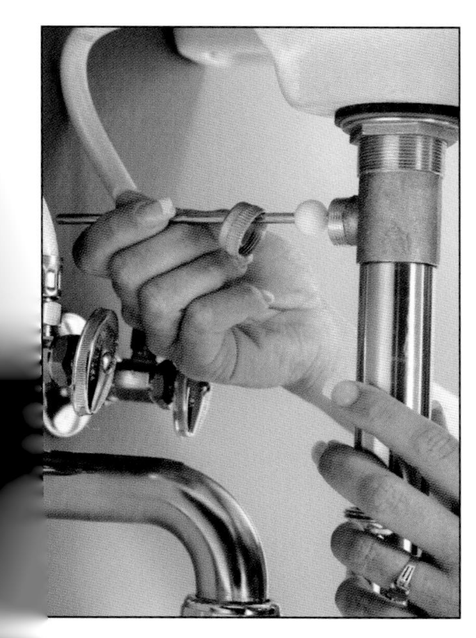

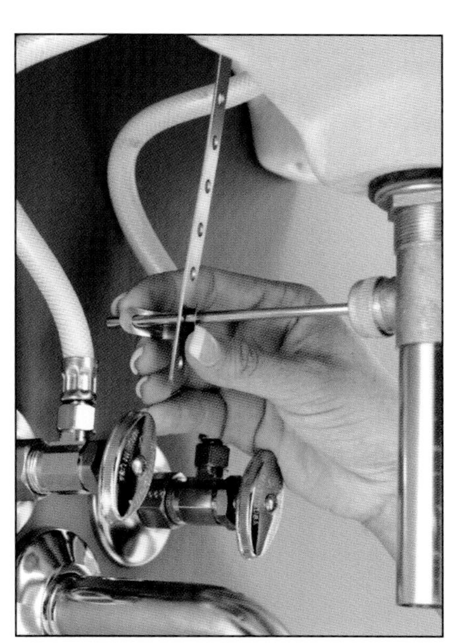

7 INSERT THE BALL INTO THE OPENING. The ball should fit snugly into the opening in the drain tailpiece.

8 THREAD THE BALL NUT HAND-TIGHT. Slide the ball nut over the arm and screw it onto the threads of the drain tailpiece.

9 SET THE POP-UP ARM. Slide the arm through the nearest hole in the lever strap and fasten it with a clip. Connect the the P-trap (see pages 146–147). Test the pop-up drain. Adjust the arm if necessary.

INSTALLING A BASKET STRAINER

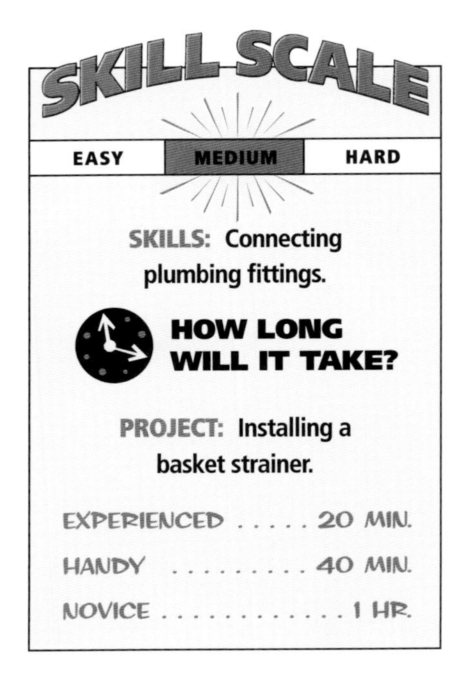

SKILL SCALE
EASY
MEDIUM
HARD
SKILLS: Connecting plumbing fittings.
HOW LONG WILL IT TAKE?
PROJECT: Installing a basket strainer.
EXPERIENCED ..... 20 MIN.
HANDY .... 40 MIN.
NOVICE .... 1 HR.

Avoid inexpensive plastic basket strainers. They will do the job in the short term, but for the long term metal is the way to go. Strainers should be replaced when they leak consistently, chip, crack, or become stained and tarnished.

✓ STUFF YOU'LL NEED

TOOLS: Basket strainer wrench or water-pump pliers

MATERIALS: Metal basket sink strainer, plumber's putty, rag

1 APPLY PUTTY TO THE SINK STRAINER HOUSING. Form a rope with plumber's putty. Wrap the rope around the underside of the strainer housing and press so it adheres to the strainer.

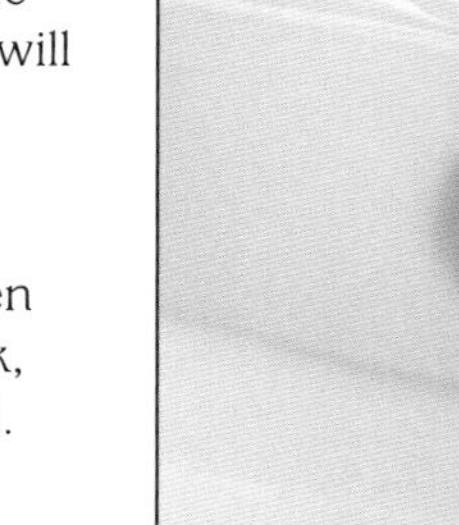

2 SET THE HOUSING INTO THE SINK DRAIN HOLE. Press the housing firmly into the drain hole.

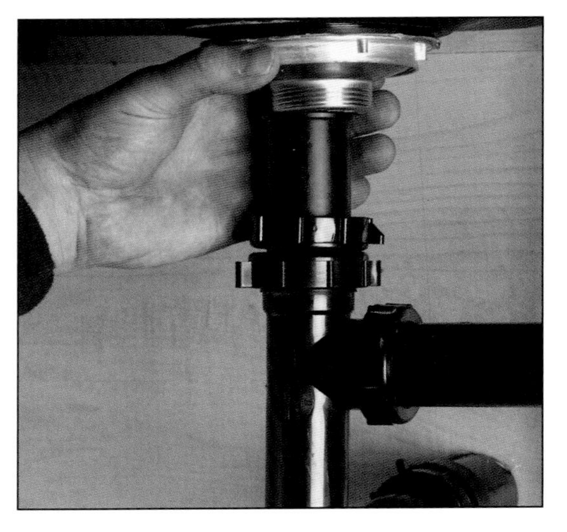

3 CONNECT TO THE SINK. Place a rubber gasket over the threads, then slide on a flat washer. Hold the gasket and washer in place while you connect the locknut hand-tight.

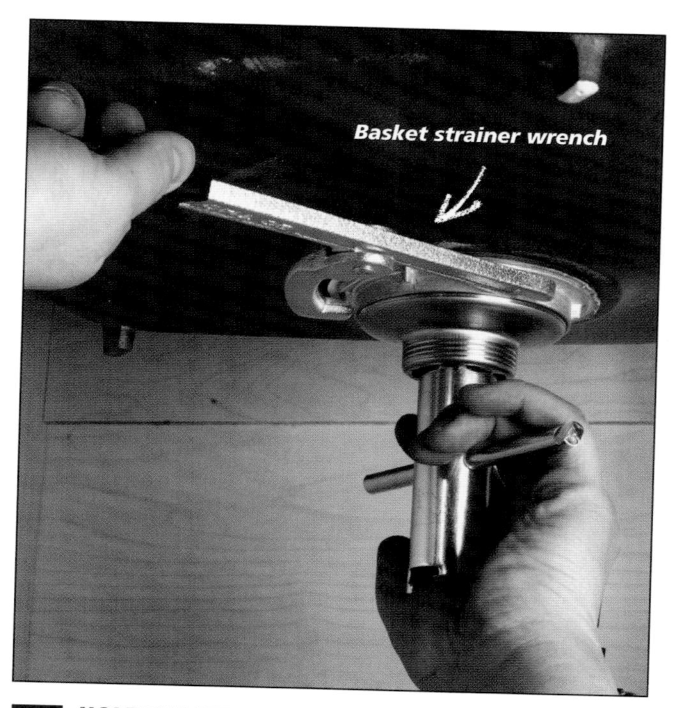

4 HOLD THE STRAINER BODY IN PLACE WHILE TIGHTENING THE LOCKNUT. Use a basket strainer wrench or water-pump pliers to tighten the locknut. DO NOT OVERTIGHTEN. Remove any excess putty that oozes out the bottom and top of the drain.

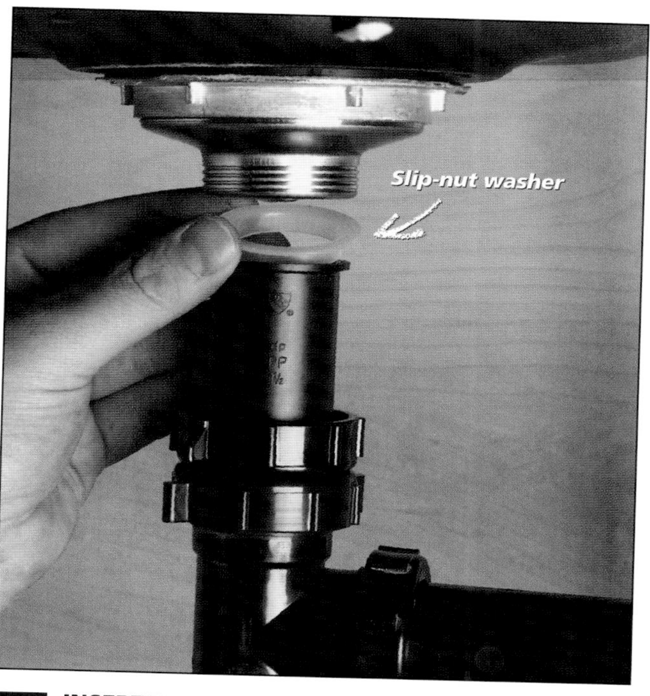

5 INSERT THE SLIP-NUT WASHER INTO THE DRAIN TAILPIECE. The washer should sit inside the tailpiece with the flange of the washer extending over the top.

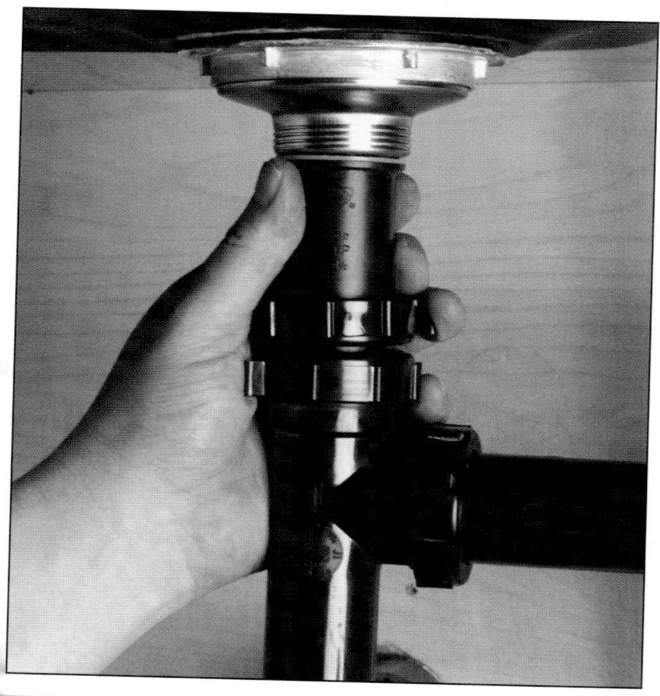

6 ADJUST THE TAILPIECE. Make sure the tailpiece is seated snugly against the strainer body.

7 CONNECT THE TAILPIECE SLIP NUT. Slide the slip nut up to the threads of the strainer body and hand-tighten. Wipe away any excess putty. Turn on the water. Inspect for leaks. Tighten if necessary.

INSTALLING A SINGLE-BOWL, ABS P-TRAP

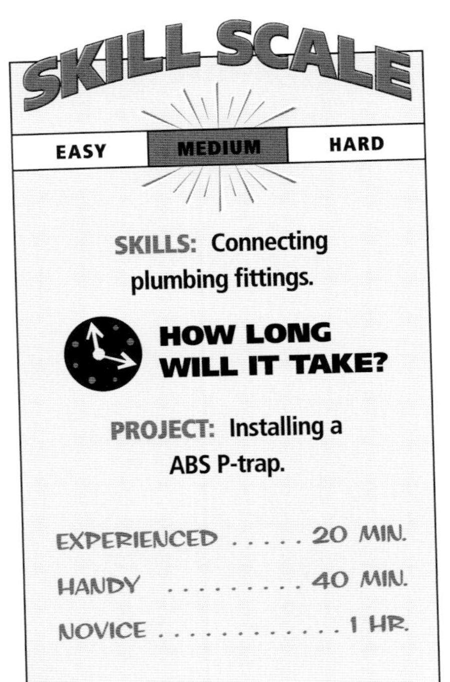

SKILL SCALE
EASY
MEDIUM
HARD
SKILLS: Connecting plumbing fittings.
HOW LONG WILL IT TAKE?
PROJECT: Installing a ABS P-trap.
EXPERIENCED ..... 20 MIN.
HANDY .... 40 MIN.
NOVICE .... 1 HR.

Drains in Canada must use a P-trap in order to meet code. The trap serves as a safety device by preventing noxious gases from backing up the sewer pipe and entering the house. Sewer gases not only pose a health hazard, they can also be explosive.

HOW THE P-TRAP WORKS

The curved portion of the trap holds standing water. Every time the drain is used, water is flushed through the trap and is replaced with fresh water. Solids will adhere to the trap over time and eventually clog the drain or possibly damage the trap—which means it's time to install a new one.

WHAT'S A SLIP JOINT ANYWAY?

Slip joints allow fixtures such as strainer baskets to be joined to drainpipes without making permanent connections. This means parts can be removed for easy replacement. A smaller pipe with slip nuts at each end is inserted into a larger pipe and the seal is made by the pressure that results from tightening the slip nuts to the threaded ends of the larger pipe.

STUFF YOU'LL NEED

TOOLS: Felt marker, tubing cutter, water-pump pliers

MATERIALS: ABS P-trap, slip-joint tailpiece

Hand-tightening is usually sufficient for a pressure connection. Use pliers gently to stop leaks if necessary.

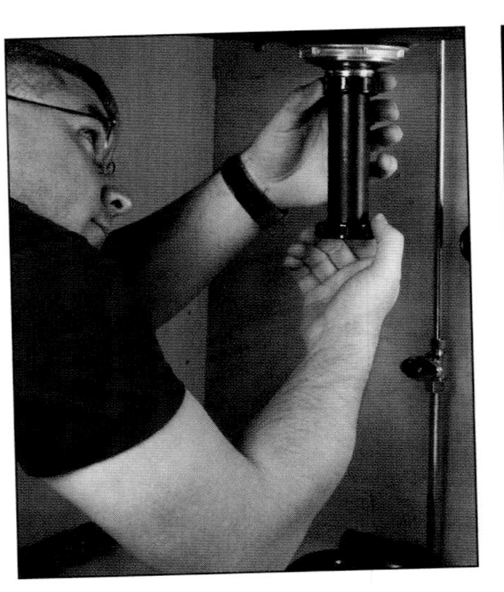

1 CONNECT THE TAILPIECE TO THE SINK DRAIN. Tighten the slip nut hand-tight.

2 MEASURE THE PIPE. Test-fit the P-trap. Mark the P-trap inlet so it will seat inside the end of the drainpipe. Remove the P-trap. Cut the inlet of the trap to length using ABS pipe cutters. Insert the inlet into the pipe socket. Slide the slip nut over the end of the inlet and tighten by hand.

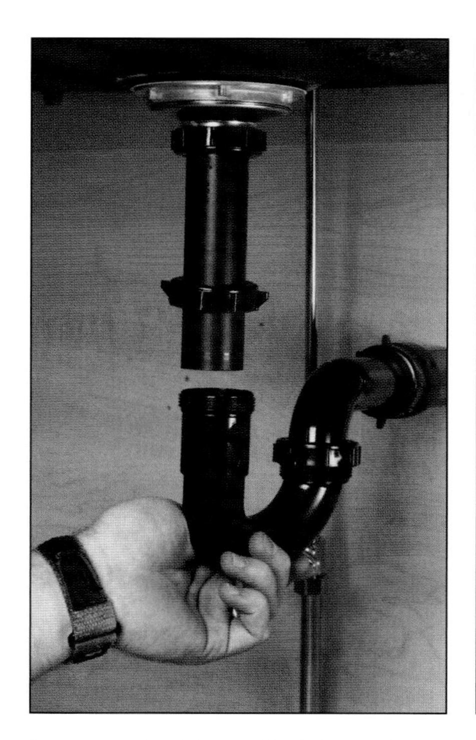

3 **ALIGN THE TRAP.** Make sure the trap will set flush against the outlet and inlet pipes. Adjust the pipe if necessary.

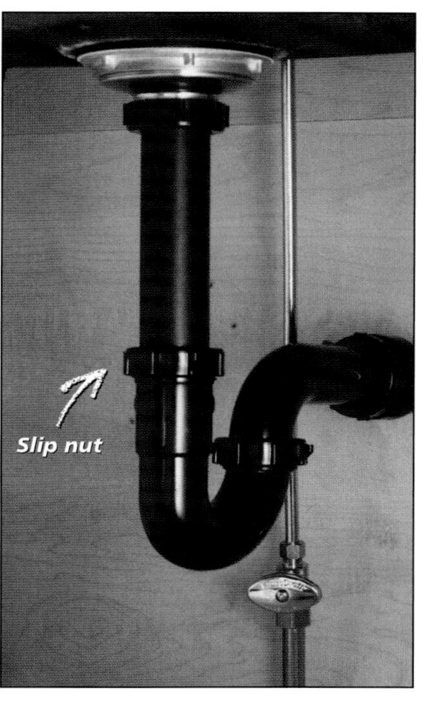

4 **TIGHTEN THE SLIP NUTS.** Turn the nuts until they are hand-tight. Turn on the water. Inspect for leaks and tighten if necessary.

ESCUTCHEON PLATES HELP TO FIGHT BUGS!

Escutcheon plates fit flush around the pipe where the drain line enters the wall or floor. They do add a finishing touch where plumbing is visable, such as under a wall-hung lavatory. Escutcheon plates also seal the drain pipe hole and can prevent drafts or unwanted insects from entering your home.

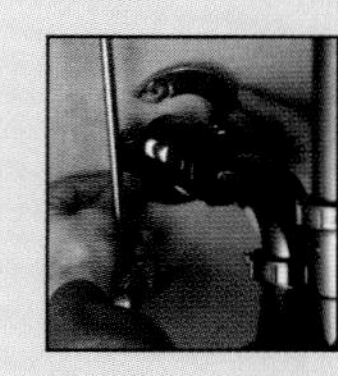

◀ **Split-ring escutcheon plates fit around a pipe that is already in place.**

◀ **Solid-ring escutcheon plates slip onto the pipe before the drain is assembled.**

CONNECTING OPTIONS FOR DOUBLE-BOWL SINKS.

Two bowls can be served by one P-trap. Conversion kits are available at your local home center. One style has the P-trap centered between the bowls with a connecting tee and separate lengths running to the sink drains. Centering the P-trap does not meet code in Canada. Another style has the P-trap aligned beneath one of the drains with a vertical tee connecting to the second bowl. Though the installation procedure is the same as for a single-bowl sink, make sure the horizontal run has a fall of ¼ inch per lineal foot toward the tee connector. Otherwise, solids will settle along the horizontal pipe and eventually stop or impede the flow of wastewater.

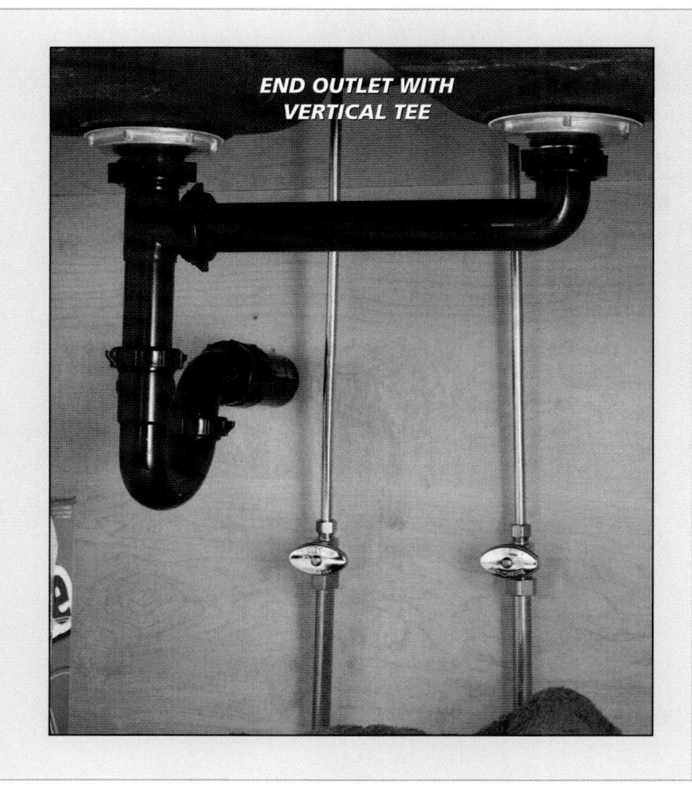

CHAPTER

9 INSTALLING TOILETS

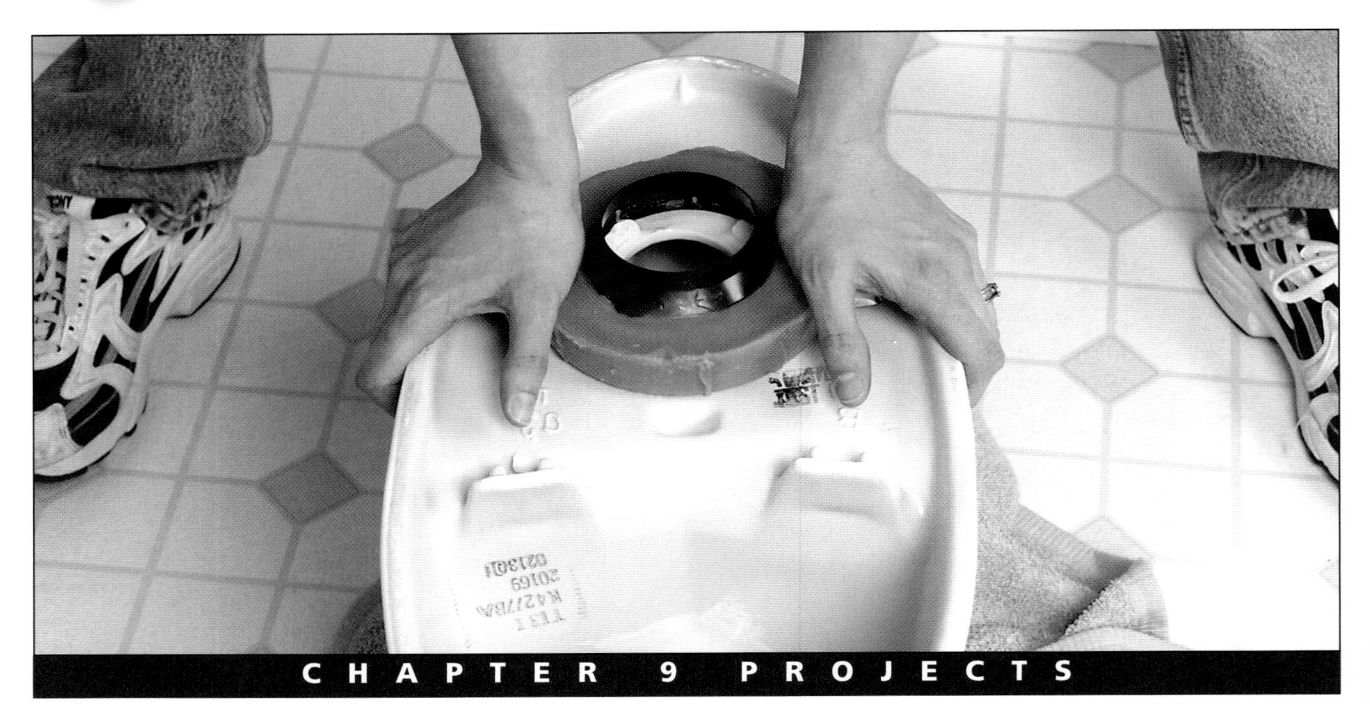

CHAPTER 9 PROJECTS

There are at least two reasons to replace your toilet. The decision can be a practical one, based on wear and tear in the form of cracks, chips, and leaks that can't be repaired. Stains will eventually develop that won't go away, no matter how hard you clean.

You may also want to replace a toilet as part of a general bathroom remodeling or to get a model in a color that fits with updated bathroom decor. Toilets are available with round or oval bowls, and in one- or two-piece models. Corner toilets are popular in some parts of the country.

You aren't limited to replacing a toilet with one of the same shape or style either. Elongated rounds and regular rounds will usually fit in the same drain hole.

WHAT ABOUT THE ENVIRONMENT?

There's an environmental issue as well. Older toilets use 13 liters of water per flush.To meet current regulations, new models must use 6 liters or less. The earliest low-flow toilets often clogged and required more than one flush, but current water-conserving toilets virtually eliminate these problems. You can also install a gravity flush or pressure-assisted flush toilet. Pressure-assisted toilets are more expensive and noisier, but are also more efficient. Gravity flush toilets are less expensive, but they may occasionally require extra flushes.

INSTALLATION ISN'T HARD

Installing a toilet isn't as hard as it may seem. While it may be a little heavy and awkward as an object, the mechanics of operation are about as basic as it gets. If your drain is in good shape and you're up to a little lifting, installing a toilet is an easy afternoon's work.

CONNECTING OLD DRAINS TO NEW TOILETS.

Before buying a new toilet, make sure the location of its drain valve is in the same place as the "rough-in" dimensions of the drain on the bathroom floor.

In newer bathrooms, the center of the floor outlet is 12 inches from the wall. Most new toilets are designed to fit these specifications. If you live in a house built before the mid-1940s, the outlet may be 10 or 14 Inches from the wall. You can get some toilets with a 10- or 14-inch rough-in, but not in all models, styles, and colors.

If you have your heart set on a toilet with a 12-inch rough-in, but your rough-in is 10 or 14 inches, it can be fixed. It's a big job, but you can install what's called an offset flange, then connect the new toilet's drain to the floor outlet.

Offset flange

GOT SOME TOUGH NUTS TO CRACK?

Nuts and bolts on toilet bowls can become so corroded that an adjustable wrench won't budge them, or the wrench will round the corners of the nut, making removal next to impossible. You may expand your vocabulary, but you won't budge the nut without applying some special techniques.

1. Before you attack the nut in the first place, assume it may be seized and squirt on some penetrating oil. Let the oil soak in thoroughly before you try removing the nut.
2. Try a mini-hacksaw. Protect the base of the toilet with masking tape and cut the nut at a slight angle (about 30 degrees) until you have a deep groove. Insert a screwdriver into the groove and twist to break the nut.
3. As a last resort (or first resort if you have one in your toolbox) you can use a nut splitter, which is a tool often found in the automotive section of your hardware store. The splitter fits over the nut. You hold it in position with an adjustable wrench and tighten it with a socket wrench that closes the jaws of the nut splitter against the nut and cuts it in half.

National Kitchen and Bath Association (NKBA) Guidelines for Bathroom Accessibility and Safety

1. Handles: Entrance doors, drawers, and faucets should have lever- or D-shaped handles. They're easier to operate than knobs, especially for people with arthritis.
2. Grab bars: People with limited mobility need grab bars next to the toilet, behind it, and in the tub and shower stall. Although the NKBA recommends lengths, locations, and installation methods, the user is always the best judge of what will—and won't—work.
3. Buy bars with a nonslip texture. Available at any medical supply outlet, they come in a variety of decorator colors and styles.
4. Avoid installing towel bars, which aren't as firmly anchored, at the same level as grab bars. In a fall, a towel bar could easily be mistaken for a grab bar. Better yet, don't use towel bars at all. Install extra grab bars to hold towels.
5. Windows: For wheelchair users, don't install double-hung or sliding windows in a barrier-free bath. Casements are easier to operate. Put windows 24 to 30 inches above the floor so wheelchair users can open and close them easily.
6. Tubs: To make a barrier-free bathtub, have enough clear space in front so a wheelchair can roll up to it. Include grab bars and a seat in the tub.

REMOVING AN OLD TOILET

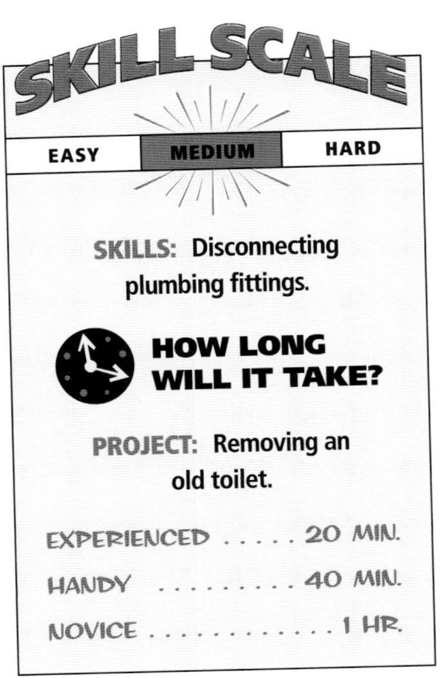

SKILL SCALE
EASY
MEDIUM
HARD
SKILLS: Disconnecting plumbing fittings.
HOW LONG WILL IT TAKE?
PROJECT: Removing an old toilet.
EXPERIENCED ..... 20 MIN.
HANDY .... 40 MIN.
NOVICE .... 1 HR.

✓ STUFF YOU'LL NEED

TOOLS: Adjustable wrench, ratchet wrench and sockets, putty knife, rubber gloves, tubing cutter

MATERIALS: Rags, sponge, bucket, towel

POTENTIAL BIOHAZARD!

The water in the toilet bowl may contain harmful bacteria. Wear rubber gloves when cleaning or removing the bowl. Wash your hands thoroughly with an antibacterial soap afterward.

Removing a toilet is basically the reverse of installation. You may have to remove the toilet to replace it, to put down a new floor, or to get it out of the way before removing the bathtub. The biggest challenges are getting the job done without creating a watery mess and without letting sewer gases seep into your home.

1. **Turn off the water supply at the shutoff valve.** Flush the toilet until the tank is empty. Wipe up any water remaining in the tank and bowl with rags and a sponge. Always wear rubber gloves when cleaning wastewater. Disconnect the supply tube with an adjustable wrench.
2. **Remove the tank bolts** with a ratcheting socket wrench.
3. **Lift the tank off the bowl.** Be careful when you remove the tank, especially if you plan to reuse it; most toilets are made of porcelain, which is easily damaged.
4. **Remove the toilet base.** Pry open the floor bolt caps at the base of the toilet. Use a socket wrench to remove the nuts from the floor bolts. Straddle the toilet and gently rock it from side to side to break the wax seal; lift it off the bolts and set it on its side on an old towel. Water may spill from the toilet trap.
5. **Scrape the old wax from the toilet flange with a putty knife.** Plug the drain opening with a rag to prevent sewer gases from escaping and entering the house while you work.

RAISING ISSUES.

If you're laying a new floor, remove the toilet so you don't have to trim around it. You'll get a cleaner and neater looking installation. The catch is that the new floor will lift the toilet by its finished thickness, but not the soil pipe. In that case, you may have to install an extension flange, which is readily available at your local home center.

INSTALLING A TOILET

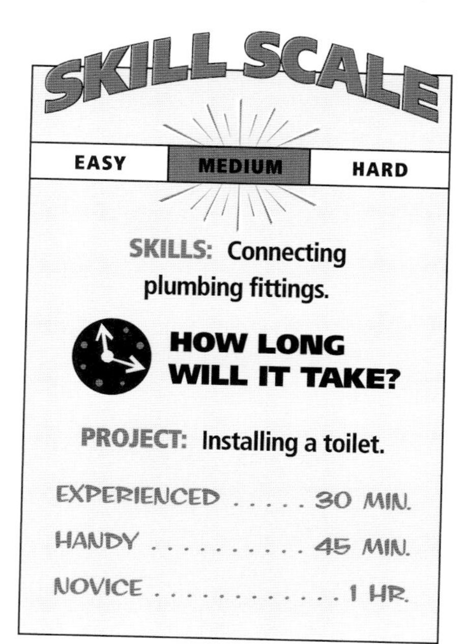

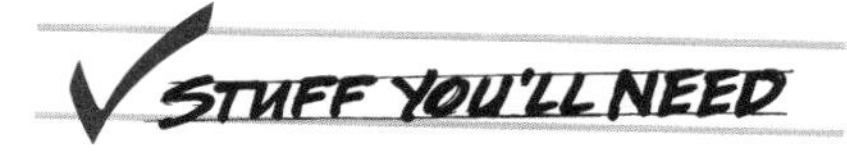

✓ STUFF YOU'LL NEED

TOOLS: Two adjustable wrenches, ratchet wrench and sockets, screwdriver, hacksaw, tubing cutter

MATERIALS: Toilet, toilet seat, "no-seep" wax ring, plumber's putty, felt marker, pipe compound

Be careful when handling the toilet. It may look strong, but it can chip or crack if dropped.

If you're installing a new toilet in a new location, you'll need to run (or have someone run) a water supply line and a drainpipe, which must be connected to the drain/vent system in compliance with code. (See Chapter 5, "The DWV System," page 92.) Even if you're just replacing an existing toilet, you should replace the water supply tubing as well. (See pages 82–85.)

SOME PARTS ARE EXTRA

Most of the parts you need to install the new toilet will come with it—except the toilet seat, which is a separate item on almost every model, with the exception of some one-piece models. The wax ring—which comes in two sizes—is also not part of the package. The manual that comes with the new toilet will tell you the correct size to buy.

BUYER'S GUIDE

SUPER-DUPER FLANGE.

Closet flanges take a lot of abuse and can crack or break. If you don't want to pull out the old one and do a new connection job, use an adapter sometimes called a super flange that fits over the old model and makes a secure connection for the toilet. There are other adapters available as well. Check them out at your local home center.

1 SET THE MOUNTING BOLTS. If you're reusing the old flange, it's a good idea to replace the 3½-inch flange bolts. Purchase two 3½-inch-long closet bolts at your local home center. If you're replacing the flange, it must be screwed into a wooden floor. Self-tapping concrete screws are used for concrete.

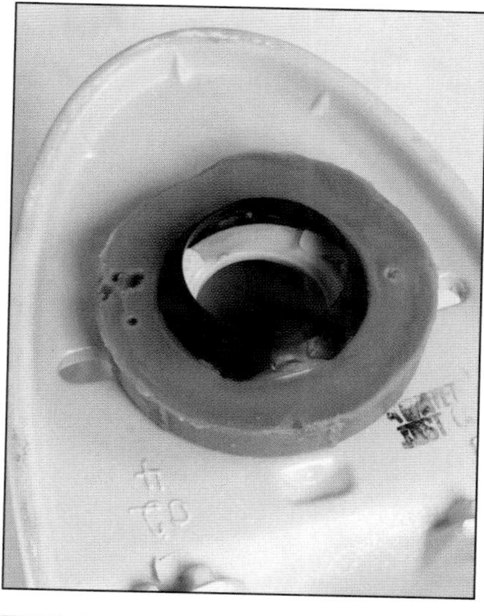

2 PLACE THE WAX RING ON THE TOILET. The "no-seep" wax ring size will vary with the size of the flange. Be sure to purchase the proper size. A 3-inch neck will fit a 3-inch closet elbow and a 4-inch neck will fit a 4-inch closet elbow. If the closet elbow is 4 inches and the neck is 3 inches in diameter, purchase a 4×3 reducer.

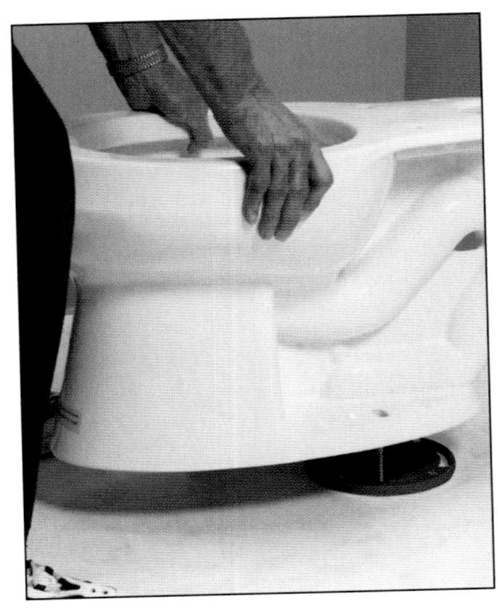

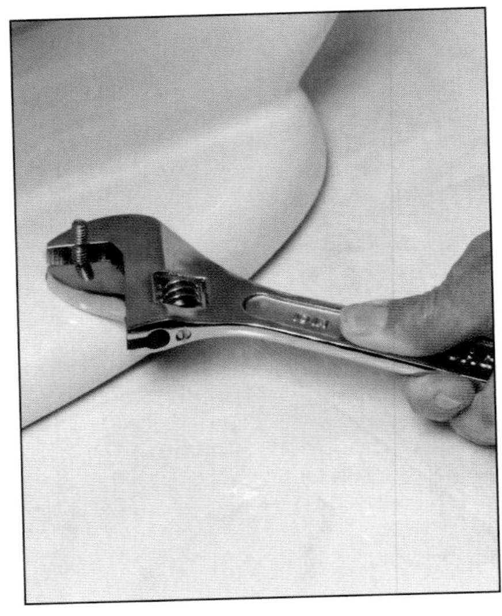

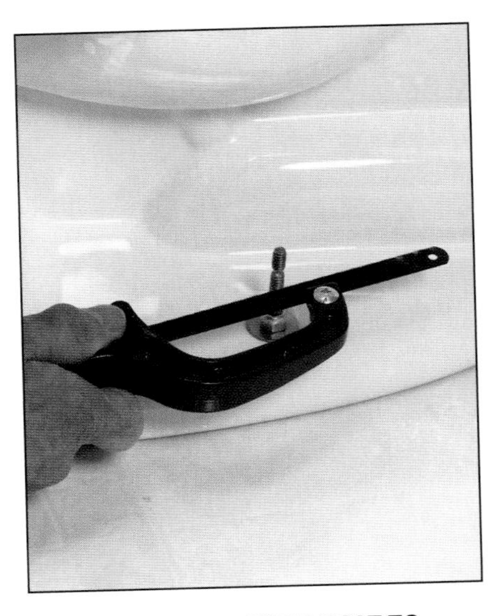

3 **SET THE TOILET BOWL.** Straddle the toilet bowl and lift, using your legs and not your back. Toilets are heavy, so get some help. Set the toilet over the anchor bolts and sit on the toilet, rocking it back and forth to seat the wax ring.

4 **TIGHTEN THE NUTS.** Draw the nuts down over both anchor bolts until they are hand-tight. Shim as necessary to maintain level. Use an adjustable wrench and—moving from one side of the toilet to the other—alternate the tightening of each nut a half-turn at a time until it's snug. Don't overtighten.

5 **CUT THE FLANGE BOLT TO SIZE.** Use a mini hacksaw to cut the flange bolt so only ¼ to ½ inch extends above the bolt. This will allow the cap to fit snugly. Most bolts have snap offs every ½ inch or so, but you should still cut through so you don't bend the bolt.

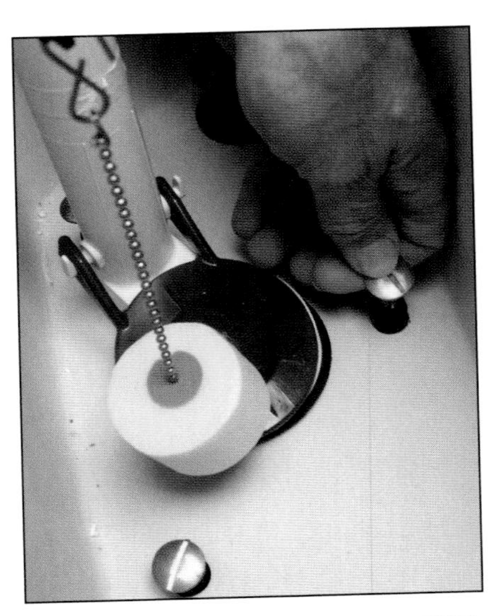

6 **INSTALL THE BOLT CAP.** Some types of caps will snap over the bolt. Others have to be filled with plumber's putty and seated over the anchor bolt.

7 **SET THE TANK ANCHOR BOLTS.** Place the tank anchor bolts in the holes of the tank to help guide the tank onto the bowl.

8 **PLACE THE TANK ON THE BOWL.** Lift the tank and place it over the bowl. You may need some help for this. Guide the tank bolts into the corresponding holes on the toilet bowl.

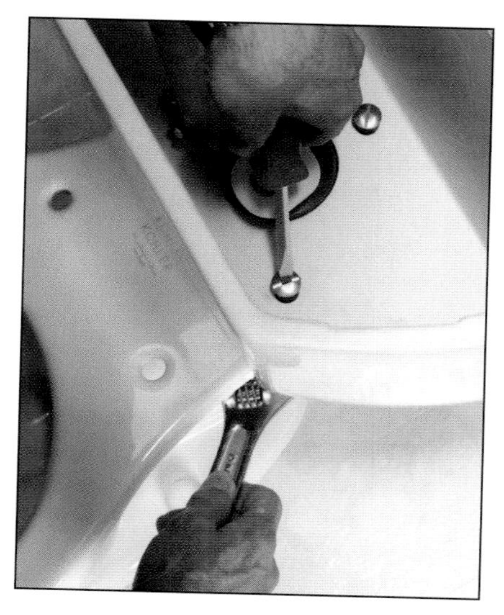

9 TIGHTEN THE TANK BOLTS. Hold an adjustable wrench over the tank bolt nut while you tighten the bolt with a screwdriver. Don't overtighten; you can crack either the tank or the toilet bowl.

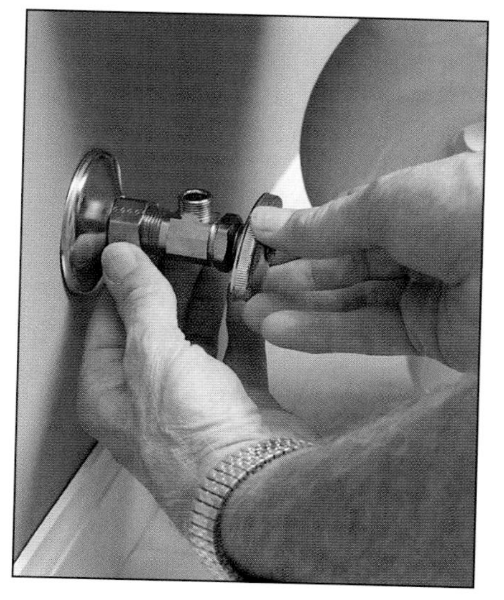

10 INSTALL THE SHUTOFF VALVE. Set the valve over the compression ring and draw the nut to it. Tighten the nut until hand-tight. Use two adjustable wrenches to tighten until snug—one to hold back the valve and the other to tighten the compression nut.

11 MEASURE THE SUPPLY TUBE. Hold the tank supply tube so it's in place. Let the extra pipe extend past the shutoff valve. Mark the pipe for cutting. Leave enough pipe so it will fit snugly inside the shutoff valve outlet. Cut the supply pipe with a tubing cutter.

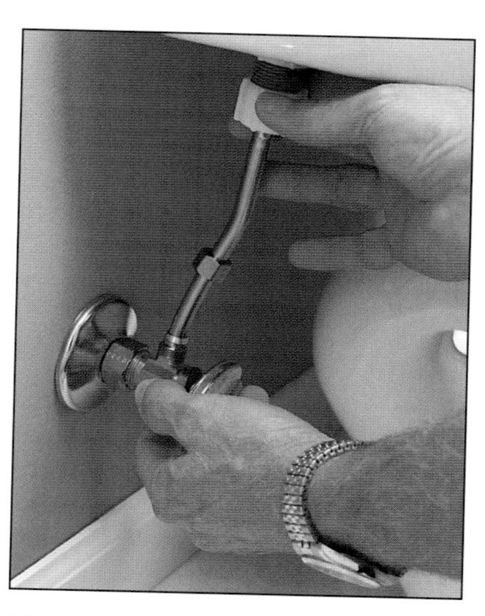

12 CONNECT THE SUPPLY PIPE TO THE TANK. Seat the end of the pipe against the tank. Draw up the tank nut. Hand-tighten until snug. Slide the compression nut over the other end, then place the compression ring over the end. Seat the end in the outlet of the shutoff valve.

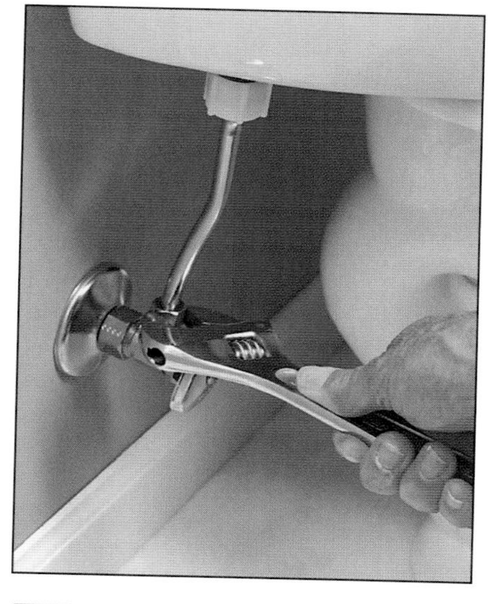

13 TIGHTEN THE COMPRESSION NUT. Use an adjustable wrench to carefully tighten the compression nut. Don't overtighten. Turn on the water supply and check for leaks along the supply line, visually and by feel. Flush the toilet and check for leaks around the base of the tank. If there is a leak, tighten the connections a half-turn.

GET THE RIGHT SIZE TOILET.

We were walking around the plumbing aisle when we saw the toilet of our dreams on clearance. We decided right then and there to take it home and install it. The problem was that it didn't fit over our closet flange. I didn't know that the distance the drain is from the wall will determine the size of toilet you buy. Ninety percent of toilets are made for a drain opening 11 to 12 inches from the wall. The rest are either 9 to 10 or 13 to 14 inches away. But those few inches make a big difference when something's on clearance and can't be returned.

CHAPTER 10

INSTALLING APPLIANCES

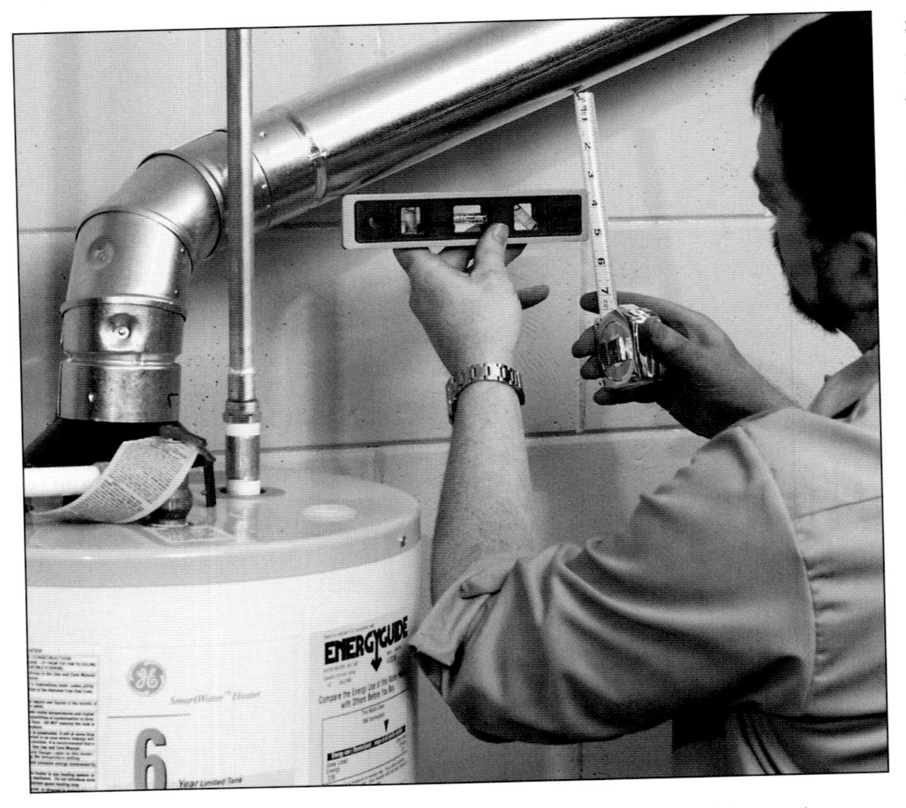

The total budget for materials, fixtures, and labor for remodeling a kitchen or bath and replacing major appliances can be high. You can lower that cost significantly by installing many of the appliances—such as the garbage disposer, dishwasher, water purifier, hot-water dispenser, and icemaker—yourself. What you save on labor and installation costs may mean the addition of specialty items you wouldn't otherwise have considered. The same logic applies to making changes in laundry rooms and bathrooms, and adding or replacing primary system features such as water heaters or water purifiers.

REVIEW WARRANTIES AND CODES

Check warranty information carefully when selecting appliances. Some warranties may be voided if the products are installed incorrectly or without manufacturer-approved materials. Also check to see if the warranty remains valid if you install the appliance yourself. Some warranties are voided if a licensed professional does not complete the installation. As with any installation, make sure the work you are doing complies with local codes and regulations.

Take your time when shopping for new or replacement appliances. Good research will save you a lot of headaches down the road.

CHAPTER 10 PROJECTS

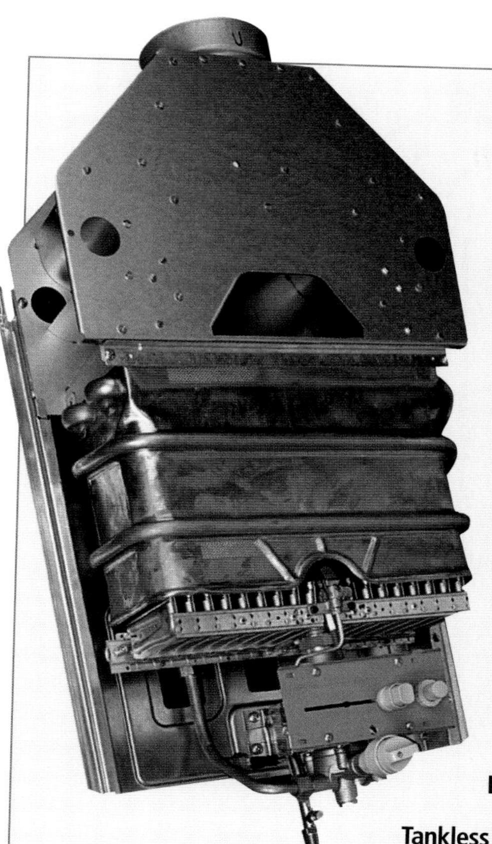

REPLACING A WATER HEATER? YOU MAY WANT TO THINK TANKLESS.

Although common in Europe, tankless (on-demand) water heaters are beginning to make their mark in Canada and the United States. If you're adding a bathtub and shower to your home or need to upgrade the size of your water heater, tankless may be the way to go. On-demand heaters come in sizes small enough to mount in a kitchen cabinet and large enough to handle almost any size home. In milder climates, some can even be placed outdoors. Tankless water heaters not only guarantee a constant supply of hot water, they also take up far less space than a conventional water heater. While they are more expensive than standard water heaters, their savings in energy and low maintenance might quickly make up the difference in cost.

HOW DO THEY WORK?

Tankless heaters flow water through a set of coiled copper pipes. The water is heated by a gas burner as it passes through the coils. Water heater tanks, on the other hand, maintain water at a constant temperature, even when hot water is not being drawn. Heat loss in the tank can be as much as 4 percent per hour. To maintain a constant water temperature in the tank, a thermostat will activate to heat the water. This requires an increase in gas and electric usage, resulting in higher energy costs.

WHAT ABOUT CAPACITY?

Tankless water heaters eliminate the need for maintaining a constant water temperature, supplying heated water only when there is a demand. They can supply continuously heated water at 40 degrees C to 46 degrees C at rates of 7.5 to 11 liters per minute. Most conventional water heaters, on the other hand, will eventually run out of hot water when demand is high, and they require time to recharge.

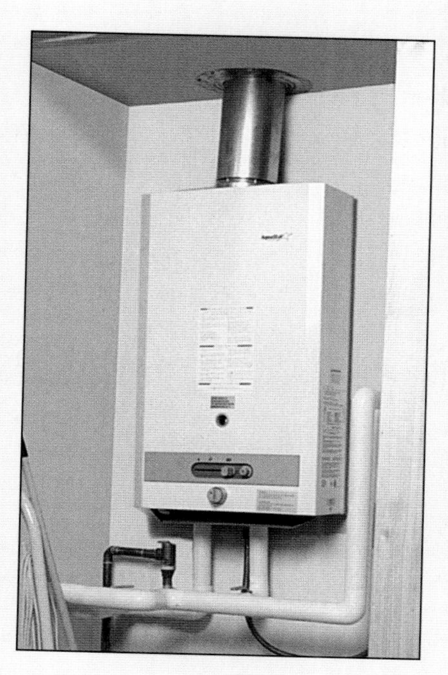

WHAT'S THE LIFE EXPECTANCY?

Because tankless water heaters are not constantly heating water, they may have a longer life expectancy than conventional water heater tanks. Conventional water heaters are constantly running and have life expectancies of 2 to 10 years. Tankless water heaters may have life expectancies of more than 20 years.

INSTALLING A GARBAGE DISPOSER

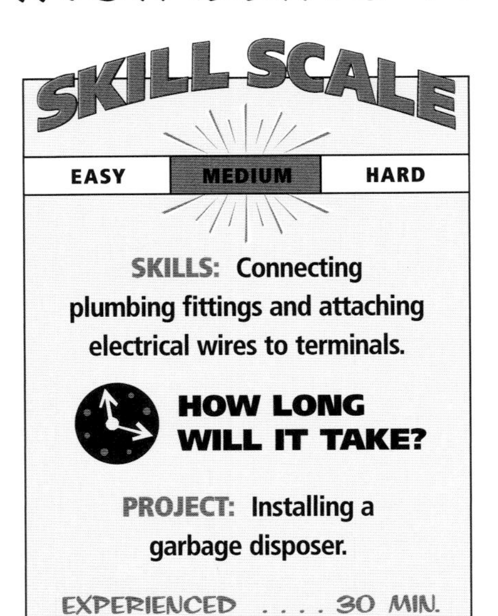

SKILLS: Connecting plumbing fittings and attaching electrical wires to terminals.

HOW LONG WILL IT TAKE?

PROJECT: Installing a garbage disposer.

EXPERIENCED .... 30 MIN.
HANDY .... 45 MIN.
NOVICE .... 1.5 HRS.

NOTE: Installation may vary depending on whether or not your drain line comes out of the wall or the floor. Know which you have and ask the sales associate about the right procedure.

A garbage disposer requires an electrical source for power. If you don't have one under the sink, you will need to install one. Check with local codes before installation. Some communities have codes that don't allow disposers because of limits on sewer capacity. They may also require an air gap for a disposer and a dishwasher.

If you have a septic system, install a disposer specifically designed for use with a septic tank. Too much food waste can interfere with the normal decomposition of septic waste.

HOW MUCH HORSEPOWER?

In-home disposers operating at less than ½ horsepower are not usually recommended for households of more than two people. A 1-horsepower disposer is an ideal choice for larger households and tougher jobs.

STUFF YOU'LL NEED

TOOLS: Screwdriver or disposer wrench, hacksaw or tubing cutters, water-pump pliers

MATERIALS: Garbage disposer, plumber's putty

SAVE THAT BRACKET. Are you replacing an existing garbage disposer with the same brand? You may be able to use the existing mounting bracket to make the job easier and quicker.

APPLIANCES

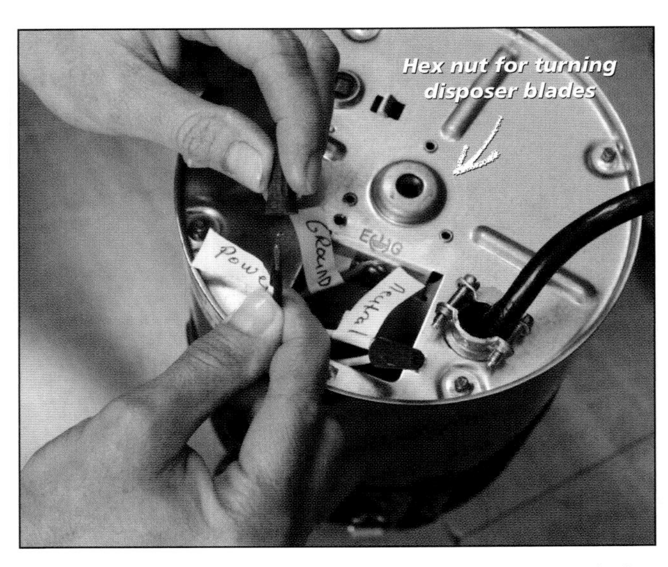

1 WIRE THE DISPOSER. The disposer may come with the appliance cord attached. If not, you will have to connect one. Remove the cover plate beneath the disposer. Most cords and disposers have the same colored wires. Connect white to white, black to black, and the green wire to the disposer's ground screw. If the colors are different, read the manufacturer's instructions for wiring.

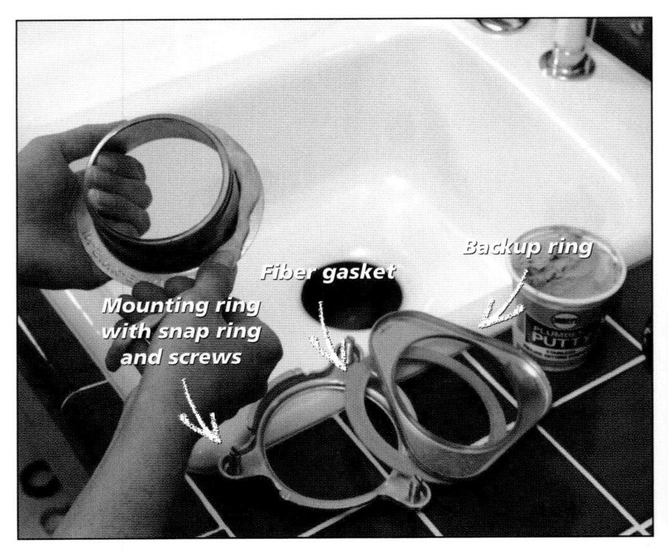

2 APPLY PLUMBER'S PUTTY. Press a rope of plumber's putty onto the underside of the drain flange. Insert the flange into the drain hole and press down evenly. Install the backup ring, fiber gasket, and mounting ring from beneath the sink.

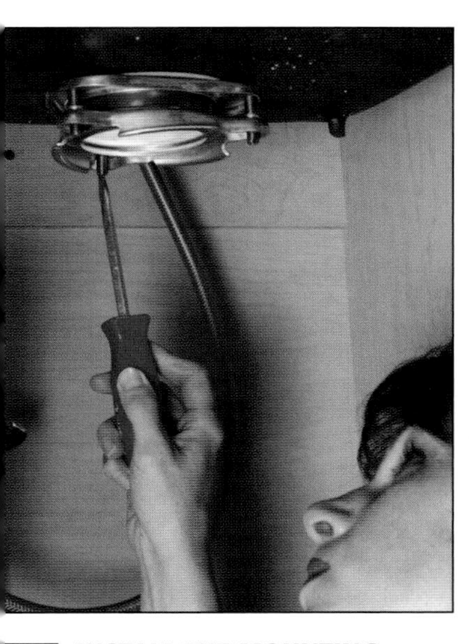

3 **INSTALL THE MOUNTING RING.** Tighten the mounting screws for the upper mounting ring. Alternate the tightening of the screws to pull the ring up evenly against the sink.

4 **MOUNT THE DISPOSER.** Place the disposer into the mounting ring. Make sure the outlet of the disposer is facing the drain pipe connection. Turn the lower ring clockwise until the disposer is supported by the mounting assembly.

5 **CONNECT THE OUTLET TO THE P-TRAP.** Use a hacksaw or tubing cutters to measure the discharge pipe and cut it to length. Install the discharge pipe to the outlet of the disposer. Attach to the drain line with slip nuts.

6 **TIGHTEN THE MOUNTING LUG.** Insert a screwdriver or disposer wrench into the mounting lug on the lower mounting ring. Turn clockwise until the disposer is locked into place. Tighten all slip nuts snug using water-pump pliers. Run water into the sink. Turn on the disposer and check for leaks. Tighten fittings if necessary.

CONNECT A DISHWASHER.

Be sure the disposer you purchase has a knockout for a dishwasher. Remove the knockout for the dishwasher connection on the disposer. Connect the dishwasher discharge line to the disposer using hose clamps. If local codes require an air gap between the dishwasher and the disposer, see page 159 on how to install one.

GETTING SOME LEVERAGE.

Sometimes the trickiest part of mounting the garbage disposer is the moment when you lift it up and lock it into the mounting rings. The pros will stack a couple of thick telephone books under the unit so they won't have so far to lift. You can do the same thing with scrap lumber or a toolbox.

INSTALLING A DISHWASHER

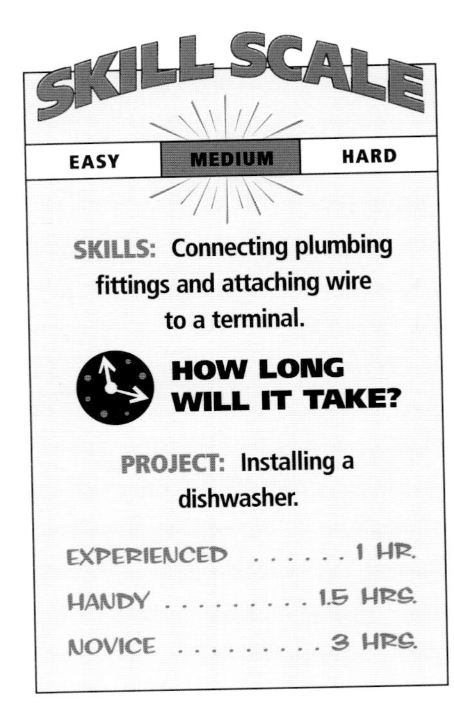

Plan ahead before installing a dishwasher. Measure the space to find out what size you should buy. Most dishwashers are 18 or 24 inches wide, which will hold about 12 place settings.

There are many options available with today's dishwashers. Dishwashers have light cycles for washing delicate or lightly soiled dishes, normal cycles for everyday usage, rinse cycles to remove food, and energy-saver cycles for drying.

Energy-conscious buyers should inspect the yellow Energy Guide label for the efficiency rating. The lower the number, the less energy the dishwasher will use over a one-year period.

✓ STUFF YOU'LL NEED

TOOLS: Power drill, hole saw, level, screwdriver, adjustable wrench, tubing cutter

MATERIALS: Wire nuts, hose clamps, drain hose, flexible copper tubing and compression fittings, drain tailpiece with inlet, air gap

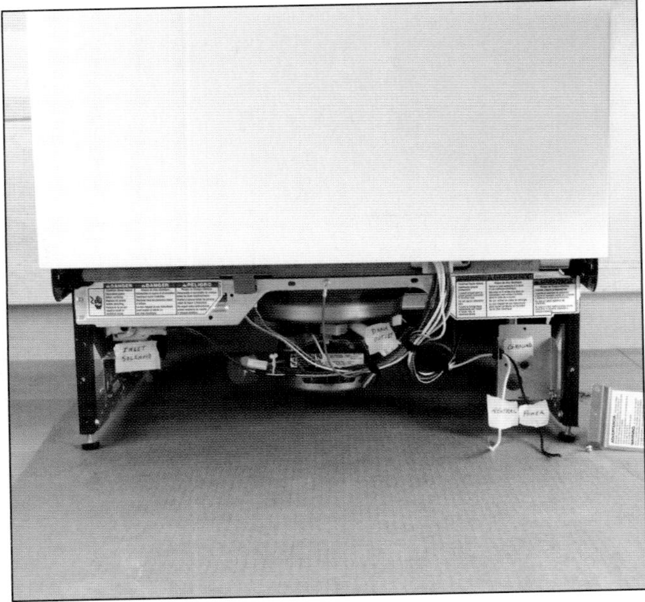

1 PREPARE THE DISHWASHER FOR INSTALLATION. Read the manufacturer's instructions for the location of the power supply and drainage lines—some will have separate holes for each line. Remove the bottom access panel to find the power cord connection, inlet solenoid valve, and drain outlet. Measure and mark the location to cut a 1½-inch hole in the wall of the sink cabinet. Slide the dishwasher into place and verify the location. Adjust the mark if necessary. Slide the dishwasher back out and set it aside to make more room to work beneath the cabinet.

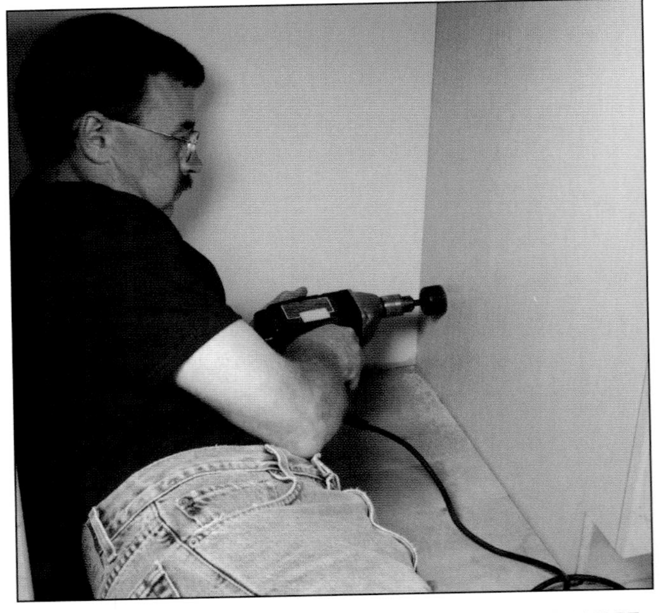

2 CUT THE HOLE FOR THE SUPPLY AND DISCHARGE LINES. Cut the hole in the cabinet wall with a power drill and hole saw.

3 **CONNECT THE LINES.** Run the drain tubing, supply line, and power cord. Slide the dishwasher into place. Level the dishwasher by adjusting the threaded feet. A good way to check for level is to open and close the door—it will operate smoothly if the dishwasher is level. When level, tighten the locknuts. The mounting brackets for the dishwasher should also be aligned with the underside of the countertop and the side of the cabinet. Install the compression elbow on the dishwasher solenoid. Bend the supply line so it lines up with the elbow. Slide the compression nut over the tubing, then place the compression ring over the end. Connect the tubing to the solenoid using an adjustable wrench. Install a tee on the hot water supply. Measure and cut tubing to the tee. Connect with compression fittings. Turn on the water. Inspect for leaks, then tighten connections if necessary.

HOW TO INSTALL AN AIR GAP.

Some local codes require an air gap between the dishwasher and the disposer. Mount the air gap in the countertop; if there is an extra hole available in the sink, mount it there. Connect a 5/8-inch drain hose to the 1/2-inch leg of the air gap with a hose clamp. Attach a 7/8-inch hose to the 3/4-inch leg of the air gap. Make sure there are no low spots or kinks in either hose.

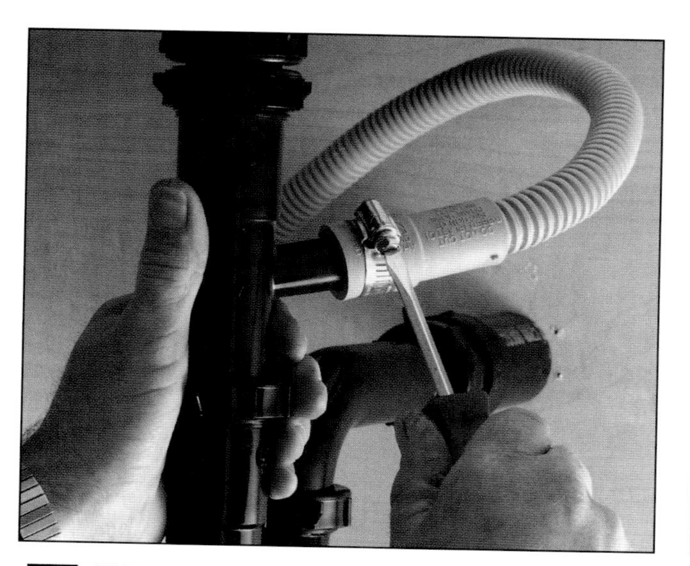

4 **INSTALL THE DRAIN LINE.** Replace the regular sink drain tailpiece with one that has a side inlet. Connect the drain line to the dishwasher outlet using hose clamps. Measure the hose and cut it to length so it connects to the inlet of the tailpiece. Connect the discharge hose to the inlet with another hose clamp. If local codes require an air gap, read the instructions to install one (see "Work Smarter" on this page).

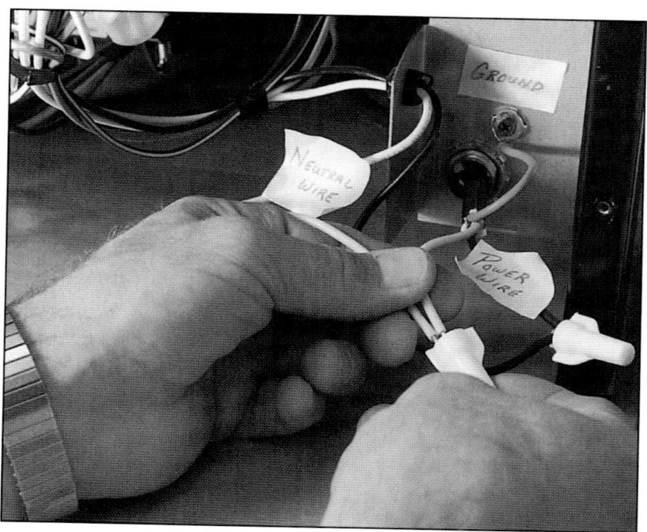

5 **HOOK UP THE POWER.** Match the wires: white to white, black to black, and connect the green to the ground screw. If the colors are different, read the manufacturer's instructions for connecting the wires. Connect the wires with wire nuts. Plug the power supply cord into an electrical outlet installed under the sink cabinet. Run the dishwasher through a test cycle to make sure it works properly.

INSTALLING A WATER PURIFIER

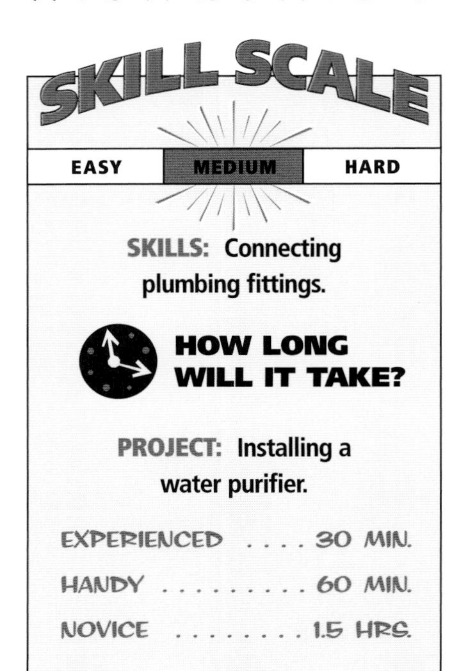

SKILLS: Connecting plumbing fittings.

HOW LONG WILL IT TAKE?

PROJECT: Installing a water purifier.

EXPERIENCED .... 30 MIN.
HANDY .... 60 MIN.
NOVICE .... 1.5 HRS.

Water purifiers remove impurities or sediment from tap water. There are two types of water purifiers: faucet-mounted and in-line. Faucet-mounted water purifiers are easy to install, but the filter has to be changed frequently. The in-line system may be more costly, but it is more comprehensive and requires only an occasional filter change. In-line systems are not difficult to install.

GET THE RIGHT FILTER

There are specific filters to do specific jobs. Some remove sediment, but most filter systems are carbon-based and are installed to improve the taste of the water while removing impurities. Talk to the sales associates at your home center or hardware store to find the one that will do the job for you.

STUFF YOU'LL NEED

TOOLS: Power drill and bits, screwdriver, two adjustable wrenches, tubing cutter

MATERIALS: Water purifier, compression fittings, bucket, wood screws

KEEP A SPARE NEARBY.
The filters need to be changed regularly in order for the unit to be effective. The kit usually comes with a special wrench to remove the cartridge. Hang that wrench near the unit so you don't have to hunt for it when it's time for a new filter.

GIVE YOURSELF SOME ROOM TO WORK!

If the space under your cabinet looks like mine, working under it is about as impossible as finding anything. Clear out the area so you have room to work. When you finish, take a second to organize under the sink. Install trays so it will be easier to find items. If you have young children, install childproof locks on the doors and keep poisonous or caustic items in another area.

1 **POSITION THE WATER PURIFIER.** Set the bracket so the bottom of the purifier hangs at least 3 inches above the cabinet floor. This will allow you to remove the cartridge easily. Predrill the holes, then attach the bracket to the wall with wood screws. You may find it easier to attach the bracket with the cartridge removed.

2 **TAP INTO THE WATER SUPPLY LINE.** Place a bucket under the water line. Turn off the shutoff valve for the line. Use a tubing cutter to cut into the supply line at least 3 inches above the shutoff. Drain the line into the bucket.

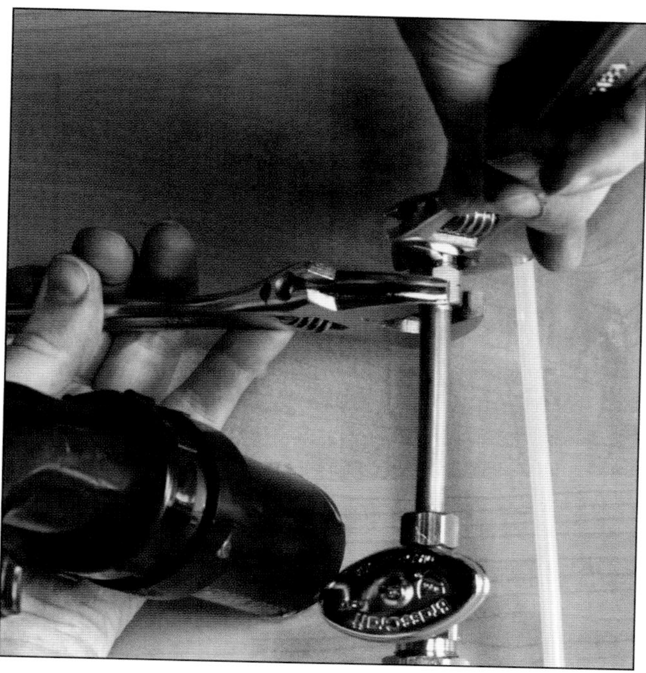

3 **INSTALL A COMPRESSION FITTING ON EACH END OF THE CUT LINE.** Insert the lines to and from the water purifier into the compression fitting. (See pages 140–141.) Tighten the compression nuts for the fittings using two adjustable wrenches.

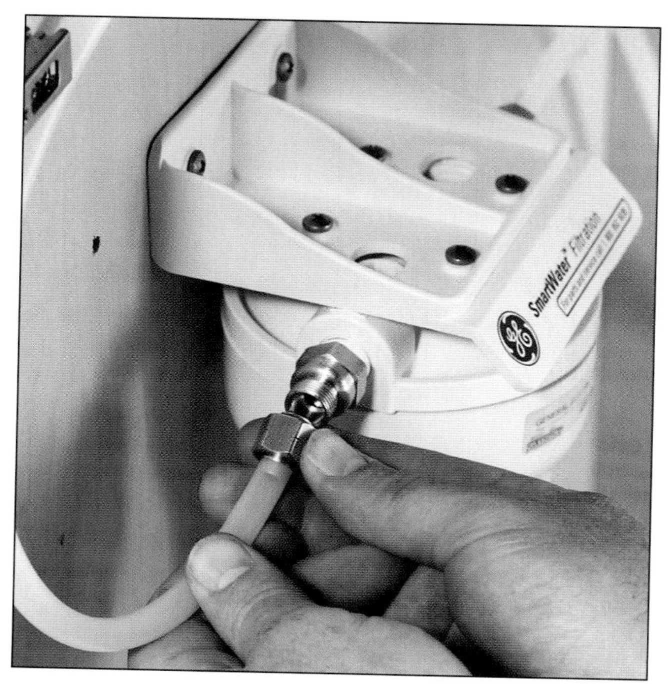

4 **CONNECT THE INLET AND OUTLET LINES TO THE WATER PURIFIER.** Slide the compression nuts over each line. Place compression rings over each end. Insert the lines into the water purifier and tighten the nuts using an adjustable wrench.

5 **FLUSH THE SYSTEM.** Install the cartridge and turn the water back on. While you're checking for leaks, flush the system until the water runs clear.

INSTALLING A HOT-WATER DISPENSER

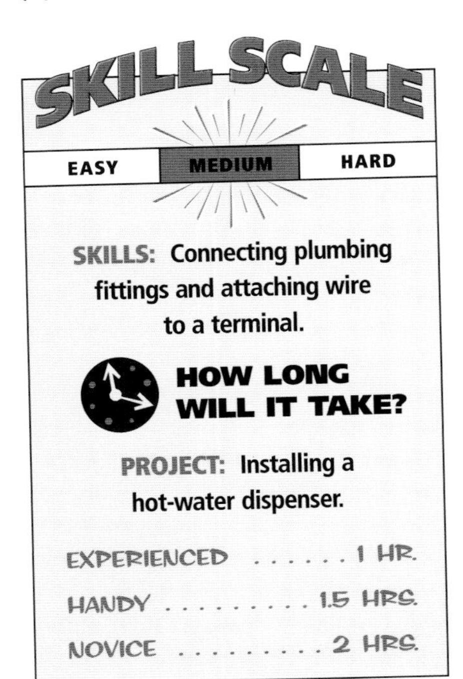

SKILL SCALE
EASY
MEDIUM
HARD
SKILLS: Connecting plumbing fittings and attaching wire to a terminal.
HOW LONG WILL IT TAKE?
PROJECT: Installing a hot-water dispenser.
EXPERIENCED .... 1 HR.
HANDY .... 1.5 HRS.
NOVICE .... 2 HRS.

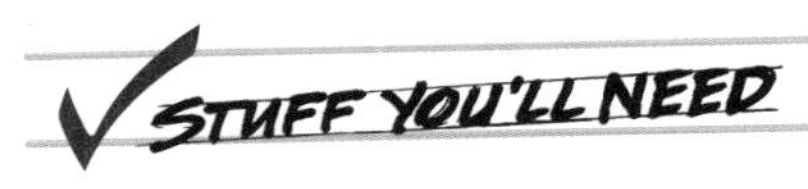

STUFF YOU'LL NEED

TOOLS: Power drill and bits, center punch, hammer, knockout punch, adjustable wrench, screwdriver

MATERIALS: Hot-water dispenser, saddle tee or compression tee valve, wood screws

Hot-water dispensers supply 88° C water so making soup, coffee, or tea is a snap. Instant hot-water dispensers are small electric water heaters that supply a single tap. The system connects directly to the cold water supply under the sink. Water is heated in the dispenser by an electric coil.

Check local codes before installing a hot-water dispenser. Some communities require the electrical outlet to have a ground fault circuit interrupter. Most codes allow the dispenser to be connected to the same outlet that supplies power to the disposer as long as the receptacle is split.

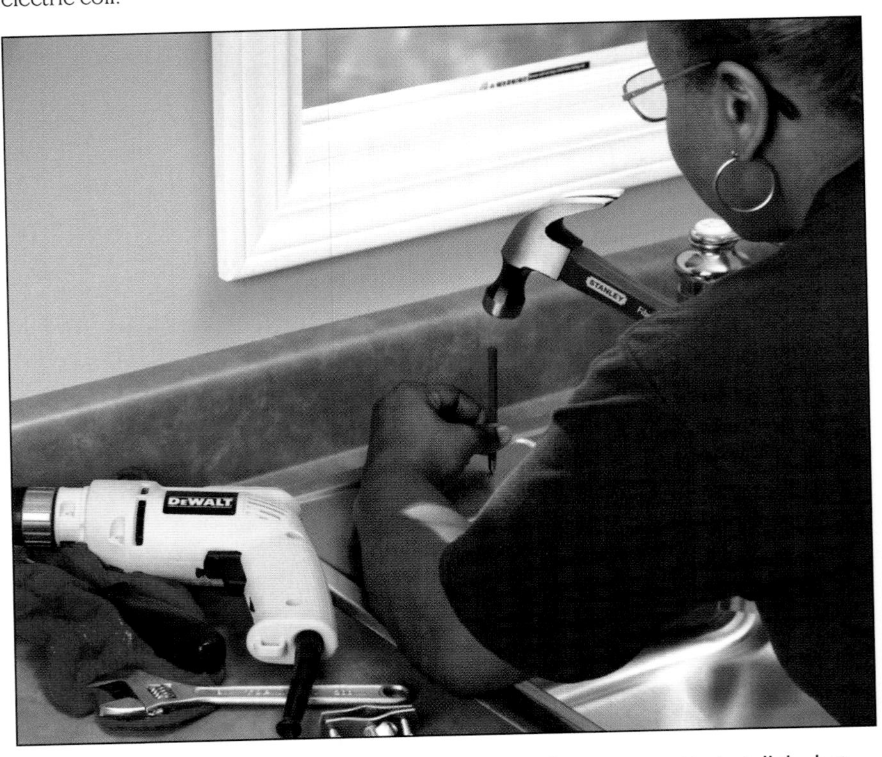

1 MARK THE DISPENSER LOCATION. Decide where you want to install the hot-water dispenser, making sure the area is flat and large enough. Measure the area for the center and mark with a center punch. Drill a hole at the mark.

BUYER'S GUIDE

SADDLE TEES AND LOCAL CODE

Even though they often come with the kits, saddle tees may not meet code requirements in your area. In that case you will have to install a compression tee valve on the supply line to get water to the hot-water dispenser.

WORK SMARTER

NOT ENOUGH HOLES IN THE SINK?

Install a one-touch faucet so one hole can be used for the hot-water dispenser. Plug the other hole or install a sprayer or drinking water dispenser.

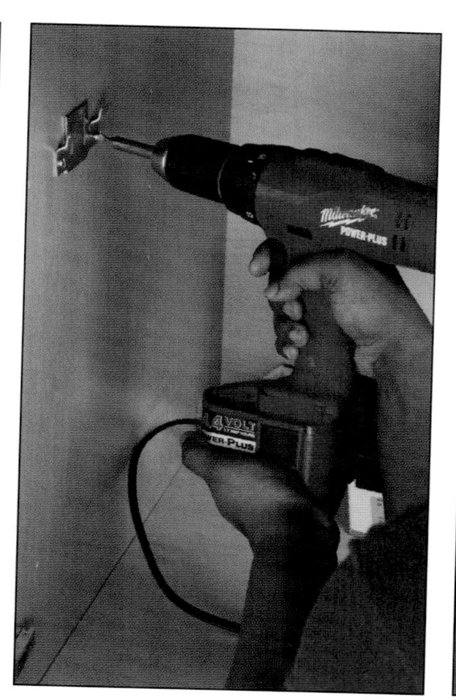

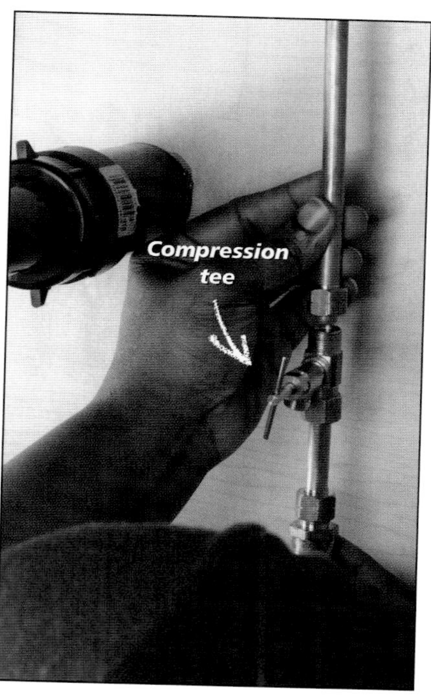

2 CUT A HOLE IN THE SINK. Connect the knockout punch through the hole. Tighten the bolt, using an adjustable wrench. Continue to tighten until the hole is cut. Remove the knockout, insert the hot-water dispenser spout through the hole, and connect.

3 MOUNT THE DISPENSER BRACKET. Set the bracket so the bottom of the dispenser hangs at least 3 inches above the cabinet floor. Predrill the holes, attach the bracket to the wall with wood screws, then attach the dispenser.

4 CONNECT TO THE WATER SUPPLY. INSTALL THE VALVE. The kit may come with a saddle tee valve but they don't always meet code so you should use a compression tee valve instead. (See "Buyer's Guide," page 162.) Installing the compression tee follows the same steps as installing a shutoff valve on pages 140-141.

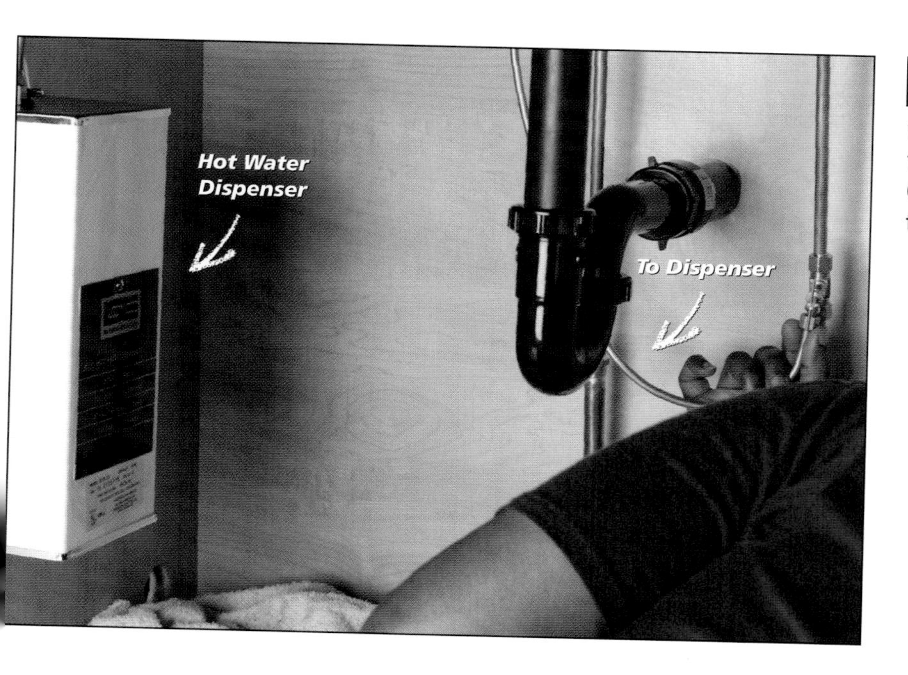

5 INSTALL THE TUBING. Connect the supply tubing from the supply valve to the dispenser. Then, connect the tubing from the dispenser to the spout. Open the valve and inspect for leaks. Plug the dispenser into an electrical outlet.

INSTALLING AN ICEMAKER

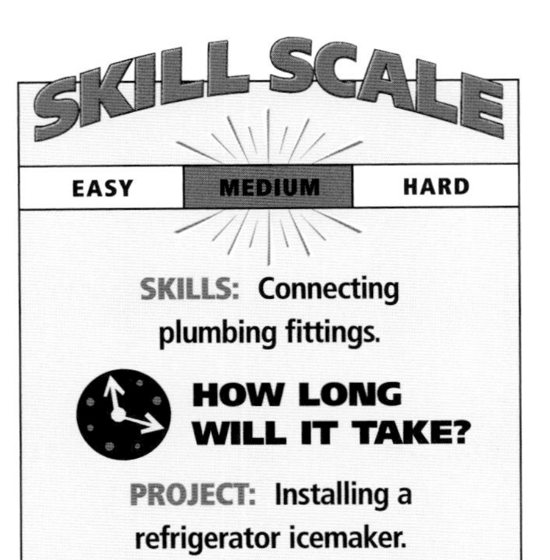

SKILLS: Connecting plumbing fittings.

HOW LONG WILL IT TAKE?

PROJECT: Installing a refrigerator icemaker.

EXPERIENCED .... 20 MIN.
HANDY .... 40 MIN.
NOVICE .... 1 HR.

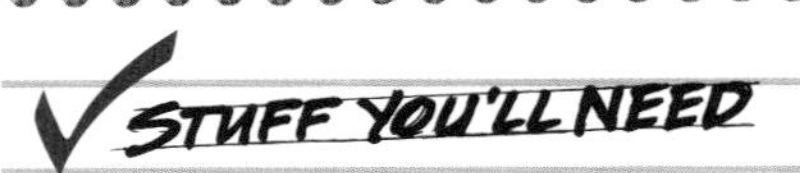

STUFF YOU'LL NEED

TOOLS: Power drill and bits, screwdriver, adjustable wrench

MATERIALS: Icemaker kit, saddle tee or compression tee valve

BUYER'S GUIDE

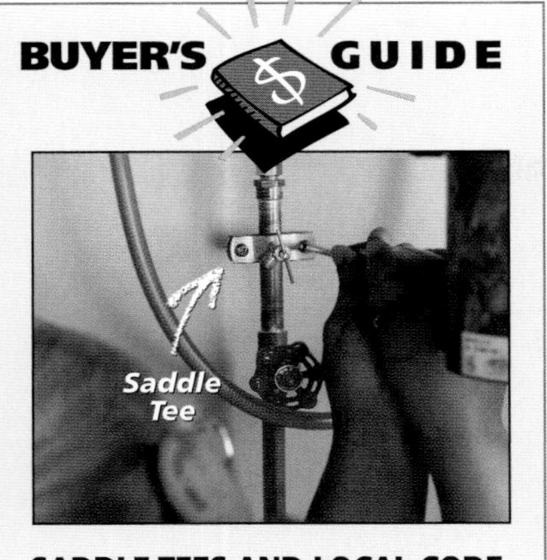

SADDLE TEES AND LOCAL CODE. Even though they often come with the kits, saddle tees may not meet code in your area. You may have to install a compression tee valve to meet local requirements.

If you want a new refrigerator that includes an icemaker you don't have to pay someone to hook it up. Installation is an easy do-it-yourself project.

The only thing you need to be careful about is installing the copper tubing. Uncoil only the length of tubing you need. Be careful not to kink the tubing, as you will then need to replace it. Leave enough coiled tubing behind the refrigerator to allow you to move it away from the wall without stressing or crimping the tubing.

GOOD IDEA

For pure, clear, ice, purchase an in-line water filter for your icemaker. Locate it in an accessible place so you can change the filter on the recommended schedule. Follow the instructions that come with the kit.

Water filter

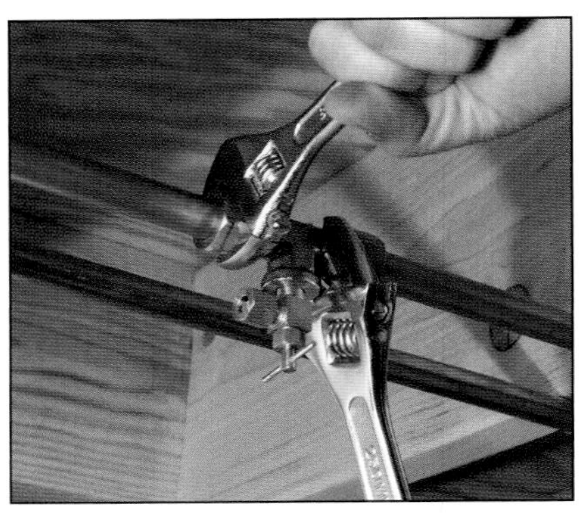

1 INSTALL THE VALVE. The kit may come with a saddle tee but they don't always meet code so you should use a compression tee instead. (See "Buyer's Guide" below left.) Installing the compression tee follows the same steps as installing a shutoff valve. (See pages 140–141.)

2 DRILL A HOLE IN THE FLOOR. The hole should be large enough for the tubing. Use a spade bit with a diameter of at least ⅜ inch and a power drill. From above, carefully uncoil the tubing and feed it through the hole. Feed enough tubing to run to the valve.

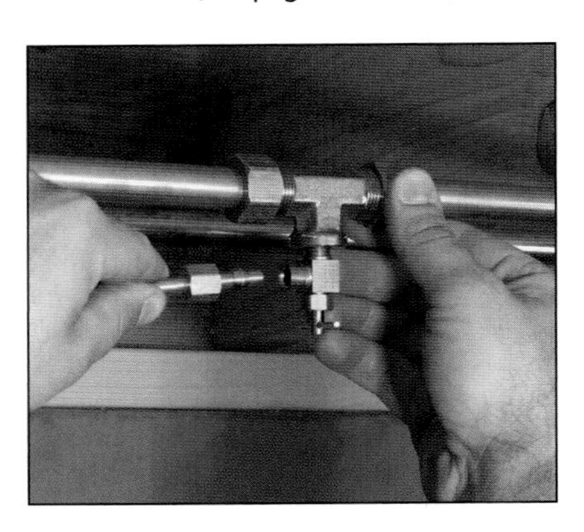

3 CONNECT THE TUBING TO THE VALVE. Insert the tubing into the valve and tighten the nut. Use an adjustable wrench to connect the compression fitting. (See pages 140–141 for connecting compression fittings.)

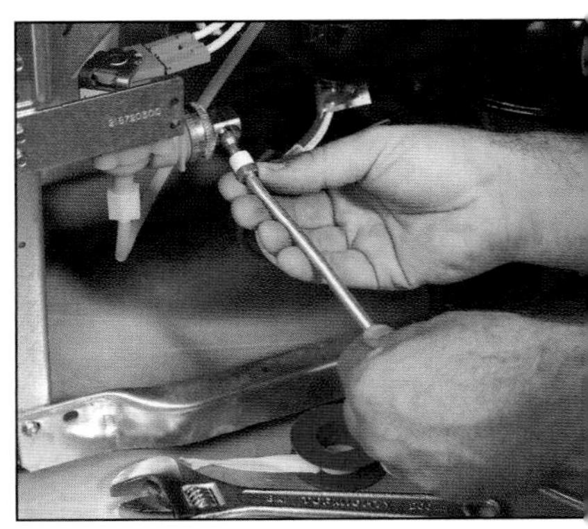

4 HOOK UP THE ICEMAKER. Remove the access panel at the back of the refrigerator. Unwind enough tubing to reach the supply valve; leave the rest coiled. Insert the tubing into the valve. Connect the fitting, turn on the water supply, and inspect for leaks.

WATER HEATERS

Water heaters are easy to forget about. They do their work in the dark corners of basements or supply closets and aren't given much thought until a rush of cold water ruins a morning shower. Even with proper maintenance, your water heater will eventually need to be replaced. Here are some considerations when looking for a replacement:

GAS OR ELECTRIC?

Either a gas or electric water heater will supply your hot-water needs efficiently. Choosing one over the other is a matter of what kind of service is available to your home and what makes the most economic sense. Make a list of the different appliances and fixtures in your home that are supplied with hot water, then talk it over with a salesperson at your local home center.

- If you are adding a dishwasher, jetted tub, or other new appliance, you may need a heater with a larger capacity.
- Make sure your electrical system can handle the wattage requirements of your heater.

INSULATION IS A FACTOR

Insulation is installed to maintain water temperature inside the water heater, reducing energy costs. Choose a water heater insulated with polyfoam over fiberglass batting. Thicknesses vary from 2.54 to 7.62 cm in the higher-efficiency water heaters. Insulation type and thickness are key factors determining the length of the warranty.

DON'T IGNORE THE WARRANTY

Warranties vary in length. Six to 12 years on the tank is the usual life span. Check the limitations of the warranties of different manufacturers. The life of the water heater will depend on environmental factors; hard water shortens its life.

THE ENERGY FACTOR

Water heaters bear an Energy Guide sticker. The sticker will display the units of kilowatt-per-year usage for electric models, therms-per-year usage for natural gas, gallons per year for propane, and first-hour rating.

Rate of recovery is not listed, but should be considered when purchasing a water heater, because it determines the amount of rise in temperature per liter of output (how quickly it heats the water).

SIZE MATTERS—ELECTRICITY

A family of up to four in a home with two full bathrooms, a clothes washer, and a dishwasher should have a heater with at least a 175-liter capacity.

SIZE MATTERS—GAS

A family of up to four in a home with two full bathrooms, a clothes washer, and a dishwasher needs a 189-liter tank with a 45,000-BTU burner unit.

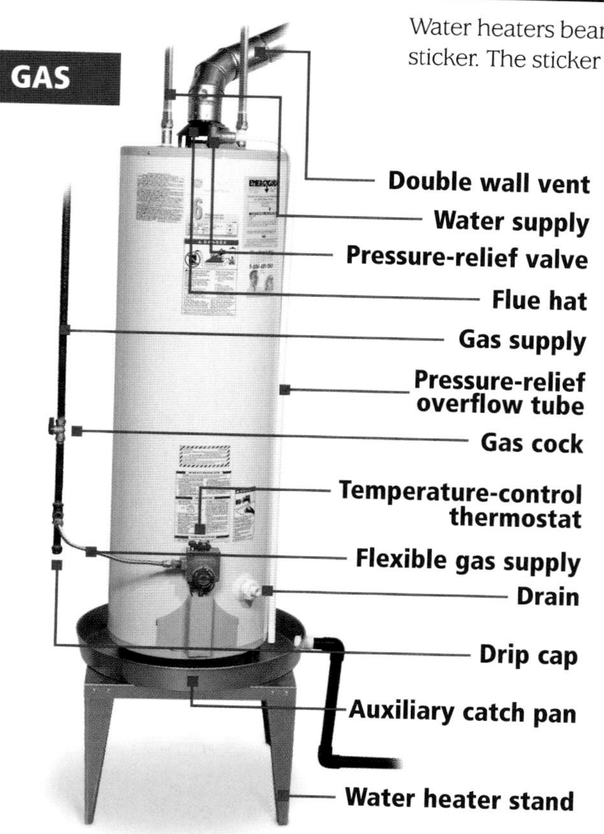

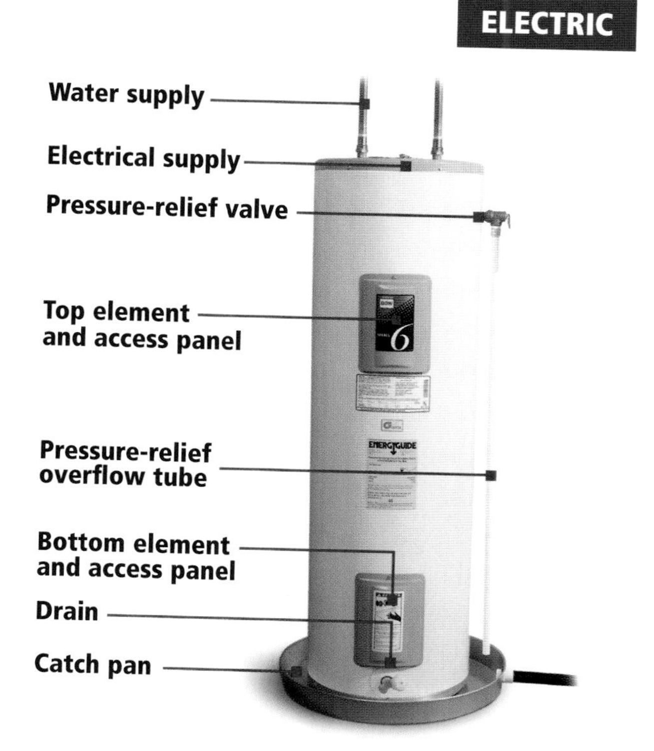

INSTALLING AN ELECTRIC WATER HEATER

SKILLS: Connecting plumbing fittings, stripping wire, and attaching wire to a terminal.

HOW LONG WILL IT TAKE?

PROJECT: Installing an electric water heater.

EXPERIENCED . . . . 45 MIN.
HANDY . . . . 1 HR.
NOVICE . . . . 2 HRS.

STUFF YOU'LL NEED

TOOLS: Carpenter's level, propane torch, two adjustable wrenches, screwdriver, hacksaw or tubing cutter

MATERIALS: Electric water heater, acid-free flux, shims, Teflon tape or pipe compound, wire nuts, water heater heat trap fittings, masking tape, 2x4 support

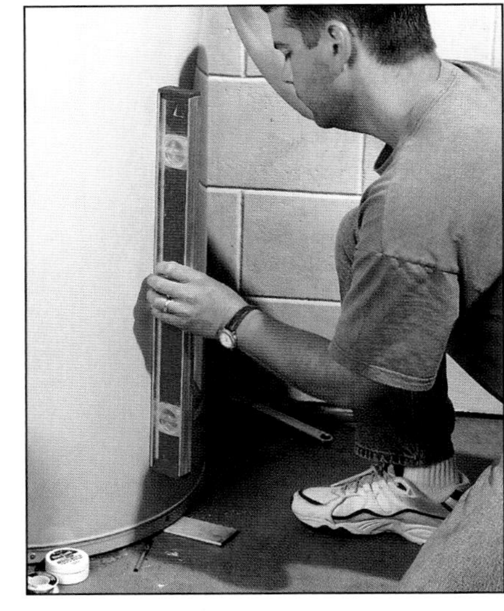

1 SET THE WATER HEATER IN PLACE. Use a carpenter's level along the side of the water heater to level it with wooden or plastic shims.

Hauling out an old water heater is usually the hardest part of the installation process.

WATER AND ELECTRICITY ARE A DANGEROUS MIX.
The electricity should always be the first thing off and the last thing on.

REMOVING THE OLD WATER HEATER: IT'S AS EASY AS 1-2-3

1 Turn off the electrical power to the water heater at the main circuit breaker; if you have fuses, remove them. Remove the access panel to the thermostat. Use a voltage tester across the terminal connections to make sure there is no power to the water heater. Turn off the water supply at the main valve.

2 Connect a garden hose to the tank discharge valve; place the end of the hose in the floor drain or a bucket. Open the valve and drain the water out of the tank and supply lines. Remove the hose. Disconnect the piping. If the piping has been soldered into place, use a hacksaw or tubing cutter to cut the pipe. Be sure to make straight cuts.

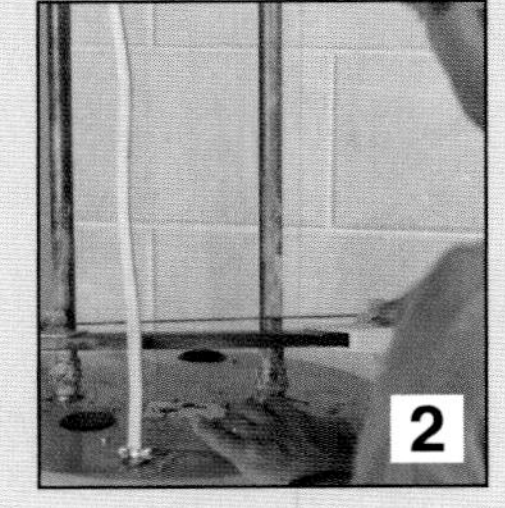

3 Remove the electrical supply access plate at the top of the water heater. Disconnect the wires one at a time; label the supply wires as you go. Get some help removing the water heater. Sediment builds up in the body over time, so it could be even heavier than the new one.

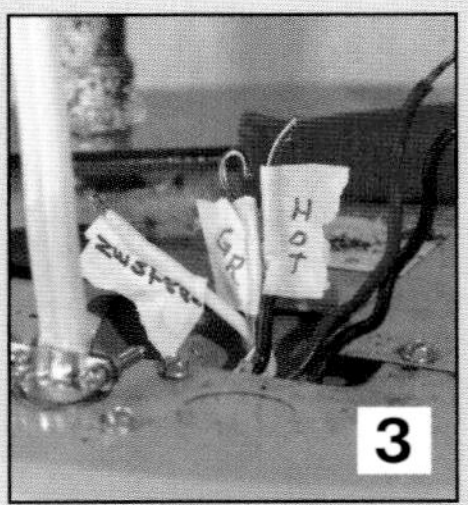

Cut a 2x4 to support the shutoff valve while soldering

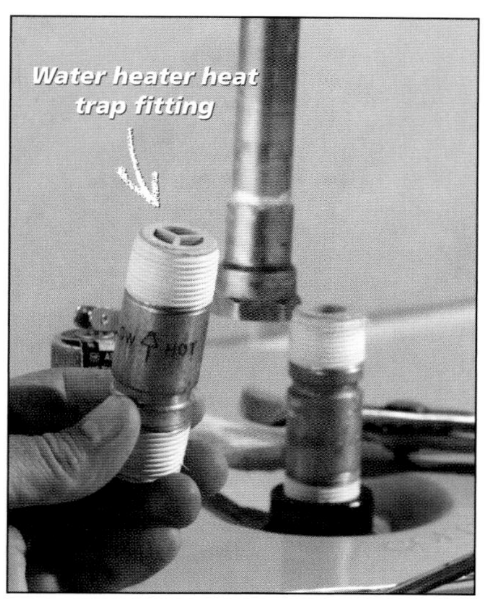

2 **SWEAT SOLDER THE SHUTOFF VALVE TO THE END OF THE COLD WATER SUPPLY PIPE.** Use a propane torch and lead-free solder to connect the valve to the supply line. The valve must be in the open position when you solder. (See pages 86–88.)

3 **WRAP THE THREADS OF THE WATER HEATER HEAT TRAP FITTING WITH TEFLON TAPE OR COAT WITH PIPE COMPOUND.** Attach the blue-coded fitting to the cold water inlet and the red-coded fitting to the hot-water outlet. These fittings are directional; they have arrows that show the proper installation. Tighten with two adjustable wrenches.

4 **INSTALL THE WATER SUPPLY.** Use two adjustable wrenches to connect the pipe from the pipe run to the water heater. Turn the main shutoff on and open all line valves. Open all the faucets in the house and run the water until it flows steadily from the faucets. Close the faucets.

SAVE AT THE METER.
The water heater heat trap fittings in step 3 may have to be purchased separately, but they are highly recommended to conserve energy.

The tank must be full of water before you restore power. The elements will burn out in seconds when exposed to air.

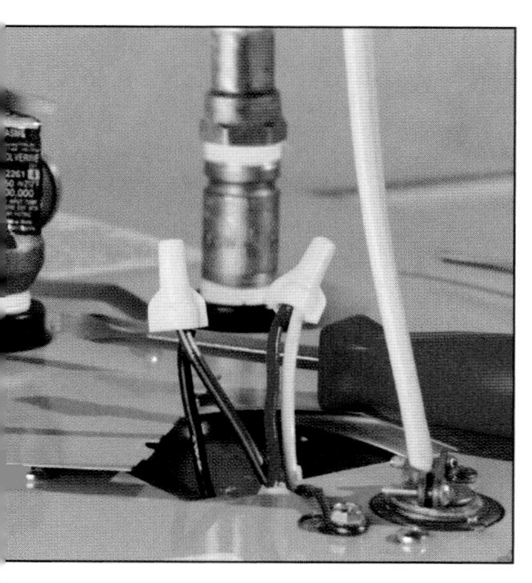

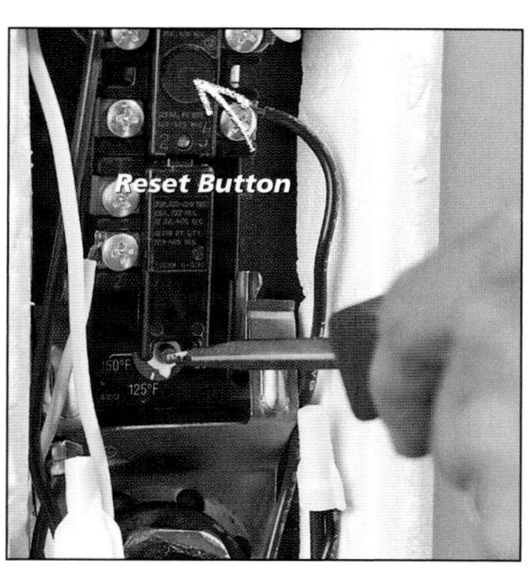

5 **REMOVE THE ELECTRICAL ACCESS PLATE.** **Always turn off the power to the unit before you do any electrical work.** Connect the electrical supply according to the manufacturer's instructions using wire nuts. Connect the bare copper or ground wire to the ground screw. Replace the electrical access plate. Remove the thermostat access plate.

6 **ADJUST THE THERMOSTAT.** Recommended settings are 48 to 51 degrees C. Now open a faucet near the heater, turn on the water supply, and fill the tank until the faucet is flowing. The tank must be full before you restore power. When the tank is full, restore power and press the reset button on the panel.

BUYER'S GUIDE

INSULATION IS THE KEY.

A new water heater is lined with rigid foam insulation to conserve heat so you won't need an insulating blanket to keep the water hot. The newest models also have LED lights wired to the upper and lower elements so you can see if the element is working.

INSTALLING A GAS WATER HEATER

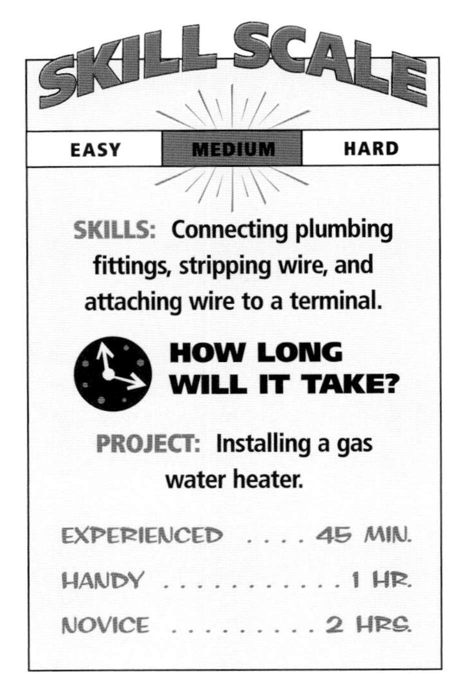

SKILL SCALE
EASY
MEDIUM
HARD
SKILLS: Connecting plumbing fittings, stripping wire, and attaching wire to a terminal.
HOW LONG WILL IT TAKE?
PROJECT: Installing a gas water heater.
EXPERIENCED .... 45 MIN.
HANDY .... 1 HR.
NOVICE .... 2 HRS.

Gas water heaters require a gas supply, a cold water supply, and a flue connection to carry away harmful gases. Today's homes tend to be airtight, so make sure you have adequate venting. Carbon monoxide—a byproduct of combustion—can be fatal when inadequate venting allows it to build up in the home. Call your local utility immediately if you suspect the water heater is not working properly.

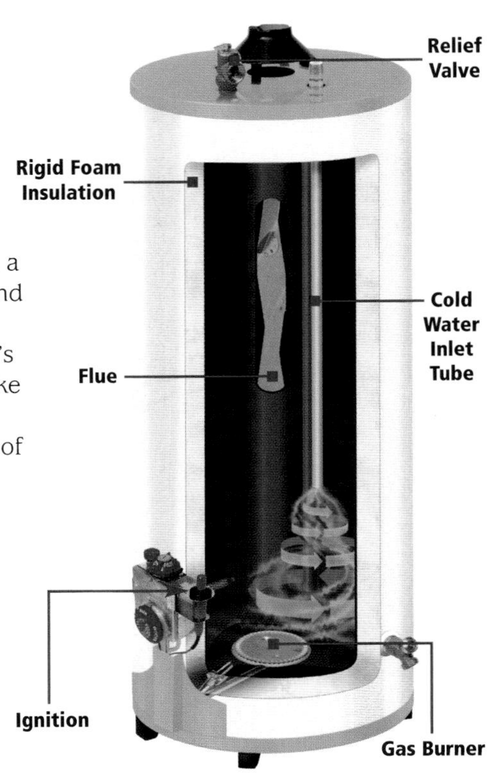

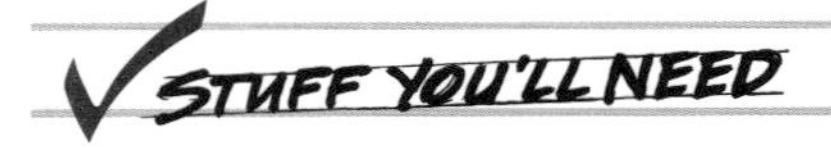

STUFF YOU'LL NEED

TOOLS: Continuity tester, carpenter's level, propane torch, two adjustable wrenches, two pipe wrenches, hacksaw or tubing cutter, screwdriver, garden hose, fireplace match or grill igniter

MATERIALS: Gas water heater, acid-free flux, plastic shims, Teflon tape, water heater heat trap fittings, rigid copper supply lines, liquid soap, sponge, bucket

Disconnect the supply wires one at a time and label each one so you can rewire without guessing.

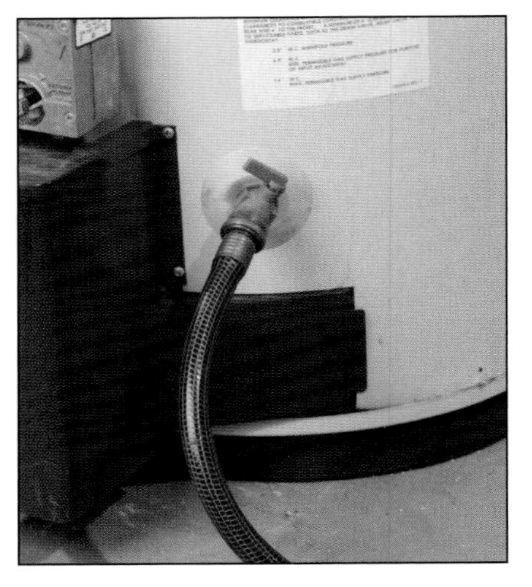

1 REMOVE THE OLD WATER HEATER. Turn off the water and gas supply valves. Turn off the power. Remove the electrical supply access plate at the top of the water heater. Check the connections with a continuity tester to make sure the power is off.

DRAIN THE TANK. Attach a garden hose to the drain valve and empty the tank.

DISCONNECT THE GAS LINE. Using two pipe wrenches, disconnect the gas line at the union fitting if the pipe is galvanized, or at the flare fitting if the gas supply line is copper.

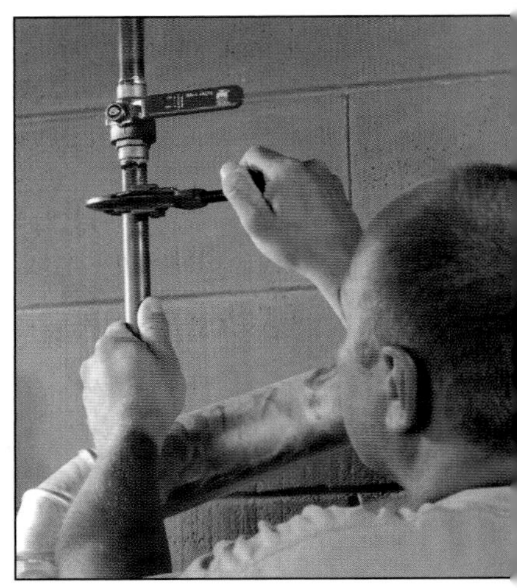

2 DISCONNECT THE WATER LINES. Using two adjustable wrenches or p wrenches, disconnect the piping above the tank. If the piping has been soldered into p use a hacksaw or tubing cutter to cut the p Make sure the cuts are straight.

SAVE AT THE METER.
The water heater heat trap fittings in step 5 may have to be purchased separately, but they are highly recommended to conserve energy.

3 DETACH THE GAS EXHAUST VENT FROM THE FLUE HAT. Remove the screws connecting the vent to the water heater using a screwdriver or nut driver. Save your back—get some help removing the old water heater. They fill with sediment, so the old one may be heavier than the new one.

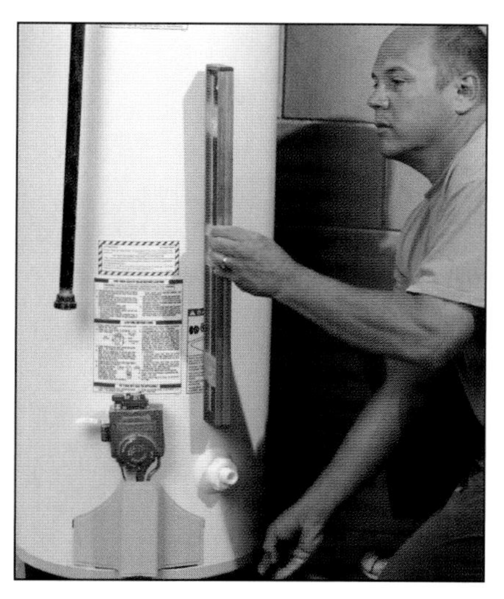

4 SET THE NEW WATER HEATER. Install the water heater in an area where it won't be cramped. There should be at least 6 inches of clearance around it for ventilation. Don't set it next to flammables. Turn the water heater so there is unobstructed access to the burner and controls. Place a carpenter's level on the side of the water heater and level it with plastic shims. Connect the flue hat to the gas exhaust vent. (See page 171.)

PILOTLESS IGNITION IS THE NEWEST THING.

The latest in water heater technology is a pilotless ignition system that requires no more than the push of a button to fire up the burner. It means no more searching for matches or the barbecue lighter to get the heater up and running.

5 WRAP THE WATER HEATER HEAT TRAP FITTING THREADS. Use Teflon tape on the pipe threads. These fittings are directional and must be installed properly. Both have arrows showing the correct direction for installation. Attach the blue-coded fitting to the cold water inlet with the arrow facing into the water heater. Attach the red fitting to the hot water outlet with the arrow pointing away from the water heater. Tighten using two pipe wrenches or adjustable pliers.

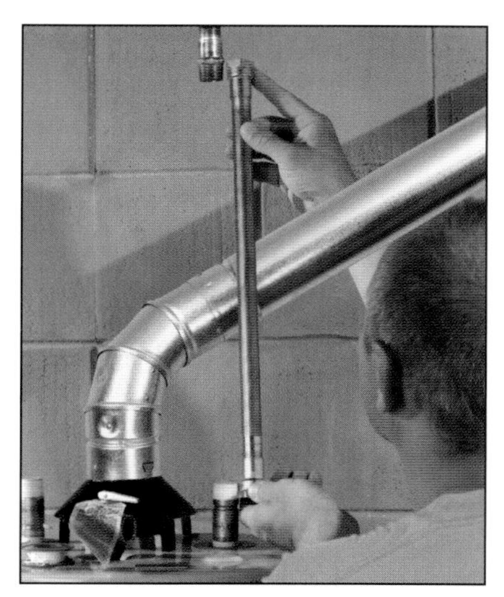

6 CONNECT THE PIPE LENGTHS. Measure and cut the water line connections to length. Precut lengths of rigid copper pipe are available at your local home center.

7 CONNECT THE GAS SUPPLY LINE. Clean all threads with a wire brush and rag. Apply piping compound to the threads of the galvanized pipes as you connect them. Assemble and tighten each fitting with two pipe wrenches. The union fitting should be the last fitting you install because it connects the new line to the existing line. Once finished, open the gas supply valve.

8 TEST THE GAS LINE FOR LEAKS. Fill a sponge with liquid dishwashing soap and water. Apply it to the new fitting and look for bubbles. It's the same process used for finding a leak in a car tire. If there's a leak, bubbles will form on the surface and you'll have to refit the joint. Test all connections.

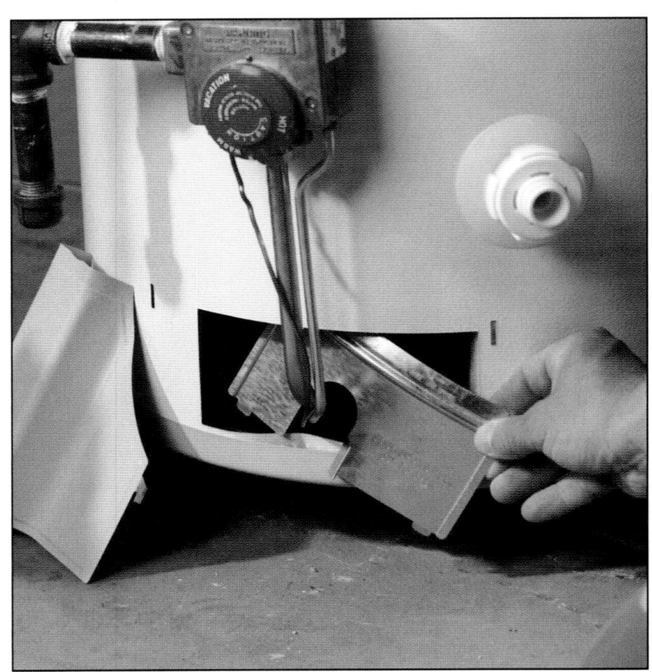

9 REMOVE THE BURNER ACCESS PANEL. This panel covers the burner chamber and prevents debris from entering.

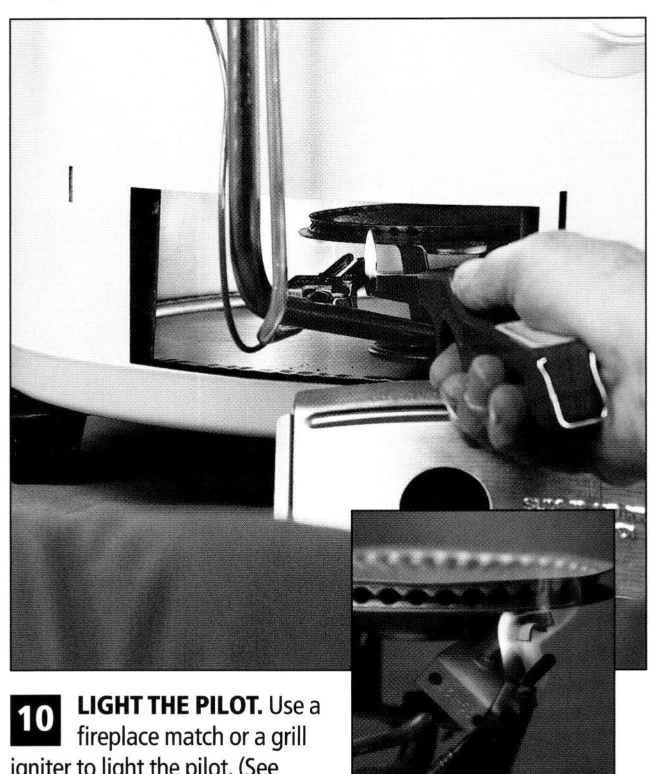

10 LIGHT THE PILOT. Use a fireplace match or a grill igniter to light the pilot. (See page 61.) Replace the burner access panel and set the control at 48 to 51 degrees C.

11 TURN THE CONTROL KNOB TO THE "ON" POSITION. You should hear the burner ignite. If it doesn't, remove the access panel and relight the pilot. Then adjust the thermostat to the desired setting

INSTALLING A FLUE HAT

SKILL SCALE
EASY
MEDIUM
HARD
SKILLS: Connecting ductwork.
HOW LONG WILL IT TAKE?
PROJECT: Installing a flue hat.
EXPERIENCED .... 30 MIN.
HANDY .... 45 MIN.
NOVICE .... 1 HR.

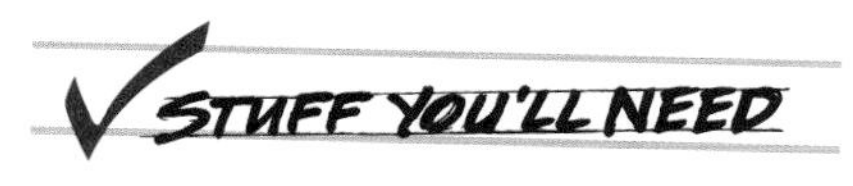

STUFF YOU'LL NEED

TOOLS: Tape measure, level, screwdriver

MATERIALS: Flue hat, vent pipe, sheet metal screws

The combustion of natural gas produces harmful fumes. Without a means of removing these gases, they would build up in the house. The flue hat serves as a funnel to collect these gases and carry them away from the house through metal duct pipe.

VENTS VARY

Vent duct for flues comes in three varieties: "A," which is single wall, "B," which is double wall, and "C," which is triple wall. Check local codes for the one that is approved in your area.

1 INSTALL THE VENT LINE. Measure, cut, and assemble the single-walled vent from the flue hat to the roof vent stack. Make sure horizontal sections have a slope of ¼ inch of rise for every foot of length to efficiently carry fumes away from the house.

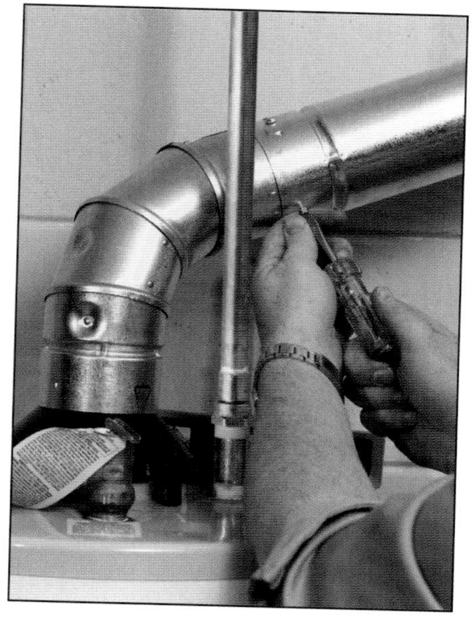

2 ATTACH THE DUCTWORK. Connect the ductwork by driving ⅜-inch sheet metal screws into the vent every 3 to 4 inches around the duct. You will need at least three screws per joint.

CO IS DANGEROUS.
Of the gases the combustion process produces, carbon monoxide is one of the most dangerous. Colorless and odorless, it is responsible for about 10,000 injuries a year. Symptoms of gas poisoning include headaches, fatigue, and nausea. Install a carbon monoxide detector and have gas appliances such as the furnace and water heater inspected once a year.

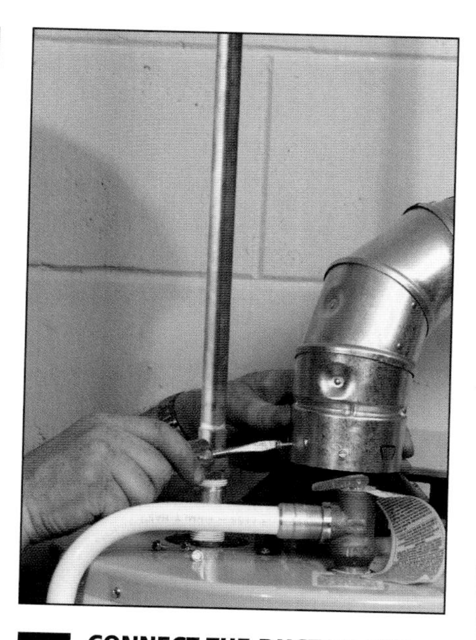

3 CONNECT THE DUCT TO THE FLUE HAT. Drive a ⅜-inch sheet metal screw through the duct connecting it to the flue hat. Some codes require copper pipe for the pressure-relief valve. (See page 172.)

INSTALLING A PRESSURE-RELIEF VALVE

SKILL SCALE
EASY
MEDIUM
HARD
SKILLS: Connecting plumbing fittings and CPVC pipe.
HOW LONG WILL IT TAKE?
PROJECT: Installing a relief valve and drain line.
EXPERIENCED .... 20 MIN.
HANDY .... 40 MIN.
NOVICE .... 1 HR.

When water is heated, pressure builds up in the tank. If this pressure is too great, a rupture could occur. A pressure-relief valve serves as a check valve to automatically open and relieve the pressure. Most water heaters now come with one, but if yours doesn't, you'll have to install one, and occasionally they need to be replaced. Consult your local home center to find the valve that will fit your water heater.

✓ STUFF YOU'LL NEED

TOOLS: Pipe or adjustable wrench

MATERIALS: Pressure-relief valve, Teflon tape or piping compound

1 PREPARE THE FITTING THREADS. Wrap the threads of the pressure-relief valve clockwise with Teflon tape or apply piping compound.

Never use more than two Teflon wraps; more will get in the way.

2 ATTACH THE VALVE TO THE WATER HEATER. Use a pipe wrench or an adjustable wrench to connect the relief valve to the opening in the water heater. Measure the distance from the relief valve to the floor. Cut lengths of CPVC pipe to match the ones you're replacing, as long as they extend to within 3 inches of the floor. Connect the pipe and elbow to the pressure-relief valve. (See pages 89–90.)

TAKE EXTRA PRESSURE OFF THE SYSTEM.

Water heater safety tanks provide relief from pressure that can build up in a contained system like your water heater. Pressure can damage fittings and fixtures, as well as shorten the life of appliances. Symptoms are water dripping from the pressure-relief valve, frequent dishwasher repairs or breakdowns, dripping faucets, periodic hot water pressure surges, and puddles of water at the base of the water heater. Rapidly heated water can also cause scalding.

Safety tanks are designed to be installed in-line on the outlet pipe of the water heater. Typically, they are attached at a location along the line with a tap saddle that supports the weight of the tank. You will have to provide support for the pipe to accommodate the extra weight.

CLOTHES WASHERS AND DRYERS

Washers and dryers are regulated by standards set by the U.S. Department of Energy and the U.S. Environmental Protection Agency. These standards define requirements for energy efficiency and water conservation. New washer designs use half the water of older models, while requiring 30 to 40 percent less energy.

TOP- OR FRONT-LOADING

Choose from top-loading and front-loading washers. Front loaders tend to perform better, are gentler on clothes, and use less water and energy than top loaders. But top loaders are easier to load and unload.

WHAT SIZE, PLEASE?

Your first decision will probably be about the size of drum you need. Sizes of washers and dryers are measured in cubic feet. Large families may want to have a 3-cubic-foot drum or larger. Capacities of 2.5 to 3 cubic feet will meet most needs. Extra-large dryers have capacities of 5 to 7 cubic feet. Smaller-capacity models are available for homes and apartments that have space limitations. The compact models may use less energy and require less space, but you'll have to do more loads of laundry to get the same amount of clothes done.

Stainless-steel drums won't chip or rust like porcelain-coated drums. A plastic drum won't either, and it costs less than the other two.

ACCESSORIES

Manufacturers offer a number of features. Some have automatic detergent dispensers and automatic temperature controls. Three spin and wash speed cycles are common, but you can choose as many as five.

People who are sensitive to detergents may want multiple-rinse cycle options such as warm rinses and extra-rinse cycles. Stackable units save space, and cost-conscious people will probably want a time-delay feature to take advantage of off-peak utility rates.

Moisture detectors in dryers help save energy. A sensor in the dryer detects the moisture level in the load and turns the dryer off when the clothes are dry. Some allow you to set for various degrees of dampness so you can remove clothes for ironing.

IF IN DOUBT, UPGRADE.
When I was installing our new washer and dryer in my old house, I took a quick peek at the electrical outlet and decided everything seemed in pretty good shape. There was even a ground wire, so I went ahead with the installation. It was a bad move because the washer drew more amps than the circuit breaker could handle and the gauge of the wire was too small to handle the current load. The result was that the breaker was constantly tripping and I had to upgrade both the wire and the breaker. Wiring in older homes is always suspect because electrical codes may not have been strictly followed. Also, deterioration of the wire and receptacle over time can create potential fire hazards, so you should make a habit of upgrading whenever you're unsure.

INSTALLING A WASHING MACHINE

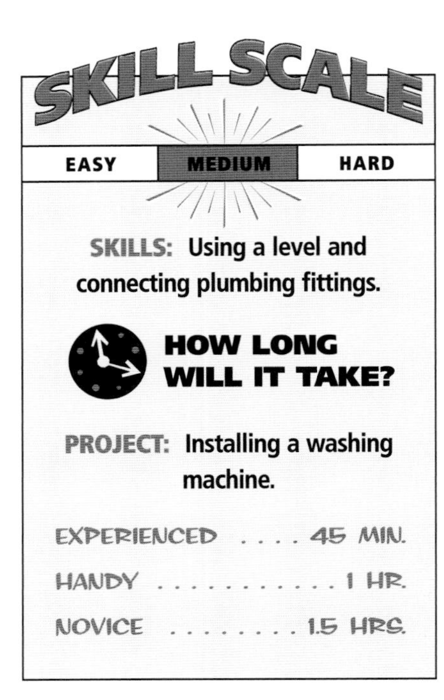

SKILL SCALE
EASY
MEDIUM
HARD
SKILLS: Using a level and connecting plumbing fittings.
HOW LONG WILL IT TAKE?
PROJECT: Installing a washing machine.
EXPERIENCED .... 45 MIN.
HANDY .... 1 HR.
NOVICE .... 1.5 HRS.

Washing machines are easy to install, especially if you're replacing the old unit with a new one. All you have to do is hook up the drain and supply lines, level it, plug it in, and wash a load of laundry. If you're installing in a new location, you'll have to run plumbing pipes and set up an electrical outlet.

Washing machines aren't particularly heavy, but they are awkward, bulky, and very likely to be installed in a basement or some other out-of-the-way corner. You'll want to have some helpers around or make sure the delivery people get the unit into its final position. You can rent a furniture truck or a wheeled dolly to make the work a little easier on your back.

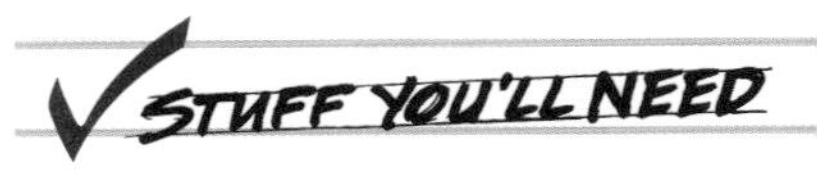

TOOLS: Carpenter's level, water-pump pliers, hacksaw

MATERIALS: Washing machine, washer hoses, ABS Cement, ABS cleaner

WALK THE WALK.
If you're installing a washing machine in a new location, make sure you've walked the delivery route and measured openings so there won't be any surprises when the unit is delivered.

1 CONNECT THE STANDPIPE. Install a standpipe to a P-trap connected to the drainage and vent system. (See pages 146–147.) The standpipe must have a greater diameter than the hose, and the top should be above the washer's water level to prevent overflow.

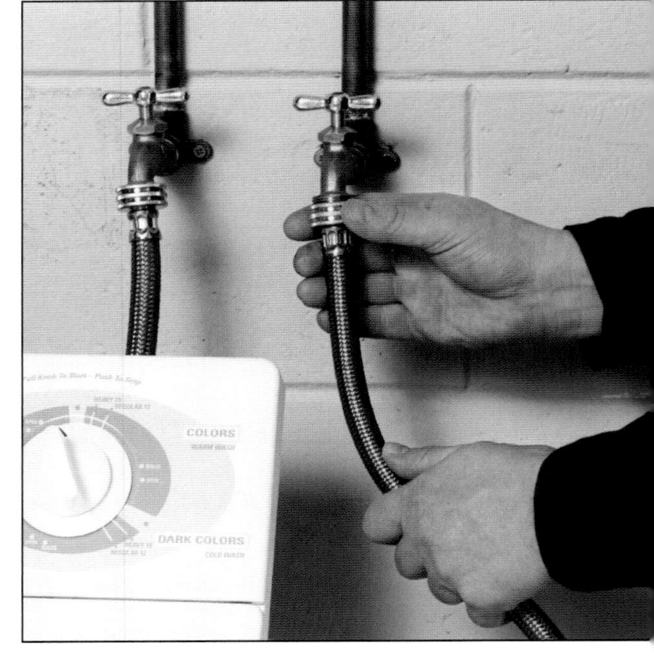

2 HOOK UP THE WASHER SUPPLY HOSES. Connect the hot and cold water lines using water-pump pliers. Turn on the water valves and run the clothes washer. If there are any leaks, tighten the connections.

3 LEVEL THE CLOTHES WASHER. Once the machine is in place, you'll want to make sure it's seated securely on the floor and is perfectly level to keep it from "walking" and banging loudly while you're doing the family wash. Place a carpenter's level or a bubble level on top of the washer. Level the unit by adjusting the legs and securing the locknuts against the frame.

Do all the work at the back of the machine (hooking up the waste line and attaching the hoses) before you set and level the washer.

INSTALL WATER HAMMER SHOCK ABSORBERS ON NOISY PIPES.

Turn off the hot- and cold-water supply valves. Disconnect the supply hoses from the valves with water-pump pliers. Connect water hammer shock absorbers to the valves and attach the supply hoses to the shock absorbers. To make your own shock absorbers, cut two 12-inch lengths of copper pipe, solder caps to one end, solder adapters with threads to the other end, and screw them onto the tees. Turn on the water supply valves and enjoy living with pipes that don't bang when water is turned on or off.

INSTALLING A DRYER VENT

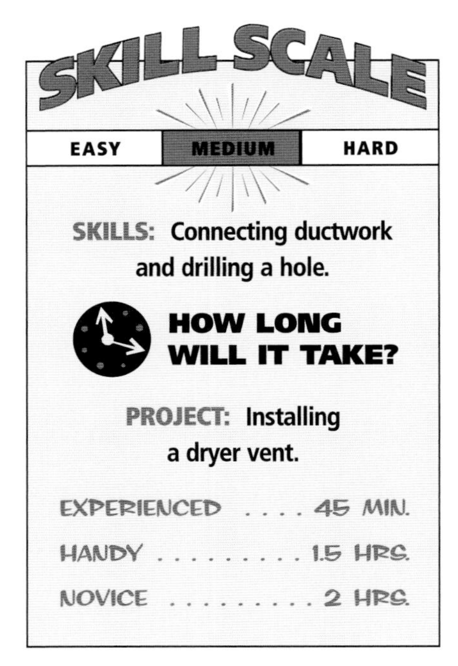

Forced heat, whether produced by natural gas, propane, or electricity, dries the clothes. It's important to remove this moist air from the house because it may be mixed with hazardous gases such as carbon monoxide, a byproduct of combustion. Vent pipe is normally made of 4-inch diameter, rigid sheet metal.

DON'T USE FLEXIBLE DUCTS

Flexible vinyl cannot be used because it doesn't support its weight and lint that collects in low spots presents a fire hazard. Seal joints with foil duct tape. Never use sheet metal screws; they will catch lint.

✓ STUFF YOU'LL NEED

TOOLS: Power drill, hole saw, tape measure, screwdriver, caulking gun, level

MATERIALS: Dryer vent, silicone caulk, wood screws, hanger straps, foil tape

Don't run a duct into a chimney, a crawlspace, or under a floor. The lint will build up and create a potential fire hazard.

1 CUT THE VENT HOOD OPENING. The hood will need to be where it's accessible from inside. Measure the location and find out where the hood will be from the inside. Translate it to its outside location. Drill a pilot hole and check inside to make sure the hole is in the right place. Plug and redrill if necessary. Use a 4¼-inch hole saw to cut the hood opening.

2 INSTALL THE VENT HOOD. Insert the duct pipe through the hole. Attach the hood to the siding with wood screws. Caulk around the edges of the hood to seal against the elements.

3 CONNECT TO THE HOOD DUCT. The location may require an elbow to the hood duct. You may have one or two elbows back-to-back to the run. Attach the duct lengths to the elbow.

4 SECURE THE DUCT. Attach straps to support the duct. Apply foil tape around the joints to seal them and help support the duct. Use a carpenter's level and set horizontal sections with a fall of ¼ inch per foot to prevent moisture from collecting.

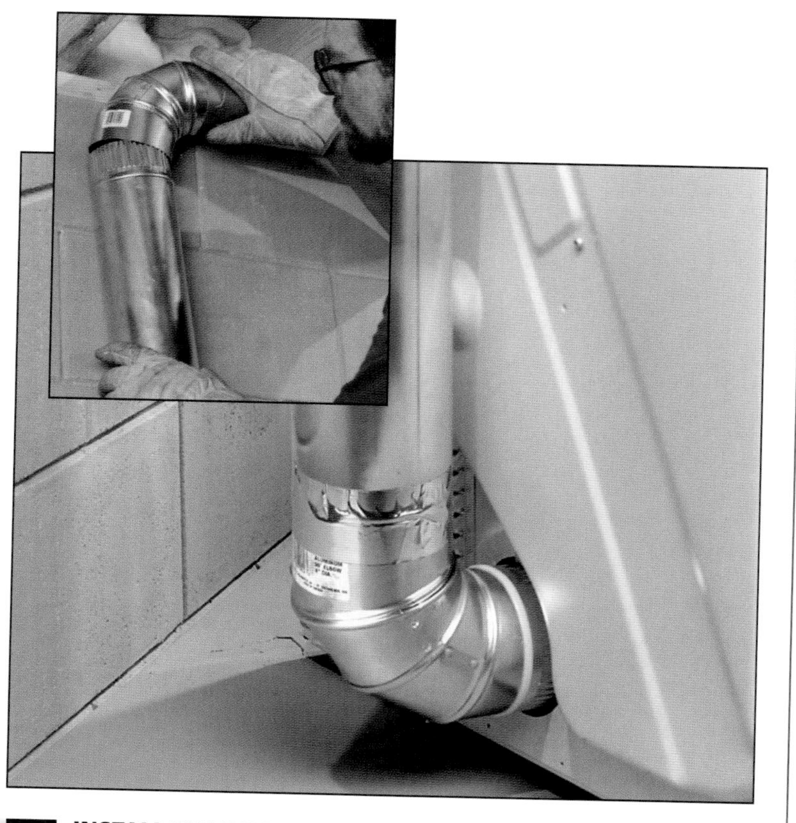

5 INSTALL THE DUCT INTO THE DRYER VENT OUTLET. Insert an elbow over the dryer outlet. Connect the duct pipe to the elbow. (See inset.) Slide the dryer back into place. Level using a carpenter's or bubble level. Adjust the legs and lock into place by tightening the locknut against the dryer.

There are indoor dryer vents on the market that may work as effectively. They're worth checking out.

WHY YOU CAN'T USE SCREWS!

Sheet metal screws are not approved by code for use in areas where dust or lint collects, so cannot be used to connect the duct pipe and fittings. The ends of the screws will stick into the duct and catch lint. The lint will build up over time and block the flow of harmful gases. Without anywhere to go, gases such as carbon monoxide will vent into the house. The back pressure created by the blockage can also shorten the life of the motor.

CHAPTER 11 OUTDOOR SYSTEMS

The earth may be three-quarters water, but it still seems the wet stuff is never exactly where you want it. If you're tired of dragging a hose and sprinkler around to water the far corners of the yard or mopping up standing water from seepage in the basement, it may be time to consider some outdoor plumbing solutions.

Hose bibs installed in more convenient locations will make general watering easier. A more efficient solution is to install underground irrigation or perhaps a drip irrigation system. Deal with soggy basements by putting in a sump pump to remove water and keep the rec room dry.

Indoors or out, the same plumbing techniques apply—good planning, proper tool use, and following code.

CHAPTER 11 PROJECTS

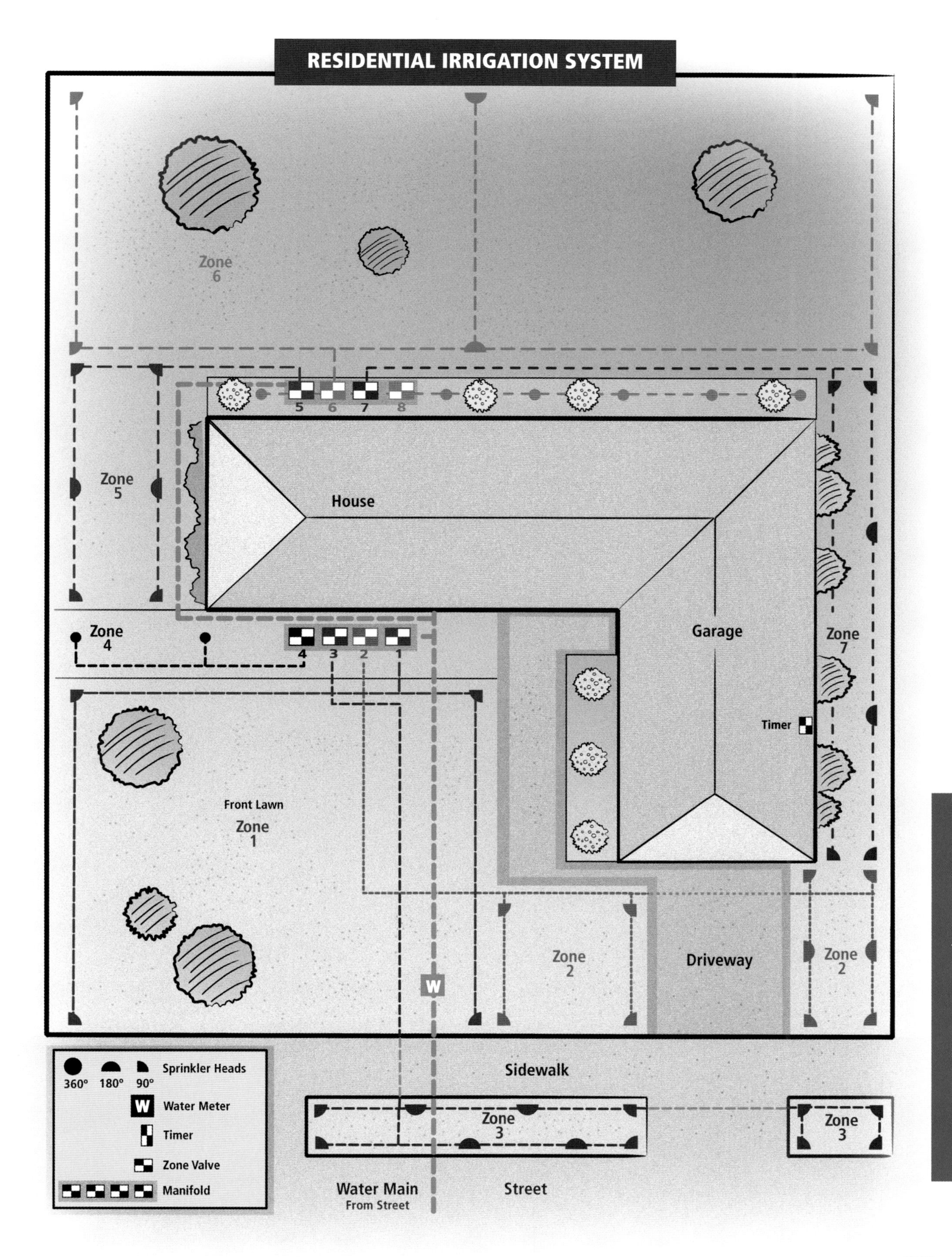
RESIDENTIAL IRRIGATION SYSTEM
Zone 6
Zone 5
5 6 7 8
House
Zone 4
4 3 2 1
Garage
Zone 7
Timer
Front Lawn
Zone 1
Zone 2
Driveway
Zone 2
W
Sprinkler Heads
360° 180° 90°
Water Meter
Timer
Zone Valve
Manifold
Sidewalk
Zone 3
Zone 3
Water Main
From Street
Street

INSTALLING AN IRRIGATION SYSTEM

EASY | MEDIUM | **HARD**

SKILLS: Connecting plumbing fittings, digging, stripping wire, and attaching wire to a terminal.

HOW LONG WILL IT TAKE?

PROJECT: Installing an irrigation system.

INSTALLATION TIME WILL VARY DEPENDING ON THE COMPLEXITY OF THE SYSTEM AND HOW MUCH YOU CHOOSE TO INSTALL YOURSELF.

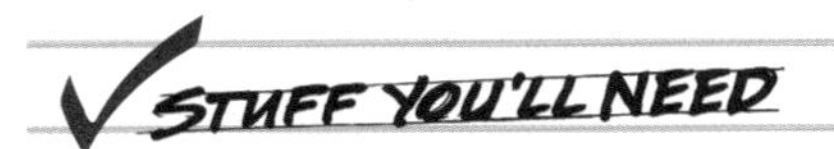

STUFF YOU'LL NEED

TOOLS: Water-pressure gauge, 5-gallon bucket, garden hose, stopwatch, 50-foot tape measure, plastic tubing cutter, shovel or trenching tool, level, keyhole saw, copper tubing cutter, 4-in-1 tool, propane torch, spark striker, adjustable wrench, pipe wrench, caulking gun, gloves, flags

MATERIALS: Graph paper, pencil, compass, ruler, copper pipe and fittings, lead-free flux, lead-free solder, silicone sealant, PVC, or Poly pipe and fittings, primer and cement (if required), valve manifolds and manifold box, sprinkler heads, sprinkler timer, rag, Teflon tape, flour or spray paint

Manufacturers have simplified the installation of underground irrigation systems to the point that all you need are a good plan that meets local code requirements, basic plumbing skills, and a little elbow grease. Whether you are adding on to an existing system or starting from scratch, there are many options for installation. This project is designed to be hooked up to a city water supply using PVC pipe, but systems can also draw service from private wells and lakes and many installations will use polyethylene pipe and fittings for the service. Check local codes.

Consult your home center to get help with the layout and for recommendations for a system that will work for you.

Some areas require the use of "reclaimed" water for irrigation systems. It is not potable and requires specially coded pipes and fittings.

Homeowners install irrigation systems to get lush, green lawns.

1 CHECK THE WATER PRESSURE AND THE SIZE OF THE WATER METER. Screw a pressure gauge onto an outside faucet. Turn off all inside and outside faucets. Fully open the valve with the pressure gauge. Read the gauge in pounds-per-square-inch (PSI) units of measure and record it.

2 DETERMINE THE WATER CAPACITY. Place a 5-gallon bucket near an outside faucet. Turn off all faucets inside and outside. Turn on the faucet, and use a stopwatch to determine how long it takes to fill the bucket. Calculate the capacity. Divide the number of gallons (5) by the number of seconds needed to fill the bucket, then divide the result by 60 seconds. Round down to the nearest whole number to calculate the gallons per minute (GPM).

WARNING! Cutting through buried utility and electric lines can cost you time and money! Call your local utility company and have them locate any buried pipes or power lines before you turn the trencher loose.

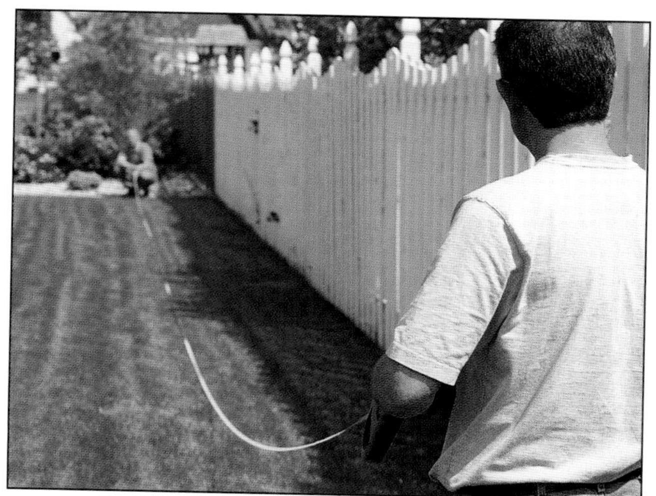

3 MEASURE THE PROPERTY. Find a fixed starting point such as a corner of a fence or house so you can tie all the measurements together. First measure the size of the lot and record it. Outline the house, garage, and any structures on the property. Locate walks, drives, slabs, decks, and surface areas. Record tree and shrub locations. Identify areas of groundcover, grass, flowerbeds, and landscaping. Note the location of the water meter.

4 SKETCH OUT THE PROPERTY. Use graph paper and set a scale of one square for 1 foot or one square for 10 feet. Show all structures. Locate the manifold position. Place sprinkler heads along the perimeter and work toward the center. Don't mix sprinkler sizes within zones. When placing stationary heads, use one-fourth sprays in the corners, one-half sprays along the edges, and full sprays for interior areas. Avoid setting sprinklers that spray the side of the house, walls, fences, drives, sidewalks, and into the street.

5 DETERMINE COVERAGE. Use a compass to draw the areas of coverage for each sprinkler head. Overlap areas to compensate for wind and pressure fluctuations. Unless you have a very small yard or unlimited water pressure, break up the system into zones. A zone is a group of similar-type sprinklers such as standard impact that operate together and are supplied by one common valve. (See "Select the Right Sprinkler," page 183.) Don't mix different types of sprinklers in the same zone. A zone's capacity must not exceed that of the water supply system. Record the GPM usage of each sprinkler head. The total is the GPM for each zone. If the total flow exceeds the water capacity, split the zone. Zones are also split based on area of coverage, slope, sun, and shade. The number of zones equals the number of manifold valves required.

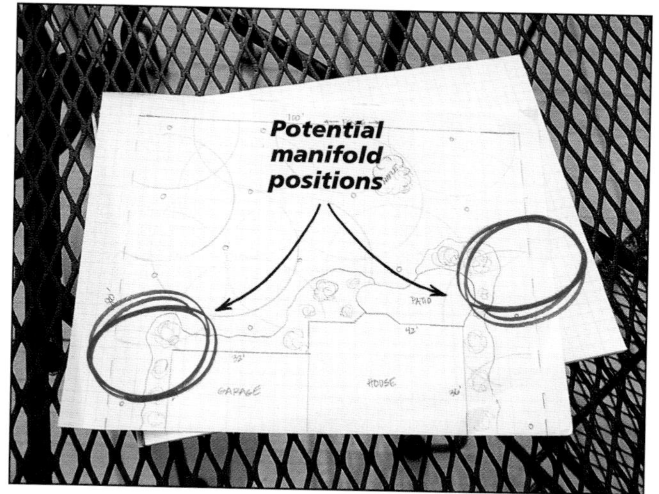

6 GROUP THE MANIFOLD VALVES. Separate valve locations or manifolds will be needed to cover all the zones on a large property. The actual number of valves should be equal to or less than the number of zones available on the automatic timer. Locate the valves as close to the water supply as possible. Lay out the pipe runs with different kinds of lines to represent different runs—a heavy dashed line for the main line and a short dashed line for the branch lines. Branch line connections should be at right angles to the main line. Plan straight runs and avoid turns that will cause a loss of pressure. Once you're satisfied with the layout, it's time to tap into the main for the irrigation system supply and to install the valve manifold. (See "Supply and Manifold" on page 182.)

Requirements for proper installation of an irrigation system vary around the country according to climate conditions, water sources, and local codes. Common to every area, however, is tapping into the water supply correctly and grouping the approved system of control valves into a manifold or series of manifolds to create centrally located controls for the system.

Before you tap into the supply system and start digging trenches, you need to know the source of the supply and its operating pressure. Installing the right manifold system depends on the climate in your area (frost zone or frost free), the construction of your home (basement or slab), and local code requirements for operation. To help you get the right answers, check local codes and get installation advice directly from the manufacturer or from your local home center. The following is an overview of the different options and some of the choices you will have to make.

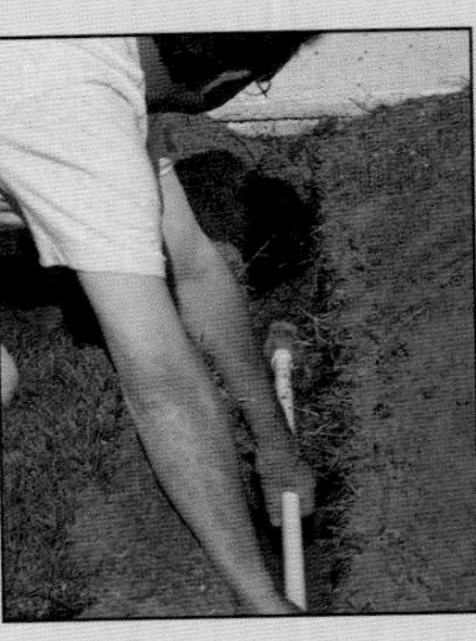

TAPPING THE WATER SUPPLY SYSTEM. The water for your irrigation system can come from several sources—potable water from the municipal system, reclaimed (nonpotable) water from the municipal system, water from a private well, or water from a lake. If your water comes from the municipality, you will need to make sure you have enough pressure to run the system. If your water comes from a well or a lake, a separate or more powerful pump may need to be installed to provide the proper operating pressure.

In colder climates the supply pipes will be buried below the frost line and you will probably tap into the system from your basement. In warmer climates where slab construction is more common and the water meter is buried outside in the yard, the tap will be located in the main supply after the meter and before it enters the house.

In either case you will break into the line and install a soldered or compression tee, a shutoff valve for the irrigation system, and a drain cap so you can flush or drain the system. You will run the supply line from the main to the position you've chosen for the manifold, which will contain the valves that operate the watering zones.

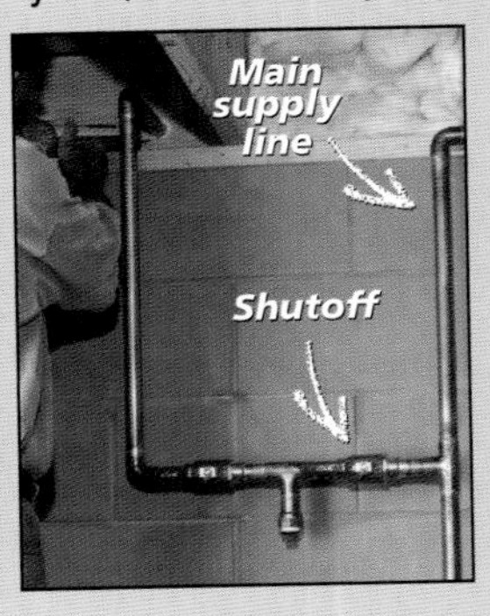

INSTALLING THE MANIFOLDS. The location you choose for the manifold should be easily accessible for maintenance and placed where you're not likely to get a shower while you're adjusting the system. A large system may actually require several manifold locations to provide the most efficient operation. There are two basic options for installing the manifold—above ground (inside or outside), or below ground in a manifold box. Local codes will also help determine the above or below question as well as the kind of manifold valves you should install. Wiring for an automated system will run back to the timer location in the same trench as the manifold supply line. Check with local codes and the timer manufacturer to determine the correct type of wire and method of installation.

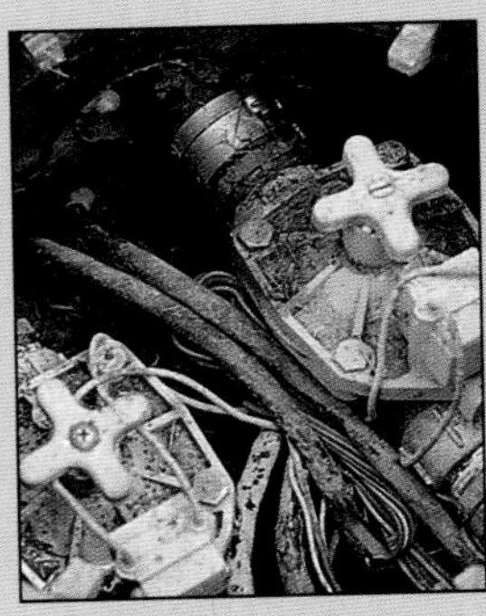

VALVE BASICS. There are basically three types of valves designed for use in irrigation systems. Some valves offer either electronic or manual operation or both. Which ones can be used in your area will probably be defined by local code. Electric valves are operated by a solenoid.

- **In-Line Valves.** In-line valves are usually installed below ground level in a square, or in a round box to allow access for maintenance and inspection.
- **Piston Valves.** A piston opens and closes the valve. These are installed below ground. Piston valves are not recommended for use with pumps, wells, or where dirty water is a factor.
- **Anti-Siphon Valves.** Anti-siphon valves are installed above ground 8 to 12 inches above the highest sprinkler head in a location where water can flow away from the valve. This allows the system to drain if there is a loss of pressure from the main supply line and prevents wastewater from flowing back into the system. Used where codes do not require a pressure vacuum valve.

Local codes often require the installation of a device called a **pressure vacuum breaker** or backflow preventer which prevents contaminated water from flowing back into the potable supply system. A pressure vacuum breaker is installed online a minimum of 12 inches above the head of the highest sprinkler head.

INTO THE TRENCHES. Once you've tapped the supply and installed the manifolds, you're ready to lay out the sprinkler system.

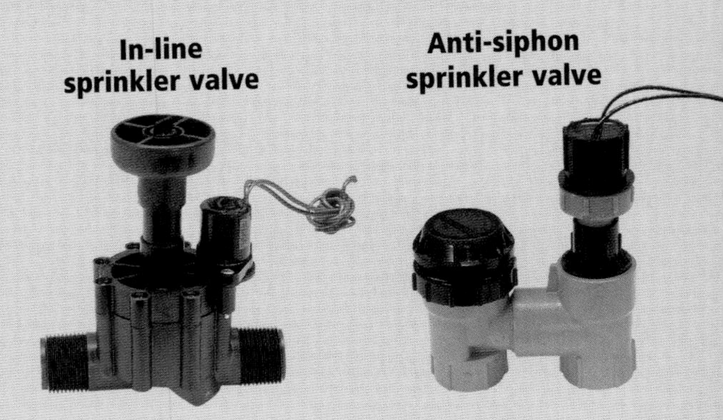

7 **MARK THE TRENCH LINES.** Measure and mark the lines with flour or spray paint to make them easy to follow. Flag the sprinkler and valve locations. Pull up the sod, set aside, and dig the trenches 8 to 12 inches deep or below the frost line—the depth should allow a ⅛-inch-per-foot slope in the lines for draining. Place removed soil on one side of the trench and sod on the other so it will be easier to backfill the trenches and to replace the sod.

8 **TEST-FIT THE LINES.** Start from the valve set and move outward, cutting, fitting, and laying the connecting pipe along the bottom of the trench. Place tees and elbows as needed. (Depending on local codes, the pipe will be PVC or Poly.) If you want to drain the system in the winter, install drain valves at the ends of lines and set a valve box to allow access to the valve. Maintain a ⅛-inch-per-foot slope toward the drain.

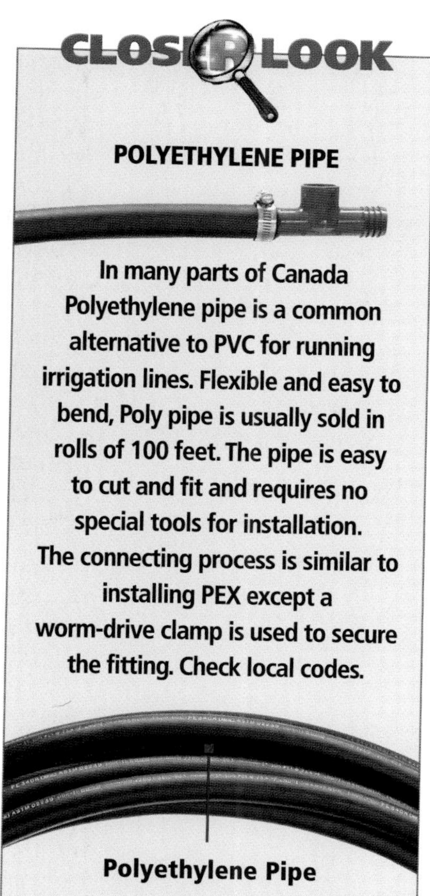

CLOSER LOOK

POLYETHYLENE PIPE

In many parts of Canada Polyethylene pipe is a common alternative to PVC for running irrigation lines. Flexible and easy to bend, Poly pipe is usually sold in rolls of 100 feet. The pipe is easy to cut and fit and requires no special tools for installation. The connecting process is similar to installing PEX except a worm-drive clamp is used to secure the fitting. Check local codes.

Polyethylene Pipe

BUYER'S GUIDE

SELECT THE RIGHT SPRINKLER

No matter which type of sprinkler head you choose, it must cover the area adequately, sprinkling water only where you want it. Ranges of coverage vary from full 360-degree circles to custom settings for tighter areas such as side yards and sidewalks. Throws vary from very short to as much as 45 feet. Heads are usually purchased with a fixed or adjustable spraying radius.

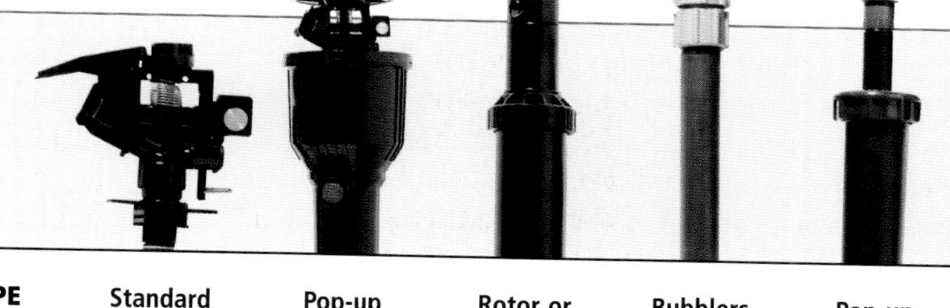

TYPE	Standard Impact Heads	Pop-up Impact Heads	Rotor or Gear Heads	Bubblers	Pop-up Rotor
COMMON USE	Large lawn areas	Medium lawn areas and shrubs	Large lawn areas	Shrubs and groundcover	Large lawn areas
PLACEMENT	Up to 45 feet apart	Up to 15 feet apart	Up to 45 feet apart	3 to 5 feet apart	Up to 45 feet apart
SPRAY DISTANCE	20 to 45 feet apart	5 to 15 feet with an average radius of 12 feet	20 to 45 feet apart	Not applicable	20 to 45 feet apart
FEATURES	Above the ground	Pop-up or flush style. Level, quiet	Above ground for plants, shrubs, and trees	Used for spot watering of flowerbeds	Below ground in a canister

9 MAKE THE FITTINGS. If you are running PVC, apply the proper primer and cement. Connect the pieces and hold until they are set. Wipe excess cement away with a rag. After the connections are set, place the runs in the trench.

10 SET THE SPRINKLER HEIGHT. Block the elbow into place. Position the sprinkler body in the trench so the top of the sprinkler is at the proper height per manufacturer's instructions. Cut an extension of PVC with a threaded end or attach a flexible extension (part of the manufacturer's system) to receive the sprinkler body, wrap the threads with Teflon tape, and install.

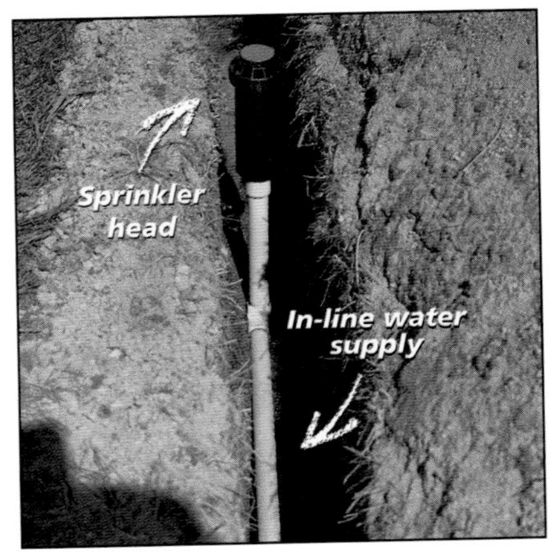

11 SET THE SPRINKLER HEADS. Install sprinkler heads one zone at a time per your layout. Turn on the water and flush the lines per the manufacturer's instructions one zone at a time until the water runs clear. Check for leaks and repair as necessary. Have materials to make repairs on hand when you test the system.

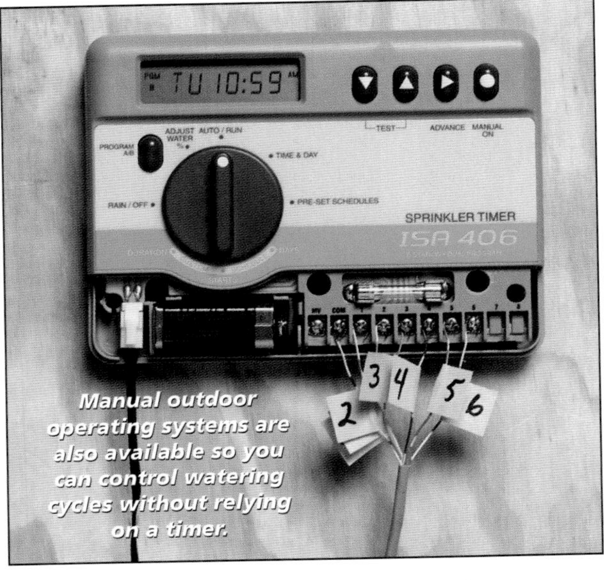

Manual outdoor operating systems are also available so you can control watering cycles without relying on a timer.

12 CONNECT THE TIMER. Put the timer where it's convenient to use, such as in the garage or the basement. If you have to locate it outside, plan on building a cabinet to protect the timer from the elements. Make sure adequate power is available—timers require a standard AC outlet. Follow the manufacturer's instructions for hooking it up. When the installation is complete, try the system out. Slowly turn on the sprinkler system. Turn on one zone at a time. Adjust the screws on the sprinkler heads to fine-tune the spray patterns. Backfill the trenches and replace the sod. Program the timer.

SELECT THE RIGHT SIZE TIMER FOR YOUR SYSTEM.

Timers control valve and sprinkler operation. Many automatic timers are available with a wide selection of features—multiple zones, multiple programs, seasonal adjustment for water conservation, rain delay, and battery backup. Select a timer that has a station for every zone on the system. You might want to buy a timer with extra stations in case you expand.

The two types of timers are mechanical and digital.

Rain switches save water by shutting down the system during storms, which also prevents overwatering the lawn. Check with your local building codes—a rain switch may be required.

Timers can be installed indoors or outdoors. If you locate it outdoors, protect the timer from the elements with an outdoor cabinet.

INSTALLING A DRIP IRRIGATION SYSTEM

EASY	MEDIUM	HARD

SKILLS: Connecting fittings.

HOW LONG WILL IT TAKE?

PROJECT: Running a drip irrigation system in your yard.

NOTE: INSTALLATION TIME VARIES WITH THE TYPE OF KIT YOU PURCHASE, THE SIZE OF YOUR LAYOUT, AND YOUR SKILL LEVEL.

Drip irrigation systems conserve water by providing low-flow irrigation to flowerbeds, vegetable gardens, and landscaped areas. They can go anywhere, in fact, where plants with similar watering requirements are grouped together. They suit themselves to oddly shaped areas and work effectively in windy locations where a regular sprinkler system would cause loss of water through evaporation.

They can run effectively off a standard hose bib, as part of a more extensive irrigation system, or attached to an existing sprinkler line. The systems can be attached to a timer or operate separately from the timer.

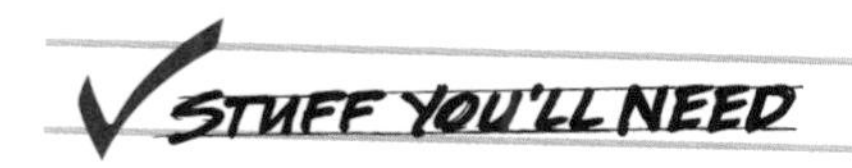

TOOLS: Utility knife or plastic tubing cutter, assembly tools provided by kit, gloves

MATERIALS: Drip irrigation kit

A backflow preventer which stops waste water from coming back into your supply system is required by many city and municipal codes. It can usually be purchased as part of the kit.

DRIP IRRIGATION SYSTEMS. Drip systems can be as simple as a perforated hose hooked to a hose bib or they can come in kits complete with fittings and pipe so you can customize the installation. The kits provide all the parts necessary for installation, and the systems are easy to set up and operate.

LAY OUT THE DRIP LINES. Sketch a layout of the pipe run before you do the actual installation so you can get the most benefit from the system. Once the system is installed, turn it on, take a tour of the run, and check for leaks.

INSTALLING A SUMP PUMP

SKILLS: Operating a jackhammer and connecting plumbing fittings.

HOW LONG WILL IT TAKE?

PROJECT: Installing a sump pump.

EXPERIENCED . . . . . 2 HRS.
HANDY . . . . 4 HRS.
NOVICE . . . . 6 HRS.

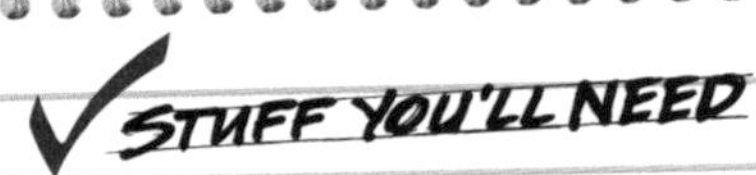

TOOLS: Electric drill, keyhole saw, jackhammer, caulking gun, water-pump pliers, trowel, screwdriver or nut driver, level, adjustable wrench, ABS cleaner and cement, gloves, eye protection

MATERIALS: Sump pump and assembly, ABS drain pipe and fittings, caulk, gravel, concrete, pit liner, electrician's tape, rubber seal

Breaking concrete to install a sump pit requires a jackhammer or a heavy-duty hammer drill and sledgehammer. If you decide to tackle a pit installation yourself, visit your local rental center.

A sump pump will carry water that gathers in a damp basement outside where it can drain away. To install the pump, you're going to have to break through the concrete slab and dig a pit. Before you attack the slab with a jackhammer, find out what's underneath. Local authorities will probably frown on shattering main sewers, branch sewers, and water service lines with a jackhammer.

LOCATE THE LINES

The sewer line should have a visible cleanout which will indicate the direction it's running. Identifying branch sewer line locations may be a little more difficult. Look for risers to help identify them. The water service line might be difficult to locate in slab construction. Most service lines enter the home 4 to 6 feet away from the sewer line. Contact the local water company; they'll tell you where service comes in.

MARK OUT THE PIT

The sump pit should be located at the lowest point on the concrete floor and near an exterior wall so the discharge piping won't be too long. Use a carpenter's level to identify the low point. Mark out the opening for the pit liner, allowing an extra 6 inches for gravel fill.

BREAK OUT THE CONCRETE

Wear heavy clothing and gloves to minimize vibration, cover your eyes with protective goggles, and give yourself plenty of light to work by. Use the electric jackhammer to cut around the perimeter and break out the center area in small chunks. (Smaller chunks are easier to haul upstairs and out of the house.)

SET THE LINER

Remove the soil from the hole to the depth of the pit liner. Set the liner in the pit and level it flush to the concrete surface. Fill around the liner with coarse gravel to within a foot of the surface. Seal around the outside of the pit with concrete. Finish the concrete surface to match the existing floor, using a trowel.

When the concrete is dry, you can begin installing the pump.

1 CUT A HOLE IN AN OUTSIDE WALL FOR THE DISCHARGE PIPE. Use a keyhole saw to drill a hole the same diameter as the discharge pipe recommended by the manufacturer. Measure and cut a section of ABS pipe to extend through the wall with 6 inches extending outside and enough extending inside to install an elbow.

2 SEAL THE OUTSIDE OF THE PIPE. Caulk around the pipe with a caulking gun. Connect an extension to the pipe long enough to carry water away from the house; if it's too close, the water will seep back through the foundation.

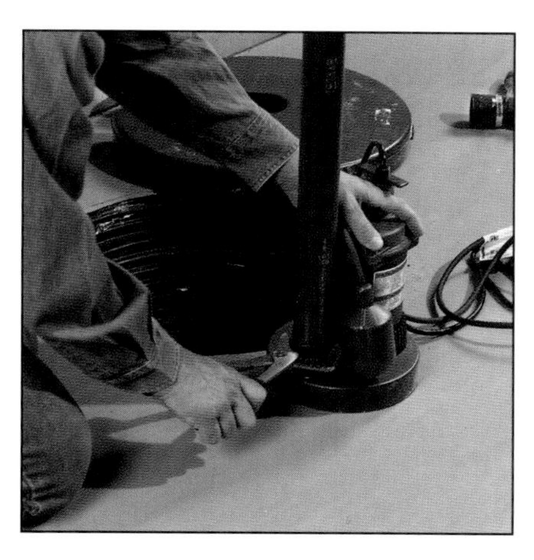

3 ATTACH THE SUMP PUMP TO THE DISCHARGE STANDPIPE. You will need to assemble the unit; the sump pump and the discharge fitting will not normally be preassembled. Connect the two.

4 INSTALL THE SUMP PUMP. Set the sump pump on the bottom of the pit.

5 SET IT ON A SOLID SURFACE. Make sure the gravel base is solid enough to keep the pump from shifting. If it is not, place a small, flat concrete block or some bricks under the pump. Set the pump so the float is several inches from the liner.

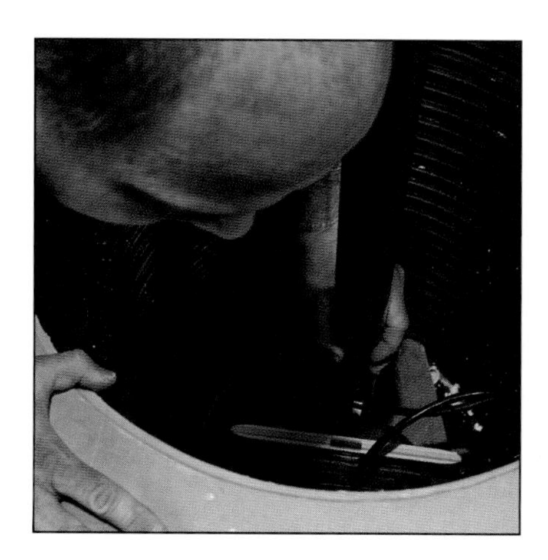

6 LEVEL THE PUMP. Place a bullet level on top of the pump housing. Level the pump using a flattened copper pipe or plastic shims. Check the discharge pipe to make sure it's also level vertically.

If your sump pump constantly turns on and off when there's no water to pump, the water left in the discharge pipe may be flowing back into the pit. The solution is a check valve that prevents the backflow. Check valves are available at your local home center.

7 CONNECT THE DISCHARGE PIPE. Install the check valve with the arrow pointing up. Slide the rubber coupling over the end of the check valve. Place a hose clamp over it and tighten, using a screwdriver or a nut driver. Measure and cut ABS pipe that extends to the pipe through the basement ceiling joist or the wall. Test-fit the line and fittings first. Make sure all the connections are snug. Make adjustments as needed. Use hangers to support longer runs. Cement the ABS (pages 98–99). Once the cement has dried, slide the rubber seal over the end and tighten it.

8 POWER IT UP. Run the power cord along the discharge pipe, securing it with electrician's tape. Adjust the float to the level recommended by the manufacturer. Plug the pump cord into a nearby GFCI-protected receptacle. Test the sump pump by emptying a 5-gallon bucket into the liner, turning it on, and watching it in action.

Installing a rubber seal between the pump and the discharge pipe minimizes vibration created when the sump pump is running.

INSTALLING A HOSE BIB

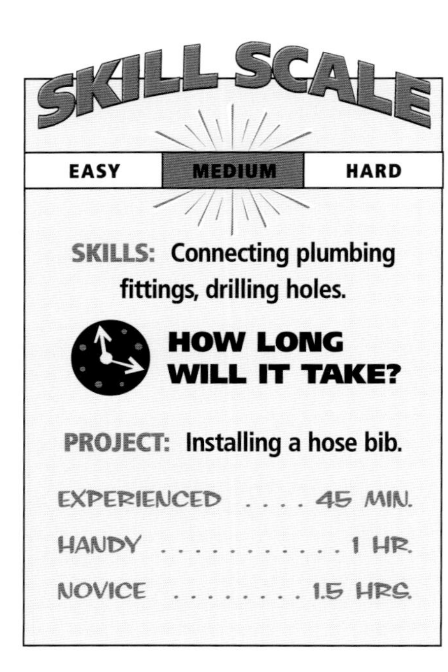

Long runs of watering hose get heavy when you're dragging them around the yard on a hot summer day. You can shorten the length of the hose by installing hose bibs in more convenient exterior locations. You'll have to tap into a water line and solder some copper pipe to connect the bib.

BIB OPTIONS

You can buy plain hose bibs, anti-siphon hose bibs that prevent wastewater from backflowing into the house water supply, "frost-free" hose bibs that won't freeze during the winter, and anti-siphon attachments that thread onto the spout of a standard hose bib valve.

✓ STUFF YOU'LL NEED

TOOLS: Hole saw, tubing cutter, 4-in-1 tool, propane torch, spark striker, adjustable wrench, pipe wrench, caulking gun, screwdriver or power screwdriver

MATERIALS: Hose bib, wood screws, copper pipe and fittings, threaded adapter fitting, union fitting , lead-free flux, lead-free solder, rags, silicone sealant

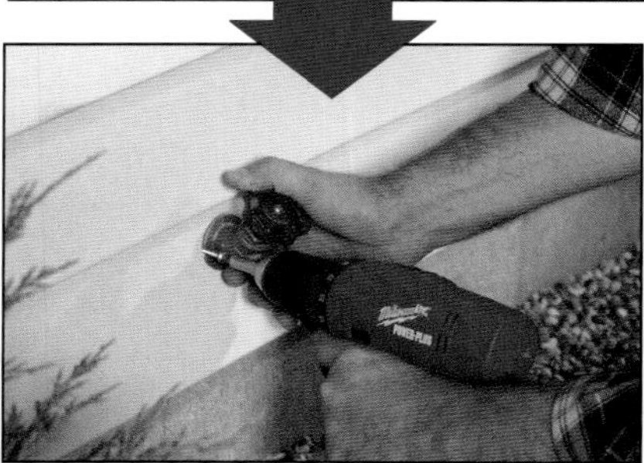

1 DRILL A HOLE FOR THE WATER LINE. Use a hole saw the diameter of the pipe and a drill to make a hole in the exterior wall. The hole needs to have a 1½ to 1¾ inch diameter, depending on the size of the supply line. Slide the hose bib shaft through the opening. Turn off the main water supply valve and drain the lines.

ATTACH THE HOSE BIB FLANGE TO THE WALL. Use a screwdriver or a power screwdriver to drive the wood screws.

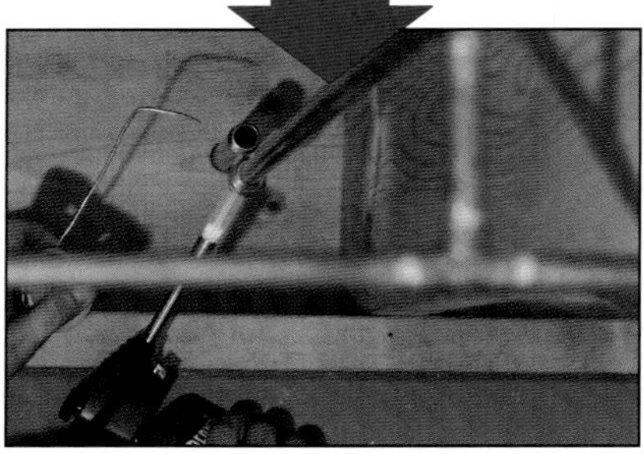

2 CUT THE WATER SUPPLY LINE WITH A TUBING CUTTER AND SOLDER THE COPPER PIPE TO THE THREADED ADAPTER FITTING. Install a copper tee and piping to the hose bib stub. Remove all flammable material or place a nonflammable object in front of them so they won't ignite. (See "Soldering Copper Pipes," page 86.)

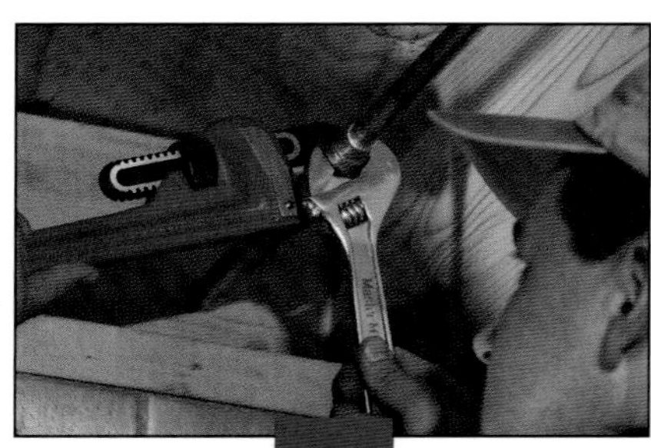

3 CONNECT THE HOSE BIB EXTENSION TO THE ADAPTER WITH A UNION FITTING. Use an adjustable wrench to hold the adapter fitting and a pipe wrench to tighten the union. You may want to consider installing a shutoff here.

CAULK AROUND THE FLANGE TO SEAL IT. Turn on the water at the main valve. Inspect the copper fittings for leaks. Turn on inside faucets until a steady stream is reestablished, then turn it off. Turn on the hose bib to test it.

PLASTIC VS. METAL: PLUMBING PIPE EXPLAINED

PLASTIC PIPES.

1. **PVC (polyvinyl chloride)**—Rigid; resistant to damage from heat and chemicals; used in drain systems; 98 percent failure rate at joints when subjected to the high water pressure in supply lines.
2. **CPVC (chlorinated polyvinyl chloride)**—Rigid; resistant to damage from heat and chemicals; withstands the high pressure of supply systems.
3. **PEX (cross-linked polyethylene)**—Flexible; resistant to damage from heat and chemicals; withstands the high pressure of supply systems. Compression fittings make PEX very easy to connect. Restricted in some areas; check local codes before using.
4. **ABS (acrylonitrile butadiene styrene)**—Rigid; first plastic developed for use in drainage systems. (Commonly used for drain systems in Canada and the western United States. Check codes in other areas.)

METAL PIPES.

5. **FLEXIBLE COPPER TUBING**—Used in gas and water supply; ability to bend around obstacles means fewer fittings are necessary when installing.
6. **RIGID COPPER**—Often the first choice for supply; durable; smooth; resists corrosion and deterioration.
7. **CAST IRON**—Old standby for drainage systems; strong; durable but difficult to work with; commonly replaced by PVC where code allows.
8. **CHROMED BRASS**—Used for drain applications where plumbing is visible.
9. **GALVANIZED IRON**—Used in supply and drain systems; strong but susceptible to corrosion and difficult to work with.
10. **BLACK IRON**—Used for gas applications exclusively; professional installation recommended.

- **CHROMED COPPER (Not shown)**—Used for supply in applications where plumbing is visible.

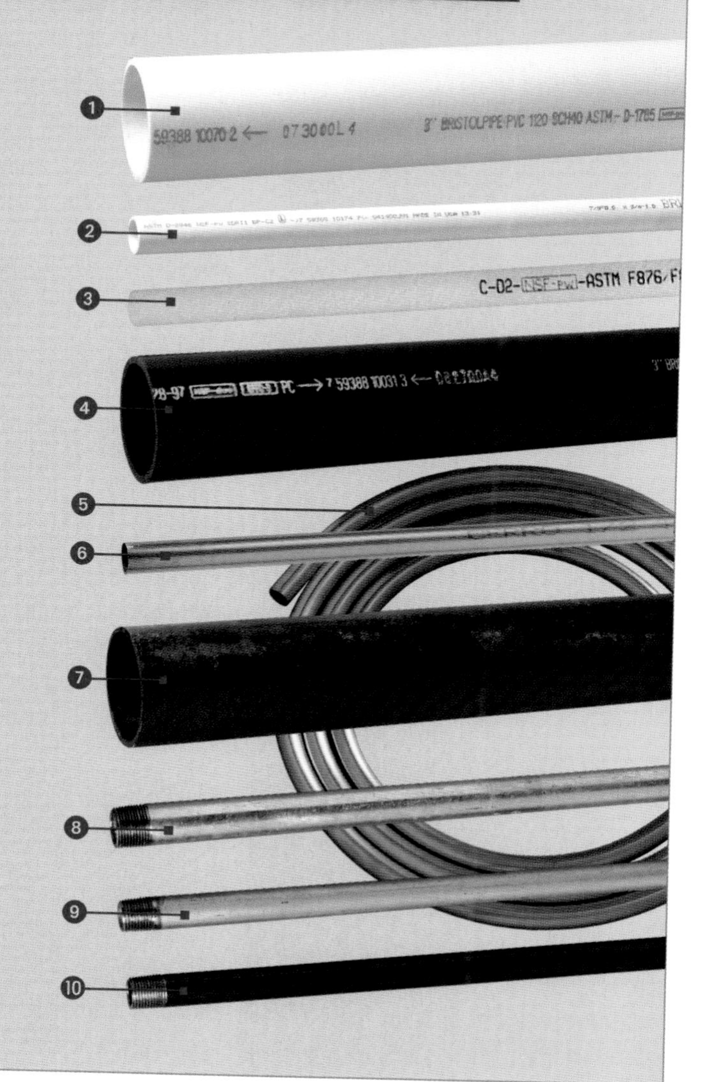

METRIC CONVERSIONS

U.S. Units to Metric Equivalents

To Convert From	Multiply By	To Get
Inches	25.4	Millimeters
Inches	2.54	Centimeters
Feet	30.48	Centimeters
Feet	0.3048	Meters
Yards	0.9144	Meters
Square inches	6.4516	Square centimeters
Square feet	0.0929	Square meters
Square yards	0.8361	Square meters
Acres	0.4047	Hectares
Cubic inches	16.387	Cubic centimeters
Cubic feet	0.0283	Cubic meters
Cubic feet	28.316	Liters
Cubic yards	0.7646	Cubic meters
Cubic yards	764.55	Liters

To convert from degrees Fahrenheit (F) to degrees Celsius (C), first subtract 32, then multiply by $5/9$.

Metric Units to U.S. Equivalents

To Convert From	Multiply By	To Get
Millimeters	0.0394	Inches
Centimeters	0.3937	Inches
Centimeters	0.0328	Feet
Meters	3.2808	Feet
Meters	1.0936	Yards
Square centimeters	0.1550	Square inches
Square meters	10.764	Square feet
Square meters	1.1960	Square yards
Hectares	2.4711	Acres
Cubic centimeters	0.0610	Cubic inches
Cubic meters	35.315	Cubic feet
Liters	0.0353	Cubic feet
Cubic meters	1.308	Cubic yards
Liters	0.0013	Cubic yards

To convert from degrees Celsius to degrees Fahrenheit, multiply by $9/5$, then add 32.

INDEX

G

H

I

J

L

M

N

O

P

Q

R

S

T

U

V

W

ENERGY-SAVING PLUMBING PROJECTS

When many of us think of saving energy, we think of turning off the lights, insulating the walls of our homes, or buying a car that gets better mileage. There are opportunities to conserve electricity and gas when it comes to plumbing, as well. Below are six sources of energy loss that are associated with your plumbing system and where you can get the information to fix them in the pages of ***Plumbing 1-2-3***™.

PLUMBING USES ENERGY, TOO!

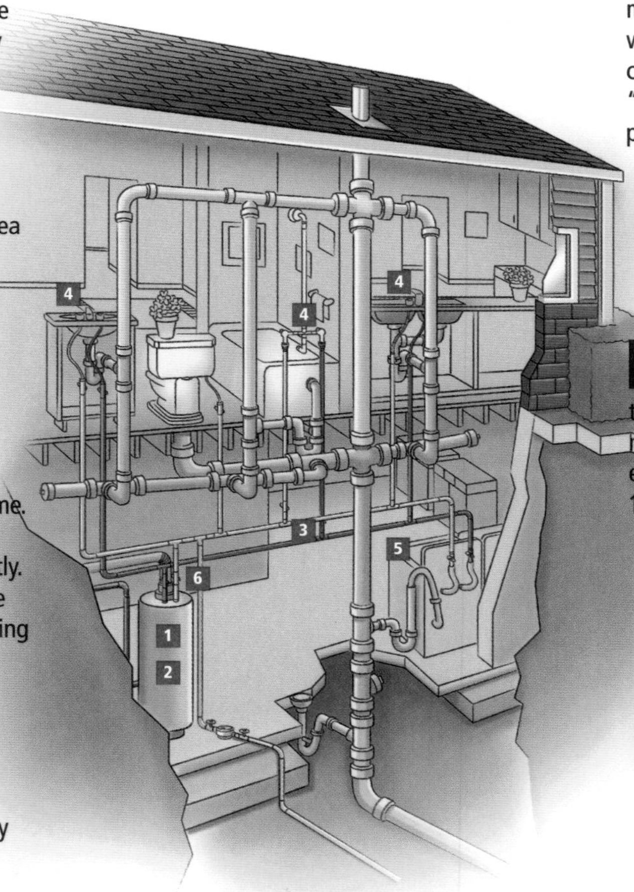

1 INSULATE THE WATER HEATER. Newer water heaters come with efficient insulation already built into the tank, but that's not necessarily the case with older models. Home Centers sell insulation blanket kits that slip over the heater to conserve heat loss. Even with newer models an insulation blanket is a good idea in very cold basements. (See "Good Idea," page 58.)

2 MAINTAIN ALL APPLIANCES, ESPECIALLY THE WATER HEATER. Water heaters are the second largest energy users in the home. Tanks that are full of sediment and debris don't work efficiently. Drain them regularly and make sure all the elements are working properly. (See page 52.)

3 INSULATE PIPES. Uninsulated pipes, especially the hot supply lines, cause significant energy loss by placing higher demand on the water heater. There are several options for insulation. (See pages 54 and 55.)

4 FIX DRIPPY FAUCETS. Leaks and drips from the faucets, especially the hot side, mean that the water heater is working overtime to stay charged. (See Chapter Three: "Repair and Maintenance," page 14.)

5 BUY ENERGY-EFFICIENT APPLIANCES. Appliances that use gas or electricity should meet government approved efficiency standards. (See pages 158, 165, and 173.)

6 INSTALL WATER HEATER HEAT TRAP FITTINGS. Some newer water heaters come with heat trap fittings included. If yours doesn't have them they are easily installed and will make your water heater more energy efficient. (See pages 165-169.)